Current Developments in Numerical Analysis

Current Developments in Numerical Analysis

Edited by **Rushel Davis**

CLANRYE
INTERNATIONAL

New Jersey

Published by Clanrye International,
55 Van Reypen Street,
Jersey City, NJ 07306, USA
www.clanryeinternational.com

Current Developments in Numerical Analysis
Edited by Rushel Davis

International Standard Book Number: 978-1-63240-125-0 (Hardback)

Printed in the United States of America.

Contents

Preface   VII

Chapter 1   **Convergence Analysis of a Fully Discrete Family of Iterated Deconvolution Methods for Turbulence Modeling with Time Relaxation**   1
R. Ingram, C. C. Manica, N. Mays, and I. Stanculescu

Chapter 2   **Solution of Nonlinear Volterra-Fredholm Integrodifferential Equations via Hybrid of Block-Pulse Functions and Lagrange Interpolating Polynomials**   33
Hamid Reza Marzban and Sayyed Mohammad Hoseini

Chapter 3   **The Optimal L^2 Error Estimate of Stabilized Finite Volume Method for the Stationary Navier-Stokes Problem**   47
Guoliang He, Jian Su, and Wenqiang Dai

Chapter 4   **An Efficient Family of Root-Finding Methods with Optimal Eighth-Order Convergence**   64
Rajni Sharma and Janak Raj Sharma

Chapter 5   **Two-Level Stabilized Finite Volume Methods for Stationary Navier-Stokes Equations**   82
Anas Rachid, Mohamed Bahaj, and Noureddine Ayoub

Chapter 6   **New Approach for Solving a Class of Doubly Singular Two-Point Boundary Value Problems Using Adomian Decomposition Method**   96
Randhir Singh, Jitendra Kumar, and Gnaneshwar Nelakanti

Chapter 7   **A Note on Fourth Order Method for Doubly Singular Boundary Value Problems**   118
R. K. Pandey and G. K. Gupta

Chapter 8   **A Class of Numerical Methods for the Solution of Fourth-Order Ordinary Differential Equations in Polar Coordinates**   139
Jyoti Talwar and R. K. Mohanty

Chapter 9   **Interpreting the Phase Spectrum in Fourier Analysis
of Partial Ranking Data**   159
Ramakrishna Kakarala

Chapter 10   **Signorini Cylindrical Waves and Shannon Wavelets**   174
Carlo Cattani

Permissions

List of Contributors

Preface

Numerical analysis is the study in the area of mathematics and computer science that uses numerical estimation and approximation to solve the problems of mathematical analysis. It creates, analyzes, and implements algorithms for numerically deciphering the problems of continuous mathematics. Numerical analysis is also responsible for continuing a long tradition of practical mathematical calculations. Numerical analysis is largely focused on obtaining approximate solutions while sustaining reasonable restraints on errors. Such problems usually originate from practical applications of geometry, algebra, and calculus, and they also involve variables which change constantly. The scientific discipline and field of study of numerical analysis varies from highly theoretical mathematical studies to computer science issues involving the effects of computer hardware and software on the implementation of specific algorithms. These problems occur throughout the social sciences, natural sciences, engineering, business and medicine. Since the 20th century, the evolution and growth in power and accessibility of digital computers has led to an escalating use of realistic mathematical models in engineering and science, and numerical analysis of increasing complexity is needed to solve these more complicated and intricate models of the world. Such real world applications have increased the demand for skilled graduates in this field.

This book is an attempt to compile and collate all available research on numerical analysis under one project. I am grateful to those who put their hard work, effort and expertise into these researches as well as those who were supportive in this endeavour.

Editor

Convergence Analysis of a Fully Discrete Family of Iterated Deconvolution Methods for Turbulence Modeling with Time Relaxation

R. Ingram,[1] C. C. Manica,[2] N. Mays,[3] and I. Stanculescu[4]

[1] Department of Mathematics, University of Pittsburgh, PA 15260, USA
[2] Departmento de Matemática Pura e Aplicada, Universidade Federal do Rio Grande do Sul, Porto Alegre 91509-900, RS, Brazil
[3] Department of Mathematics, Wheeling Jesuit University, WV 26003, USA
[4] Farquhar College of Arts and Sciences, Nova Southeastern University, FL 33314, USA

Correspondence should be addressed to N. Mays, nmays@wju.edu

Academic Editor: William J. Layton

We present a general theory for regularization models of the Navier-Stokes equations based on the Leray deconvolution model with a general deconvolution operator designed to fit a few important key properties. We provide examples of this type of operator, such as the (modified) Tikhonov-Lavrentiev and (modified) Iterated Tikhonov-Lavrentiev operators, and study their mathematical properties. An existence theory is derived for the family of models and a rigorous convergence theory is derived for the resulting algorithms. Our theoretical results are supported by numerical testing with the Taylor-Green vortex problem, presented for the special operator cases mentioned above.

1. Approximate Deconvolution for Turbulence Modeling

Numerical simulations of complex flows present many challenges. The Navier-Stokes (NS) equations (NSE), given by the following

$$\mathbf{u}_t + \mathbf{u} \cdot \nabla \mathbf{u} - \nu \Delta \mathbf{u} + \nabla p = \mathbf{f}, \qquad \nabla \cdot \mathbf{u} = 0, \text{ in } \Omega \times (0, T) \tag{1.1}$$

is an exact model for the flow of a viscous, incompressible fluid, [1]. For turbulent flows (characterized by Reynold's number Re $\gg$ 1), it is infeasible to properly resolve all significant scales above the Kolmogorov length scale $O(\text{Re}^{-3/4})$ by direct numerical simulation. Thus, numerical simulations are often based on various regularizations of NSE, rather than NSE themselves. Accordingly, regularization methods provide a (computationally) *efficient* and (algorithmically) *simple* family of turbulence models. Several of the most commonly applied regularization methods include:

$$(\text{Leray}) \qquad \mathbf{w}_t + \overline{\mathbf{w}} \cdot \nabla \mathbf{w} - \text{Re}^{-1}\Delta \mathbf{w} + \nabla p = \mathbf{f}, \qquad \nabla \cdot \mathbf{w} = 0, \qquad (1.2)$$

$$(\text{NS-}\alpha) \qquad \mathbf{w}_t + \overline{\mathbf{w}} \times (\nabla \times \mathbf{w}) - \text{Re}^{-1}\Delta \mathbf{w} + \nabla P = \mathbf{f}, \qquad (1.3)$$

$$(\text{NS-}\overline{\omega}) \qquad \mathbf{w}_t + \mathbf{w} \times (\nabla \times \overline{\mathbf{w}}) - \text{Re}^{-1}\Delta \mathbf{w} + \nabla P = \mathbf{f}, \qquad (1.4)$$

$$(\text{time relaxation}) \qquad \mathbf{w}_t + \mathbf{w} \cdot \nabla \mathbf{w} - \text{Re}^{-1}\Delta \mathbf{w} + \nabla p + \chi(\mathbf{w} - \overline{\mathbf{w}}) = \mathbf{f}, \qquad (1.5)$$

where $\nabla \cdot \mathbf{w} = 0$ in each case and $\overline{\mathbf{w}}$ is an averaged velocity field $\mathbf{w}$, p is pressure, and P is the Bernoulli pressure. More details about these models can be found for instance in [2–5] and references therein. Although these regularization methods achieve high theoretical accuracy and perform well in select practical tests, those models do not provide a fully-developed numerical solution for decoupling the scales in a turbulent flow. In fact, results show that only time relaxation regularization truncates scales sufficiently for practical computations. Indeed, it is shown that time relaxation term $\chi(\mathbf{w} - \overline{\mathbf{w}})$ for $\chi > 0$ damps unresolved fluctuations over time [5, 6]. Note that the choice of χ is an active area of research and that solutions are very sensitive to variations in χ.

Deconvolution-based regularization is also an active area of research obtained, for example, by replacing $\overline{\mathbf{w}}$ by $D(\overline{\mathbf{w}})$ in each (1.2)–(1.5) for some deconvolution operator D. In [7], Dunca proposed the general Leray-deconvolution problem ($D(\overline{\mathbf{w}})$ instead of $\overline{\mathbf{w}}$) as a more accurate extension to Leray's model [8]. Leray used the Gaussian filter as the smoothing (averaging) filter G, denoted above by *overbar*. In [9], Germano proposed the differential filter (approximate-Gaussian) $G = (-\delta^2 \Delta + I)^{-1}$ where $\delta > 0$ is the filter length. The differential filter is easily modeled in the variational framework of the finite element (FE method (FEM)). We provide a brief overview of continuous and discrete operators (Section 3).

The deconvolution-based models have proven themselves to be very promising. However, among the very many known approximate deconvolution operators from image processing, for example, [10], so far only few have been studied for turbulence modeling, for example, the van Cittert and the modified Tikhonov-Lavrentiev deconvolution operators. Their success suggests that it is time to develop a general theory for regularization models of the NSE as a guide to development of models based on other, possibly better, deconvolution operators and refinement of existing ones.

Herein, we present a general theory for regularization models of the NSE based on the Leray deconvolution model with a general deconvolution operator. We prove energetic stability (and hence existence) and convergence of an FE (in space) and Crank-Nicolson (CN) (in time) discretization of the following family of Leray deconvolution regularization models with time relaxation: find $\mathbf{w} : \Omega \times (0, T] \rightarrow \mathbb{R}^3$ and $\pi : \Omega \times (0, T] \rightarrow \mathbb{R}$ satisfying the following

$$\mathbf{w}_t + D(\overline{\mathbf{w}}) \cdot \nabla \mathbf{w} - \text{Re}^{-1}\Delta \mathbf{w} + \nabla \pi + \chi(\mathbf{w} - D(\overline{\mathbf{w}})) = \mathbf{f}, \qquad \nabla \cdot \mathbf{w} = 0, \qquad (1.6)$$

Convergence Analysis of a Fully Discrete Family of Iterated Deconvolution Methods for
Turbulence Modeling with Time Relaxation

3

for some appropriate boundary and initial conditions (the fully discrete model is presented in Problem 1 with energetic stability and well posedness proved in Theorem 4.5).

1.1. Improving Accuracy of Approximate Deconvolution Methods

The fundamental difficulties corresponding to regularization methods applied as a viable turbulence model include ensuring that:

(i) scales are appropriately truncated (model microscale = filter radius = mesh width)

(ii) smooth parts of the solution are accurately approximated ($D(\overline{\mathbf{w}}) \approx \mathbf{w}$ for smooth $\mathbf{w}$)

(iii) physical fidelity of flow is preserved.

Due to the nonlinearity in (1.6), different choices of the filter and deconvolution operator yield significant changes in the solution of the corresponding model. Implementation concerns for deconvolution methods, for example, Tikhonov-Lavrentiev regularization given by $D = (G + \alpha I)^{-1}$, include selection of deconvolution parameter $\alpha > 0$. Iterated deconvolution methods reduce approximation sensitivity relative to α-selection and, hence, allow a conservatively large α-selection for stability with updates (fixed number of iterations) used to recover higher accuracy (Section 3.3). For example we prove, under usual conditions, (Proposition 5.3),

$$\text{Modified Tikhonov } (j = 0) \qquad \text{error}(\mathbf{w} - D(\overline{\mathbf{w}})) \leq \mathcal{O}\left(\alpha\delta^2\right),$$

$$\text{Iterated Modified Tikhonov } (j > 0) \qquad \text{error}(\mathbf{w} - D(\overline{\mathbf{w}})) \leq \mathcal{O}\left(\left(\alpha\delta^2\right)^{j+1}\right), \tag{1.7}$$

so that iterated modified Tikhonov regularization gives geometric convergence with respect to the update number j. In either case, Tikhonov-Lavrentiev or iterated Tikhonov-Lavrentiev regularization, we prove, under usual conditions, (Proposition 5.4)

$$\text{error}\left(D^h\left(\overline{\mathbf{w}}^h\right) - D(\overline{\mathbf{w}})\right) \leq \mathcal{O}\left(h^k\right), \tag{1.8}$$

where D^h and $\overline{\mathbf{w}}^h$ represent the discrete deconvolution operator and discrete filter, respectively, and k is order of FE polynomial space. We propose minimal properties for a general family of deconvolution operators D and filters G (Section 3.1) satisfying (Assumptions 3.4, 3.5), for example,

(i) $\|DG\| \leq 1$, forces spectrum of DG in $[0, 1]$

(ii) $\|\nabla DG\mathbf{v}\| \leq d_1\|\nabla\mathbf{v}\|$, controls size of ∇DG

(iii) $\|(I - DG)\mathbf{v}\| \leq c_1(\alpha, \delta)\|\Delta\mathbf{v}\| \to 0$ as $\alpha, \delta \to 0$ (for smooth $\mathbf{v}$), ensures convergence of method

(iv) $\|(DG - D^hG^h)\mathbf{v}\| \leq c_2(h, \alpha, \delta) \to 0$ as $h \to 0$ (for smooth $\mathbf{v}$), ensures convergence of method.

In fact, the updates $\{\boldsymbol{w}_j\}_{j\geq 0}$ satisfying

$$\boldsymbol{w}_0 := D(\overline{\mathbf{w}}), \quad \boldsymbol{w}_j - \boldsymbol{w}_{j-1} := D\overline{(\mathbf{w} - \boldsymbol{w}_{j-1})}, \tag{1.9}$$

inherit the properties assumed for the base operator D in Assumptions 3.4, 3.5 (Propositions 3.10, 3.11). These iterates represent defect correction generalization of iterated Tikhonov regularization operator [11]. We prove that the FE-CN approximation $\mathbf{w}^h$ of the general deconvolution turbulence model (Problem 1) satisfies (Theorem 4.6)

$$\text{error}\left(\mathbf{w}^h - \mathbf{u}_{\text{NSE}}\right) \leq C\left(h^k + \Delta t^2 + c_1(\alpha, \delta, j) + c_2(h, \alpha, \delta, j)\right) \longrightarrow 0, \quad \text{as } h, \delta, \alpha \longrightarrow 0 \tag{1.10}$$

for smooth enough solutions $\mathbf{u}_{\text{NSE}}$ of the NSE (see variational formulation (2.3)–(2.5)). We show that $c_1 = C(\alpha\delta^2)^\beta$ (for any $0 \leq \beta \leq j + 1$ and $\Delta^\beta \mathbf{u}_{\text{NSE}} \in L^2(\Omega)$) (Proposition 5.3) and $c_2 = Ch^k$ (for $D(\overline{\mathbf{w}}) \in H^{k+1}(\Omega)$) (Proposition 5.4) for modified iterated Tikhonov-Lavrentiev regularization (see Corollary 5.5 for corresponding error estimate). We conclude with a numerical test that verifies the theoretical convergence rate predicted in Theorem 4.6 (Section 5.2).

1.2. Background and Overview

One of the most interesting approaches to generate turbulence models is via approximate deconvolution or approximate/asymptotic inverse of the filtering operator. Examples of such models include: Approximate Deconvolution Models (ADM) and Leray-Tikhonov Deconvolution Models. Layton and Rebholz compiled a comprehensive overview and detailed analysis of ADM [12] (see also references therein). Previous analysis of the ADM with and without the time-relaxation term used van Cittert deconvolution operators [5, 6]; although easily programmed, van Cittert schemes can be computationally expensive [5]. Tikhonov-Lavrentiev regularization is another popular regularization scheme [13]. Determining the appropriate value of α to ensure stability while preserving accuracy is challenging, see for example, [14–19]. Alternatively, *iterated* Tikhonov regularization is well known to decouple stability and accuracy from the selection of regularization parameter α, see for example, [11, 20–22]. Iterated Tikhonov regularization is one special case of the general deconvolution operator we propose herein.

2. Function Spaces and Approximations

Let the flow domain $\Omega \in \mathbb{R}^d$ for $d = 2, 3$ be a regular and bounded polyhedral. We use standard notation for Lebesgue and Sobolev spaces and their norms. Let $\|\cdot\|$ and $(\cdot, \cdot)$ be the L^2-norm and inner product, respectively. Let $\|\cdot\|_{p,k} := \|\cdot\|_{W_p^k(\Omega)}$ represent the $W_p^k(\Omega)$-norm. We write $H^k(\Omega) := W_2^k(\Omega)$ and $\|\cdot\|_k$ for the corresponding norm. Let the context determine whether $W_p^k(\Omega)$ denotes a scalar, vector, or tensor function space. For example let $\mathbf{v} : \Omega \to \mathbb{R}^d$. Then, $\mathbf{v} \in H^1(\Omega)$ implies that $\mathbf{v} \in H^1(\Omega)^d$ and $\nabla\mathbf{v} \in H^1(\Omega)$ implies that

Convergence Analysis of a Fully Discrete Family of Iterated Deconvolution Methods for
Turbulence Modeling with Time Relaxation

5

$\nabla \mathbf{v} \in H^1(\Omega)^{d \times d}$. Write $W_q^m(W_p^k(\Omega)) := W_q^m(0, T; W_p^k(\Omega))$ equipped with the standard norm. For example,

$$\|\mathbf{v}\|_{L^q(W_p^k)} := \begin{cases} \left(\int_0^T \|\mathbf{v}(\cdot, t)\|_{p,k}^q dt \right)^{1/q}, & \text{if } 1 \leq q < \infty, \\ \operatorname*{ess\,sup}_{0 < t < T} \|\mathbf{v}(\cdot, t)\|_{p,k}, & \text{if } q = \infty. \end{cases} \tag{2.1}$$

Denote the pressure and velocity spaces by $Q := L_0^2(\Omega) = \{q \in L^2(\Omega) : \int_\Omega q = 0\}$ and $X := H_0^1(\Omega) = \{\mathbf{v} \in H^1(\Omega) : \mathbf{v}|_{\partial\Omega} = 0\}$, respectively. Moreover, the dual space of X is denoted $X' := W_2^{-1}(\Omega)$ and equipped with the norm

$$\|\mathbf{f}\|_{-1} := \sup_{0 \neq \mathbf{v} \in X} \frac{\langle \mathbf{f}, \mathbf{v} \rangle_{X' \times X}}{|\mathbf{v}|_1}. \tag{2.2}$$

Fix $\mathbf{f} \in X'$ and Re > 0. In this setting, we consider strong NS solutions: find $\mathbf{u} \in L^2(X) \cap L^\infty(L^2(\Omega))$ and $p \in W^{-1,\infty}(Q)$ satisfying

$$(\mathbf{u}_t, \mathbf{v}) + (\mathbf{u} \cdot \nabla \mathbf{u}, \mathbf{v}) + \text{Re}^{-1}(\nabla \mathbf{u}, \nabla \mathbf{v}) - (p, \nabla \cdot \mathbf{v}) = (\mathbf{f}, \mathbf{v}), \quad \text{a.e. } t \in (0, T], \; \forall \mathbf{v} \in X, \tag{2.3}$$

$$(q, \nabla \cdot \mathbf{u}) = 0, \quad \text{a.e. } t \in (0, T], \; \forall q \in Q, \tag{2.4}$$

$$\mathbf{u}(\cdot, 0) = \mathbf{u}_0 \quad \text{in } L^2(\Omega). \tag{2.5}$$

Let $V := \{\mathbf{v} \in X : \nabla \cdot \mathbf{v} = 0\}$. Restricting test functions $\mathbf{v} \in V$ reduces (2.3)–(2.5) to find $\mathbf{u} : (0, T] \to V$ satisfying

$$(\mathbf{u}_t, \mathbf{v}) + (\mathbf{u} \cdot \nabla \mathbf{u}, \mathbf{v}) + \text{Re}^{-1}(\nabla \mathbf{u}, \nabla \mathbf{v}) = (\mathbf{f}, \mathbf{v}), \quad \text{a.e. } t \in (0, T], \; \forall \mathbf{v} \in V \tag{2.6}$$

and (2.5). For smooth enough solutions, solving the problem associated with (2.6), (2.5) is equivalent to (2.3)–(2.5).

Control of the nonlinear term is essential for establishing a priori estimates and convergence estimates. We state a selection of inequalities here that will be utilized later:

$$|\mathbf{u} \cdot \nabla \mathbf{v}, \mathbf{w}| \leq C(\Omega) \begin{cases} \|\mathbf{u}\|_1 |\mathbf{v}|_1 \|\mathbf{w}\|_1 \\ \|\mathbf{u}\| \|\mathbf{v}\|_2 \|\mathbf{w}\|_1 & \forall \mathbf{v} \in H^2(\Omega) \\ \|\mathbf{u}\|_2 \|\mathbf{v}\| \|\mathbf{w}\|_1 & \forall \mathbf{u} \in H^2(\Omega) \cap V. \end{cases} \tag{2.7}$$

2.1. Discrete Function Setting

Fix $h > 0$. Let $\mathcal{T}^h$ be a family of subdivisions (e.g., triangulation) of $\overline{\Omega} \subset \mathbb{R}^d$ satisfying $\overline{\Omega} = \bigcup_{E \in \mathcal{T}^h} E$ so that $diameter(E) \leq h$ and any two closed elements $E_1, E_2 \in \mathcal{T}^h$ are either disjoint or share exactly one face, side, or vertex. See Chapter II, Appendix A in [23] for more on this subject in context of Stokes problem and [24] for a more general treatment. For example, $\mathcal{T}^h$ consists of triangles for $d = 2$ or tetrahedra for $d = 3$ that are nondegenerate as $h \to 0$.

Let $X^h \subset X$ and $Q^h \subset Q$ be a conforming velocity-pressure mixed FE space. For example, let X^h and Q^h be continuous, piecewise (on each $E \in \mathcal{T}^h$) polynomial spaces. The discretely divergence-free space is given by

$$V^h = \left\{ \mathbf{v}^h \in X^h : \left(q^h, \nabla \cdot \mathbf{v}^h \right) = 0 \ \forall q^h \in Q^h \right\}. \tag{2.8}$$

Note that in general $V^h \not\subset V$ (e.g., for Taylor-Hood elements). In order to avoid stability issues arising when FE solutions are not exactly divergence free (i.e., when $V^h \not\subset V$), we introduce the explicitly skew-symmetric convective term

$$b^h(\mathbf{u}, \mathbf{v}, \mathbf{w}) := \frac{1}{2}((\mathbf{u} \cdot \nabla \mathbf{v}, \mathbf{w}) - (\mathbf{u} \cdot \nabla \mathbf{w}, \mathbf{v})), \tag{2.9}$$

so that

$$b^h(\mathbf{u}, \mathbf{v}, \mathbf{v}) = 0. \tag{2.10}$$

Note that $b^h(\mathbf{u}, \mathbf{v}, \mathbf{w}) = (\mathbf{u} \cdot \nabla \mathbf{v}, \mathbf{w})$ when $\mathbf{u} \in V$. Moreover, the trilinear from $b^h(\cdot, \cdot, \cdot)$ is continuous and skew-symmetric on $X \times X \times X$.

Lemma 2.1. *If* $\mathbf{u}, \mathbf{v}, \mathbf{w} \in X$,

$$\left| b^h(\mathbf{u}, \mathbf{v}, \mathbf{w}) \right| \leq C(\Omega) \begin{cases} (\|\mathbf{u}\| \|\mathbf{u}\|_1)^{1/2} \|\mathbf{v}\|_1 \|\mathbf{w}\|_1 \\ \|\mathbf{u}\| \|\mathbf{v}\|_2 \|\mathbf{w}\|_1 \end{cases} \quad \forall \mathbf{v} \in H^2(\Omega). \tag{2.11}$$

Proof. The proof of the first inequality can be found in [25]. The second follows from Hölder's and Poincaré's inequalities. □

For the time discretization, let $0 = t_0 < t_1 < \cdots < t_{M-1} = T < \infty$ be a discretization of the time interval $[0, T]$ for a constant time step $\Delta t = t_{n+1} - t_n$. Write $t_{n+1/2} := (t_{n+1} + t_n)/2$, $z_n = z(t_n)$ and, if $z \in C^0([t_n, t_{n+1}])$, $z_{n+1/2} = (z_{n+1} + z_n)/2$. Define

$$\|\mathbf{u}\|_{l^q(m_1, m_2; W_p^k(\Omega))} := \begin{cases} \left(\Delta t \sum_{n=m_1}^{m_2} \|\mathbf{u}_n\|_{k,p}^q \right)^{1/q}, & q \in [1, \infty), \\ \max_{m_1 \leq n \leq m_2} \|\mathbf{u}_n\|_{k,p}, & q = \infty, \end{cases} \tag{2.12}$$

for any $0 \leq n = m_1, m_1 + 1, \ldots, m_2 \leq M$. Write $\|\mathbf{u}\|_{l^q(W_p^k(\Omega))} = \|\mathbf{u}\|_{l^q(0,M;W_p^k(\Omega))}$. We say that $\mathbf{u} \in l^q(m_1, m_2; W_p^k(\Omega))$ if the associated norm defined above stays finite as $\Delta t \to 0$.

The discrete Gronwall inequality is essential to the convergence analysis in Section 4.2.

Lemma 2.2. *Let* $D \geq 0$ *and* $\kappa_n, A_n, B_n, C_n \geq 0$ *for any integer* $n \geq 0$ *and satisfy*

$$A_M + \Delta t \sum_{n=0}^{M} B_n \leq \Delta t \sum_{n=0}^{M} \kappa_n A_n + \Delta t \sum_{n=0}^{M} C_n + D, \quad \forall M \geq 0. \tag{2.13}$$

Convergence Analysis of a Fully Discrete Family of Iterated Deconvolution Methods for
Turbulence Modeling with Time Relaxation

7

Suppose that for all n, $\Delta t \kappa_n < 1$ and set $g_n = (1 - \Delta t \kappa_n)^{-1}$. Then,

$$A_M + \Delta t \sum_{n=0}^{M} B_n \leq \exp\left(\Delta t \sum_{n=0}^{M} g_n \kappa_n \right) \left[\Delta t \sum_{n=0}^{M} C_n + D \right], \quad \forall M \geq 0. \qquad (2.14)$$

Proof. The proof follows from [26]. $\qquad \square$

2.2. Approximation Theory

Let $C > 0$ be a generic constant independent of $h \to 0^+$. Preserving an abstract framework for the FE spaces, we assume that $X^h \times Q^h$ inherit several fundamental approximation properties.

Assumption 2.3. The FE spaces $X^h \times Q^h$ satisfy:

Uniform inf-sup (LBB) condition

$$\inf_{q^h \in Q^h} \sup_{\mathbf{v}^h \in X^h} \frac{(q^h, \nabla \cdot \mathbf{v}^h)}{|\mathbf{v}^h|_1 \|q\|} \geq C > 0. \qquad (2.15)$$

FE-approximation

$$\inf_{\mathbf{v}^h \in X^h} \left| \mathbf{u} - \mathbf{v}^h \right|_1 \leq C h^k \|\mathbf{u}\|_{k+1}, \quad \inf_{q^h \in Q^h} \left\| p - q^h \right\| \leq C h^{s+1} \|p\|_{s+1} \quad \text{for } k \geq 0, \ s \geq -1 \qquad (2.16)$$

when $\mathbf{u} \in H^{k+1}(\Omega) \cap X$, $p \in H^{s+1}(\Omega) \cap Q$.

Inverse-estimate

$$\left| \mathbf{v}^h \right|_1 \leq C h^{-1} \left\| \mathbf{v}^h \right\|, \quad \forall \mathbf{v}^h \in X^h. \qquad (2.17)$$

The well-known Taylor-Hood mixed FE is one such example satisfying Assumption 2.3.

Estimates in (2.18)–(2.20) stated below are used in proving error estimates for time-dependent problems: for any $n = 0, 1, \ldots, M - 1$,

$$\left\| \frac{\boldsymbol{\theta}_{n+1} - \boldsymbol{\theta}_n}{\Delta t} \right\|_k^2 \leq C \Delta t^{-1} \int_{t_n}^{t_{n+1}} \|\boldsymbol{\theta}_t(t)\|_k^2 dt, \qquad (2.18)$$

$$\|\boldsymbol{\theta}_{n+1/2} - \boldsymbol{\theta}(t_{n+1/2})\|_k^2 \leq C \Delta t^3 \int_{t_n}^{t_{n+1}} \|\boldsymbol{\theta}_{tt}(t)\|_k^2 dt, \qquad (2.19)$$

$$\left\| \frac{1}{\Delta t}(\boldsymbol{\theta}_{n+1} - \boldsymbol{\theta}_n) - (\boldsymbol{\theta}_t)_{n+1/2} \right\|_k^2 \leq C \Delta t^3 \int_{t_n}^{t_{n+1}} \|\boldsymbol{\theta}_{ttt}(t)\|_k^2 dt, \qquad (2.20)$$

where $\boldsymbol{\theta} \in H^1(H^k(\Omega))$, $\boldsymbol{\theta} \in H^2(H^k(\Omega))$, and $\boldsymbol{\theta} \in H^3(H^k(\Omega))$ is required, respectively, for some $k \geq -1$. Each estimate (2.18)–(2.20) is a result of a Taylor expansion with integral remainder.

These higher-order spatial ($k \geq 2$ or $s \geq 1$) and temporal estimates (2.18)–(2.20) require that the nonlocal compatibility condition addressed by Heywood and Rannacher in [26, 27] (and more recently, for example, by He in [28, 29] and He and Li in [30, 31]) is satisfied. Suppose, for example, that p_0 is the solution of the (well-posed) Neumann problem

$$\Delta p_0 = \nabla \cdot (\mathbf{f}_0 - \mathbf{u}_0 \cdot \nabla \mathbf{u}_0), \quad \text{in } \Omega,$$
$$\nabla p_0 \cdot \hat{\mathbf{n}}\big|_{\partial\Omega} = (\Delta \mathbf{u}_0 + \mathbf{f}_0 - \mathbf{u}_0 \cdot \nabla \mathbf{u}_0) \cdot \hat{\mathbf{n}}\big|_{\partial\Omega}. \tag{2.21}$$

In order to avoid the accompanying factor $\min\{t^{-1}, 1\}$ in the error estimates contained herein, the following compatibility condition is necessarily required (e.g., see [27, Corollary 2.1]):

$$\nabla p_0\big|_{\partial\Omega} = (\Delta \mathbf{u}_0 + \mathbf{f}_0 - \mathbf{u}_0 \cdot \nabla \mathbf{u}_0)\big|_{\partial\Omega}. \tag{2.22}$$

Replacing (2.21) with (2.21)(a), (2.22) defines an overdetermined Neumann-type problem. Condition (2.22) is a nonlocal condition relating $\mathbf{u}_0$ and $\mathbf{f}_0$. Condition (2.22) is satisfied for several practical applications including start up from rest with zero force, $\mathbf{u}_0 = 0$, $\mathbf{f}_0 = 0$. In general, however, condition (2.22) cannot be verified. In this case, it is shown that, for example, $|\mathbf{u}_t(\cdot, t)|_1, \|\mathbf{u}(\cdot, t)\|_3 \to \infty$ as $t \to 0^+$.

We finish with an approximation property of the L^2-projection. Indeed, Assumption 2.4 holds for smooth enough Ω.

Assumption 2.4. Fix $\mathbf{w} \in V$ and let $\mathbf{w}^h \in V^h$ be the unique solution satisfying $(\mathbf{w} - \mathbf{w}^h, \mathbf{v}^h)$ for all $\mathbf{v}^h \in V^h$. Then

$$\left\|\mathbf{w} - \mathbf{w}^h\right\|_{-m} \leq Ch^{m+1} \inf_{\mathbf{v}^h \in X^h} \left|\mathbf{w} - \mathbf{v}^h\right|_1 \tag{2.23}$$

for $m = -1, 0, 1$.

Note that the infimum in (2.23) is over all X^h (see intermediate estimate (1.16) of Theorem II.1.1 in [23] for the corresponding estimate relating the spaces V^h and X^h).

3. Filters and Deconvolution

We prescribe the essential properties our filter G and deconvolution operator D in this section.

Definition 3.1. Let Y be a Hilbert space and $T : Y \to Y$. Write $T \geq 0$ if T is self-adjoint $T = T'$ and $(T\mathbf{v}, \mathbf{v})_Y \geq 0$ for all $\mathbf{v} \in Y$ and call T symmetric nonnegative (snn). Write $T > 0$ if T is self-adjoint $T = T'$ and $(T\mathbf{v}, \mathbf{v})_Y > 0$ for all $0 \neq \mathbf{v} \in Y$ and call T symmetric positive definite (spd).

Let $G = G(\delta) > 0$ be a linear, bounded, compact operator on X representing a generic *smoothing filter* with filter radius $\delta > 0$:

$$G : L^2(\Omega) \longrightarrow L^2(\Omega), \qquad \bar{\phi} := G\phi. \tag{3.1}$$

Convergence Analysis of a Fully Discrete Family of Iterated Deconvolution Methods for
Turbulence Modeling with Time Relaxation

9

One example of this operator is the continuous differential filter $G = A^{-1} = (-\delta^2 \Delta + I)^{-1}$ (Definition 3.2), which is used, together with its discrete counterpart $(A^h)^{-1}$ (Definition 3.3), for implementation of our numerical scheme (Section 5.2).

Definition 3.2 (continuous differential filter). Fix $\phi \in L^2(\Omega)$. Then $\overline{\phi} \in X$ is the unique solution of $-\delta^2 \Delta \overline{\phi} + \overline{\phi} = \phi$ with corresponding weak formulation

$$\delta^2 \left(\nabla \overline{\phi}, \nabla \mathbf{v} \right) + \left(\overline{\phi}, \mathbf{v} \right) = (\phi, \mathbf{v}), \quad \forall \mathbf{v} \in X. \tag{3.2}$$

Set $A = -\delta^2 \Delta + I$ so that $A^{-1} : L^2(\Omega) \rightarrow X$, defined by $\overline{\phi} := A^{-1}\phi$, is well defined.

Definition 3.3 (discrete differential filter). Fix $\phi \in L^2(\Omega)$. Then $\overline{\phi}^h \in X^h$ is the unique solution of the following

$$\delta^2 \left(\nabla \overline{\phi}^h, \nabla \mathbf{v}^h \right) + \left(\overline{\phi}^h, \mathbf{v}^h \right) = \left(\phi, \mathbf{v}^h \right), \quad \forall \mathbf{v}^h \in X^h. \tag{3.3}$$

Set $A^h = -\delta^2 \Delta^h + \Pi^h$ so that $(A^h)^{-1} : L^2(\Omega) \rightarrow X^h$, defined by $\overline{\phi}^h := (A^h)^{-1}\phi$, is well defined. Here, $\Pi^h : L^2(\Omega) \rightarrow X^h$ is the L^2 projection and $\Delta^h : X \rightarrow X^h$ the discrete Laplace operator satisfying the following

$$\left(\Pi^h \phi - \phi, \mathbf{v}^h \right) = 0, \qquad \left(\Delta^h \phi, \mathbf{v}^h \right) = -\left(\nabla \phi, \nabla \mathbf{v}^h \right) \quad \forall \mathbf{v}^h \in X^h. \tag{3.4}$$

It is well known that A^{-1} and $(A^h)^{-1}$ are each linear and bounded, A^{-1} is compact, and the spectrum of A and A^h (on X and X^h, resp.) is contained in $[1, \infty)$ and spectrum of A^{-1} and $(A^h)^{-1}$ (on X and X^h, resp.) is contained in $(0, 1]$ so that

$$A^{-1} > 0 \quad \text{on } X, \qquad \left(A^h \right)^{-1} > 0 \quad \text{on } X^h. \tag{3.5}$$

For more detailed exposition on these operators, see [13].

3.1. A Family of Deconvolution Operators

We analyze (1.6) for stable, accurate deconvolution D of the smoothing filter G introduced in Section 3 so that $DG\mathbf{u}$ accurately approximates the smooth parts of $\mathbf{u}$.

Assumption 3.4 (continuous deconvolution operator). Suppose that $D : X \rightarrow X$ is linear, bounded, spd, and commutes with G so that

$$\|DG\| \leq 1, \qquad |D\overline{\mathbf{v}}|_1 \leq d_1 |\mathbf{v}|_1 \quad \forall \mathbf{v} \in X, \tag{3.6}$$

for some constant $d_1 > 0$. Moreover, suppose that D is parametrized by $\alpha > 0$, $\delta > 0$ so that

$$\|(I - DG)\mathbf{v}\| \leq c_1(\delta, \alpha)\|\Delta \mathbf{v}\| \longrightarrow 0, \quad \text{as } \alpha, \delta \longrightarrow 0 \tag{3.7}$$

for smooth enough $\mathbf{v} \in X$.

Note that the first estimate in (3.6) is required so that the spectral radius satisfies $\rho(DG) \leq 1$. The second estimate in (3.6) (which controls the H^1-seminorm of DG) is required for the convergence analysis in Section 4.2.

Assumption 3.5 prescribes properties of the discrete analogue $D^h : X^h \to X^h$ corresponding to the continuous deconvolution operator $D : X \to X$ (Assumptions 3.4).

Assumption 3.5 (discrete deconvolution operator). Let D satisfy Assumption 3.4. Let $G^h : X^h \to X^h$ be a discrete analogue of G that is linear, bounded, spd. Suppose that $D^h : X^h \to X^h$ is linear, bounded, spd, and commutes with G^h such that

$$\left\|D^h G^h\right\| \leq 1, \qquad \left|D^h \bar{\mathbf{v}}^h\right|_1 \leq d_1 |\mathbf{v}|_1 \quad \forall \mathbf{v} \in X, \tag{3.8}$$

for some constant $d_1 > 0$. Moreover, suppose that D^h is parametrized by $\alpha > 0$, $\delta > 0$, h such that $D = D(h, \delta, \alpha)$ and

$$\left\|\left(DG - D^h G^h\right)\mathbf{v}\right\| \leq c_2(h, \delta, \alpha)\|\mathbf{v}\|_{k+1} \longrightarrow 0, \quad \text{as } h, \delta, \alpha \longrightarrow 0, \tag{3.9}$$

for all $\mathbf{v} \in X \cap H^{k+1}(\Omega)$ for some $k \geq 0$.

The estimates in (3.8) are motivated by the continuous case of (3.6). The approximation (3.9) is required for the convergence analysis in Section 4.2 (see Theorem 4.6, Corollary 4.7).

Remark 3.6. If $D = f(G)$ for some continuous map $f : \mathbb{R} \to \mathbb{R}$, then commutativity is satisfied $DG = GD$. Tikhonov-Lavrentiev (modified) regularization with $G = A^{-1}$, $G^h = (A^h)^{-1}$ given by $D = ((1 - \alpha)A^{-1} + \alpha I)^{-1}$, $D^h = ((1 - \alpha)(A^h)^{-1} + \alpha \Pi^h)^{-1}$ is one such example with $f(x) = ((1 - \alpha)x + \alpha)^{-1}$ and $d_1 = 1$, $c_1 = \alpha \delta^2$, $c_2 = \alpha \delta^2 h^k + h^{k+1}$, see [13].

Remark 3.7. Letting $\lambda_k(\cdot)$ denote the kth (ordered) eigenvalue of a given operator, commutativity of D and G provides $\lambda_k(DG) = \lambda_k(D)\lambda_k(G)$ and similarly for the discrete operator $D^h G^h$.

We next derive several important consequences of D and D^h under Assumptions 3.4, 3.5 required in the forthcoming analysis.

Lemma 3.8. *Suppose that G, G^h, D, D^h satisfy Assumptions 3.4, 3.5. Then,*

$$\|D\bar{\mathbf{v}}\| \leq \|\mathbf{v}\|, \qquad \left\|D^h \bar{\mathbf{v}}^h\right\| \leq \|\mathbf{v}\| \quad \forall \mathbf{v} \in L^2(\Omega). \tag{3.10}$$

Proof. For the continuous operator,

$$\|D\bar{\mathbf{v}}\| \leq \|DG\|\|\mathbf{v}\|. \tag{3.11}$$

Convergence Analysis of a Fully Discrete Family of Iterated Deconvolution Methods for
Turbulence Modeling with Time Relaxation

11

Then, (3.10)(a) follows from Assumption 3.4, and (3.10)(b) is derived similarly applying Assumption 3.5 instead. □

Lemma 3.9. *Suppose that G, G^h, D, D^h satisfy Assumptions 3.4, 3.5. Then, the spectrum of both DG and $D^h G^h$ are contained in $[0, 1]$ so that*

$$\|I - DG\| \leq 1, \qquad \left\|I - D^h G^h\right\| \leq 1. \tag{3.12}$$

As a consequence,

$$
\left.
\begin{aligned}
\|\mathbf{v}\|_\star^2 &:= (\mathbf{v} - D\bar{\mathbf{v}}, \mathbf{v}) \\
\|\mathbf{v}\|_{\star h}^2 &:= \left(\mathbf{v} - D^h \bar{\mathbf{v}}^h, \mathbf{v}\right)
\end{aligned}
\right\} \geq 0, \quad \forall \mathbf{v} \in L^2(\Omega). \tag{3.13}
$$

Proof. Assumptions 3.4, 3.5 guarantee that the spectral radius $\rho(DG) \leq 1$ and $\rho(D^h G^h) \leq 1$. Also, $D > 0$ and $G > 0$ and commute so that $DG \geq 0$. Similarly, $D^h G^h \geq 0$. Therefore, the spectrum of DG, $D^h G^h \geq 0$ are each contained in $[0, 1]$. So, $I - DG$, $I - D^h G^h \geq 0$ have spectrum contained in $[0, 1]$ which ensures the non-negativity of both $\|\cdot\|_\star$ and $\|\cdot\|_{\star h}$. □

3.2. Iterated Deconvolution

One can show, by eliminating intermediate steps in the definition of the iterated regularization operator D_j in (1.9) with base operator D satisfying Assumption 3.4, that

$$D_j = D \sum_{i=0}^{j} (FD)^i, \qquad F := D^{-1} - G. \tag{3.14}$$

Similarly, the discrete iterated regularization operator D_j^h with discrete base operator D^h satisfying Assumption 3.5, is given by the following

$$D_j^h = D^h \sum_{i=0}^{j} \left(F^h D^h\right)^i, \qquad F^h := \left(D^h\right)^{-1} - G^h. \tag{3.15}$$

We next show that D_j and D_j^h for $j > 0$ inherit several important properties from D and D^h, respectively, via Assumption 3.5.

Proposition 3.10. *Fix $j \in \mathbb{N}$. Then $D_j : X \rightarrow X$ defined by (3.14) satisfies Assumption 3.4. In particular, $D_j > 0$ is linear, bounded, commutes with G and satisfies (3.6)(a). Estimate (3.6)(b) is replaced by the following*

$$\left|D_j \bar{\mathbf{v}}\right|_1 \leq d_{1,j} |\mathbf{v}|_1 \quad \forall \mathbf{v} \in X \tag{3.16}$$

for some constant $d_{1,j} > 0$. Estimate (3.7) is replaced by the following

$$\left\|(I - D_jG)\mathbf{v}\right\| \le c_{1,j}(\delta, \alpha)\|\Delta\mathbf{v}\|^2 \longrightarrow 0, \quad as \; \alpha, \delta \longrightarrow 0. \tag{3.17}$$

Moreover, $d_{1,j} \le \sum_{i=0}^{j} d_1^i$ and $c_{1,j} \le \sum_{i=0}^{j} c_1^i$.

Proof. First notice that D_j is linear and bounded since it is a linear combination of linear and bounded operators $D(FD)^i = D(I-DG)^i$, for $i = 0, 1, \ldots, j$. Moreover, since G commutes with D, it follows that G commutes with $D(I - DG)^i$ and hence with D_j. Next, D_j is a sum of spd and snn operators $D > 0$, $D(I - DG)^i \ge 0$. Hence, $D_j > 0$. Next, notice that

$$D_jG = \left(\sum_{i=0}^{j}(I - DG)^i\right)DG = \left(I + (I - DG) + \cdots + (I - DG)^j\right)DG. \tag{3.18}$$

Letting $\lambda_k(\cdot)$ denote the kth (ordered) eigenvalue of a given operator, we can characterize the spectrum of D_j by summing the resulting finite geometric series (3.18) to get

$$\lambda_k(D_jG) = \lambda_k(D)\lambda_k(G)\sum_{i=0}^{j}(1 - \lambda_k(DG))^i = \left(1 - (1 - \lambda_k(DG))^{j+1}\right). \tag{3.19}$$

Then under Assumption 3.4, Lemma 3.9 with (3.19) implies that $0 \le \lambda_k(D_jG) \le \|D_jG\| \le 1$. Hence, D_j satisfies (3.6)(a). Expanding the terms in (3.18) as powers of DG, we see that (3.18) can be written as a polynomial (with coefficients a_i) in DG, so that

$$\nabla D_j\bar{\mathbf{v}} = \sum_{i=0}^{j}a_i\nabla(DG)^i\mathbf{v}, \qquad |D_j\bar{\mathbf{v}}|_1 \le \sum_{i=0}^{j}d_1^i|\mathbf{v}|_1, \tag{3.20}$$

since $|D\bar{\mathbf{v}}|_1 \le d_1|\mathbf{v}|_1$ can be applied successfully. Therefore (3.16) follows with $d_{1,j} = \sum_{i=0}^{j} d_1^i$. Next, start with (3.18) to get

$$\left\|(I - D_jG)\mathbf{v}\right\| = \left\|\left((I - DG)\mathbf{v} + DG(I - DG)\mathbf{v} + \cdots + DG(I - DG)^j\mathbf{v}\right)\right\|$$

$$\le \|(I - DG)\mathbf{v}\| + \|DG\|\|(I - DG)\mathbf{v}\| + \cdots + \|DG\|\|I - DG\|^{j-1}\|(I - DG)\mathbf{v}\|. \tag{3.21}$$

Estimate (3.17) follows by noting $\|DG\| \le 1$, $\|I - DG\| \le 1$, and by Assumption 3.5, $\|(I - DG)\mathbf{v}\| \le c_1\|\Delta\mathbf{v}\|$. $\square$

Proposition 3.11. *Fix $j \in \mathbb{N}$. Then $D_j^h : X^h \to X^h$ defined by (3.15) satisfies Assumption 3.5. In particular, $D_j^h > 0$ is linear, bounded, commutes with G^h and satisfies (3.8)(a). Estimate (3.8)(b) is replaced by the following*

$$\left|D_j^h\bar{\mathbf{v}}^h\right|_1 \le d_{1,j}|\mathbf{v}|_1 \quad \forall\mathbf{v} \in X \tag{3.22}$$

Convergence Analysis of a Fully Discrete Family of Iterated Deconvolution Methods for
Turbulence Modeling with Time Relaxation

13

for some constant $d_{1,j} > 0$. Estimate (3.9) is replaced by the following

$$\left\|\left(D_j G - D_j^h G^h\right)\mathbf{v}\right\| \leq c_{2,j}(h,\delta,\alpha)\|\mathbf{v}\|_{k+1} \longrightarrow 0, \quad as \ h,\delta,\alpha \longrightarrow 0 \qquad (3.23)$$

for any $\mathbf{v} \in X \cap H^{k+1}(\Omega)$ for some $k \geq 0$. Moreover, $c_{2,j} \leq \beta(j)c_2$ for some constant $\beta = \beta(j) > 0$.

Proof. The first two assertions follow similarly as in the previous proof of Proposition 3.10. To prove (3.23), we start by writing

$$D_j^h G^h = \left(\sum_{i=0}^{j}\left(I - D^h G^h\right)^i\right)D^h G^h, \qquad (3.24)$$

and then subtract (3.24) from (3.18) to get

$$D_j G - D_j^h G^h = \Lambda_j\left(DG - D^h G^h\right) + \left(\Lambda_j - \Lambda_j^h\right)D^h G^h, \qquad (3.25)$$

where

$$\Lambda_j = \sum_{i=0}^{j}(I - DG)^i, \qquad \Lambda_j^h = \sum_{i=0}^{j}\left(I - D^h G^h\right)^i. \qquad (3.26)$$

Then taking norms across (3.25), we get

$$\left\|\left(D_j G - D_j^h G^h\right)\mathbf{v}\right\| = \|\Lambda_j\|\left\|\left(DG - D^h G^h\right)\mathbf{v}\right\| + \left\|D^h G^h\right\|\left\|\left(\Lambda_j - \Lambda_j^h\right)\mathbf{v}\right\|. \qquad (3.27)$$

Notice that $\|I - DG\| \leq 1$ so that $\|\Lambda_j\| \leq j + 1$. Moreover, $\|(DG - D^h G^h)\mathbf{v}\| \leq c_2\|\mathbf{v}\|_{k+1}$ via Assumption 3.5. Next, using the binomial theorem and factoring, we get

$$\begin{aligned}
\left\|\left(\Lambda_j - \Lambda_j^h\right)\mathbf{v}\right\| &= \left\|\sum_{i=0}^{j}\frac{j!}{i!(j-i)!}(-1)^i\left[(DG)^i - \left(D^h G^h\right)^i\mathbf{v}\right]\right\| \\
&= \left\|\sum_{i=0}^{j}\frac{j!}{i!(j-i)!}(-1)^i\left[\sum_{n=0}^{i}(DG)^n\left(D^h G^h\right)_{n-i}\right]\left(DG - D^h G^h\right)\mathbf{v}\right\|.
\end{aligned} \qquad (3.28)$$

Then, applying $\|DG\| \leq 1$, $\|D^h G^h\| \leq 1$ to (3.28) provides

$$\left\|\left(\Lambda_j - \Lambda_j^h\right)\mathbf{v}\right\| = \left(\sum_{i=0}^{j}\frac{j!i}{(i)!(j-i)!}\right)\left\|\left(DG - D^h G^h\right)\mathbf{v}\right\|. \qquad (3.29)$$

Again, $\|(DG - D^h G^h)\| \leq c_2\|\mathbf{v}\|_{k+1}$ via Assumption 3.5. So, we combine these above results to conclude (3.23) with $\beta(j) = \sum_{i=0}^{j} j!i/(i)!(j-i)!$. $\qquad\square$

3.3. Tikhonov-Lavrentiev Regularization

We provide two examples of discrete deconvolution operators D^h to make the abstract formulation in the previous section more concrete. The Tikhonov-Lavrentiev and modified Tikhonov-Lavrentiev operator (for linear, compact $G > 0$) is given by the following

$$\text{Tikhonov-Lavrentiev} \implies D_{\alpha,0} = (G + \alpha I)^{-1}$$

$$\text{modified Tikhonov-Lavrentiev} \implies D_{\alpha,0} = ((1 - \alpha)G + \alpha I)^{-1}.$$

$$\tag{3.30}$$

Definition 3.12 ((weak) modified Tikhonov-Lavrentiev deconvolution). Fix $\alpha > 0$. Let $G = A^{-1}$. For any $\mathbf{w} \in X$, let $\boldsymbol{w}_0 := D_{\alpha,0}\overline{\mathbf{w}} \in X$ be the unique solution of

$$\alpha\delta^2(\nabla\boldsymbol{w}_0, \nabla\mathbf{v}) + (\boldsymbol{w}_0, \mathbf{v}) = (\mathbf{w}, \mathbf{v}), \quad \forall \mathbf{v} \in X. \tag{3.31}$$

Definition 3.13 ((discrete) modified Tikhonov-Lavrentiev deconvolution). Fix $\alpha > 0$. Let $G^h = (A^h)^{-1}$ and $D^h_{\alpha,0} = ((1 - \alpha)(A^h)^{-1} + \alpha\Pi^h)^{-1}$. For any $\mathbf{w} \in X$, let $\boldsymbol{w}^h_0 := D^h_{\alpha,0}\overline{\mathbf{w}}^h \in X^h$ be the unique solution of

$$\alpha\delta^2\left(\nabla\boldsymbol{w}^h_0, \nabla\mathbf{v}^h\right) + \left(\boldsymbol{w}^h_0, \mathbf{v}^h\right) = \left(\mathbf{w}, \mathbf{v}^h\right), \quad \forall \mathbf{v}^h \in X^h. \tag{3.32}$$

The iterated modified Tikhonov-Lavrentiev operator (for linear, compact $G > 0$) is obtained from the Tikhonov-Lavrentiev operator with updates via (1.9):

$$\text{Iterated Tikhonov-Lavrentiev} \implies D_{\alpha,j} = D_{\alpha,0}\sum_{i=0}^{j}(\alpha D_{\alpha,0})^i,$$

$$\tag{3.33}$$

$$\text{Iterated modified Tikhonov-Lavrentiev} \implies D_{\alpha,j} = D_{\alpha,0}\sum_{i=0}^{j}(\alpha(I - G)D_{\alpha,0})^i.$$

Definition 3.14 (iterated modified Tikhonov-Lavrentiev deconvolution (weak)). Fix $\alpha > 0$ and $J \in \mathbb{N}$. Let $G = A^{-1}$. Define $\boldsymbol{w}_{-1} = 0$, then for any $\mathbf{w} \in X$ and $j = 0, 1, \ldots, J$, let $\boldsymbol{w}_j := D_{\alpha,j}\overline{\mathbf{w}} \in X$ be the unique solution of

$$\alpha\delta^2(\nabla\boldsymbol{w}_j, \nabla\mathbf{v}) + (\boldsymbol{w}_j, \mathbf{v}) = (\mathbf{w}, \mathbf{v}) + \alpha\delta^2(\nabla\boldsymbol{w}_{j-1}, \nabla\mathbf{v}), \quad \forall \mathbf{v} \in X. \tag{3.34}$$

Definition 3.15 (iterated modified Tikhonov-Lavrentiev deconvolution (discrete)). Fix $\alpha > 0$ and $J \in \mathbb{N}$. Let $G^h = (A^h)^{-1}$, and $D^h_{\alpha,j} = D^h_{\alpha,0}\sum_{i=0}^{j}(\alpha(\Pi^h - (A^h)^{-1})D^h_{\alpha,0})^i$. Define $\boldsymbol{w}^h_{-1} = 0$, then for any $\mathbf{w} \in X$ and $j = 0, 1, \ldots, J$, let $\boldsymbol{w}^h_j := D^h_{\alpha,j}\overline{\mathbf{w}}^h \in X^h$ be the unique solution of

$$\alpha\delta^2\left(\nabla\boldsymbol{w}^h_j, \nabla\mathbf{v}^h\right) + \left(\boldsymbol{w}^h_j, \mathbf{v}^h\right) = \left(\mathbf{w}, \mathbf{v}^h\right) + \alpha\delta^2\left(\nabla\boldsymbol{w}^h_{j-1}, \nabla\mathbf{v}^h\right), \quad \forall \mathbf{v}^h \in X^h. \tag{3.35}$$

Convergence Analysis of a Fully Discrete Family of Iterated Deconvolution Methods for
Turbulence Modeling with Time Relaxation

15

4. Well Posedness of the Fully Discrete Model

We now state the proposed algorithm.

Problem 1 (CNFE for Leray-deconvolution). Let $(\mathbf{w}_0, \pi_0) \in (X^h, Q^h)$. Then, for each $n = 0, 1, \ldots, M-1$, find $(\mathbf{w}_{n+1}^h, \pi_{n+1}^h) \in (X^h, Q^h)$ satisfying

$$\frac{1}{\Delta t}\left(\mathbf{w}_{n+1}^h - \mathbf{w}_n^h, \mathbf{v}^h\right) + b^h\left(\varphi^h\left(\mathbf{w}_{n+1/2}^h\right), \mathbf{w}_{n+1/2}^h, \mathbf{v}^h\right) - \left(\pi_{n+1/2}^h, \nabla \cdot \mathbf{v}^h\right)$$
$$+ \operatorname{Re}^{-1}\left(\nabla \mathbf{w}_{n+1/2}^h, \nabla \mathbf{v}^h\right) + \chi\left(\mathbf{w}_{n+1/2}^h - \varphi^h\left(\mathbf{w}_{n+1/2}^h\right), \mathbf{v}^h\right) = \left(\mathbf{f}_{n+1/2}, \mathbf{v}^h\right), \quad \forall \mathbf{v}^h \in X^h,$$
$$\tag{4.1}$$

$$\left(\nabla \cdot \mathbf{w}_{n+1}^h, q^h\right) = 0, \quad \forall q^h \in Q^h, \tag{4.2}$$

where $\varphi^h(\mathbf{w}_{n+1/2}^h) = \overline{D^h \mathbf{w}_{n+1/2}^h}^h$.

Notice that $(q_{n+1/2}^h, \nabla \cdot \mathbf{v}^h) = 0$ when $\mathbf{v}^h \in V^h$ so that the problem of finding $\mathbf{w}_{n+1}^h \in V^h$ satisfying

$$\frac{1}{\Delta t}\left(\mathbf{w}_{n+1}^h - \mathbf{w}_n^h, \mathbf{v}^h\right) + b^h\left(\varphi^h\left(\mathbf{w}_{n+1/2}^h\right), \mathbf{w}_{n+1/2}^h, \mathbf{v}^h\right) + \operatorname{Re}^{-1}\left(\nabla \mathbf{w}_{n+1/2}^h, \nabla \mathbf{v}^h\right)$$
$$+ \chi\left(\mathbf{w}_{n+1/2}^h - \varphi^h\left(\mathbf{w}_{n+1/2}^h\right), \mathbf{v}^h\right) = \left(\mathbf{f}_{n+1/2}, \mathbf{v}^h\right), \quad \forall \mathbf{v}^h \in V^h. \tag{4.3}$$

4.1. Well Posedness

We establish existence of $\mathbf{w}$ at each time step of (4.3) by Leray-Schauder's fixed-point theorem.

Lemma 4.1. *Let*

$$a\left(\boldsymbol{\theta}^h, \mathbf{v}^h\right) = \frac{\Delta t}{2 \operatorname{Re}}\left(\nabla \boldsymbol{\theta}^h, \nabla \mathbf{v}^h\right) + \frac{\chi \Delta t}{2}\left(\boldsymbol{\theta}^h - \varphi^h\left(\boldsymbol{\theta}^h\right), \mathbf{v}^h\right),$$
$$l_y\left(\mathbf{v}^h\right) = \left(\mathbf{y}, \mathbf{v}^h\right). \tag{4.4}$$

for any $\mathbf{y} \in X'$ *and* $\boldsymbol{\theta}^h, \mathbf{v}^h \in V^h$. *Suppose that* D^h *satisfies Assumption 3.5. Then* $a(\cdot, \cdot) : V^h \times V^h \to \mathbb{R}$ *is a continuous and coercive bilinear form and* $l_y(\cdot) : V^h \to \mathbb{R}$ *is a linear, continuous functional.*

Proof. Linearity for $l_y(\cdot)$ is obvious, and continuity follows from an application of Hölder's inequality. Continuity for $a(\cdot, \cdot)$ also follows from Hölder's inequality and Assumption 3.5. Coercivity is proven by application of (3.13). □

Lemma 4.2. *Let* $T : X' \to V^h$ *be such that, for any* $\mathbf{y} \in X'$, $\boldsymbol{\theta}^h := T(\mathbf{y})$ *solves*

$$a\left(\boldsymbol{\theta}^h, \mathbf{v}^h\right) = l_y\left(\mathbf{v}^h\right), \quad \forall \mathbf{v}^h \in V^h. \tag{4.5}$$

Then T *is a well-defined, linear, bounded operator.*

Proof. Linearity is clear. The results of Lemma 4.1, and the Lax-Milgram theorem prove the rest. □

Lemma 4.3. *Fix* $n = 0, 1, \ldots, M - 1$. *Let* $\mathbf{w}_n^h$ *be a solution of Problem 1 and let* $N : V^h \to X'$ *satisfy, for any* $\boldsymbol{\theta}^h \in V^h$,

$$
\left(N\left(\boldsymbol{\theta}^h\right), \mathbf{v}^h \right) = -\left(\boldsymbol{\theta}^h - 2\mathbf{w}_n^h, \mathbf{v}^h \right) - \frac{\Delta t}{4} b^h\left(\varphi^h\left(\boldsymbol{\theta}^h\right), \boldsymbol{\theta}^h, \mathbf{v}^h \right) + \Delta t\left(\mathbf{f}_{n+1/2}, \mathbf{v}^h \right)
$$

$$
=: c\left(\boldsymbol{\theta}^h, \mathbf{v}^h \right), \quad \forall \mathbf{v}^h \in V^h.
$$

(4.6)

Then $N(\boldsymbol{\theta}^h)$ *is well-defined, bounded, and continuous.*

Proof. For each $\boldsymbol{\theta}^h \in V^h$, the map $\mathbf{v}^h \in V^h \mapsto c(\boldsymbol{\theta}^h, \mathbf{v}^h)$ is a bounded, linear functional (apply Hölder's inequality and (2.11)). Since V^h is a Hilbert space, we conclude that $N(\boldsymbol{\theta}^h)$ is well defined, by the Riesz-Representation theorem. Moreover, $N(\boldsymbol{\theta}^h)$ is bounded on V^h and since the underlying function space is finite dimensional, continuity follows. □

Lemma 4.4. *Fix* $n \in \mathbb{N}$. *Let* $F : V^h \to V^h$ *be defined such that* $F(\boldsymbol{\theta}^h) = (T \circ N)(\boldsymbol{\theta}^h)$. *Then,* F *is a compact operator.*

Proof. $N(\cdot)$ is a compact operator (continuous on a finite dimensional function space). Thus, F is a continuous composition of a compact operator and hence compact itself. □

Theorem 4.5 (well posedness). *Fix* $n = 0, 1, 2, \ldots, M - 1 < \infty$. *There exists* $(\mathbf{w}_n^h, \pi_n^h) \in X^h \times Q^h$ *satisfying Problem 1. Moreover,*

$$
\left\| \mathbf{w}_m^h \right\|^2 + \frac{1}{2\,\mathrm{Re}} \Delta t \sum_{n=0}^{m-1} \left| \mathbf{w}_{n+1/2}^h \right|_1^2 + \chi \Delta t \sum_{n=0}^{m-1} \left\| \mathbf{w}_{n+1/2}^h \right\|_{\star h}^2 \leq \left\| \mathbf{w}_0^h \right\|^2 + \frac{\Delta t\,\mathrm{Re}}{2} \sum_{n=0}^{m-1} \|\mathbf{f}_{n+1/2}\|_{-1}^2,
$$

(4.7)

for all integers $1 \leq m \leq M$, *independent of* $\Delta t > 0$.

Proof. First, assume that $(\mathbf{w}_{n+1}^h, q_{n+1}^h)$ is a solution to (4.1), (4.2). Set $\mathbf{v}^h = \mathbf{w}_{n+1/2}^h$ in (4.1) so that skew-symmetry of the nonlinear term provides

$$
\frac{1}{2\Delta t}\left(\left\| \mathbf{w}_{n+1}^h \right\|^2 - \left\| \mathbf{w}_m^h \right\|^2 \right) + \mathrm{Re}^{-1} \left| \mathbf{w}_{n+1/2}^h \right|_1^2
$$

$$
+ \chi\left(\mathbf{w}_{n+1/2}^h - \varphi^h\left(\mathbf{w}_{n+1/2}^h \right), \mathbf{w}_{n+1/2}^h \right) = \left(\mathbf{f}_{n+1/2}, \mathbf{w}_{n+1/2}^h \right).
$$

(4.8)

Duality of $X' \times X$ with Young's inequality implies

$$
\left(\mathbf{f}_{n+1/2}, \mathbf{w}_{n+1/2}^h \right) \leq \frac{\mathrm{Re}}{2} \|\mathbf{f}_{n+1/2}\|_{-1}^2 + \frac{\mathrm{Re}}{2} \left| \mathbf{w}_{n+1/2}^h \right|_1^2.
$$

(4.9)

Convergence Analysis of a Fully Discrete Family of Iterated Deconvolution Methods for
Turbulence Modeling with Time Relaxation

17

From (3.13), we have

$$\left\| \mathbf{w}_{n+1/2}^h \right\|_{\star h}^2 = \left(\mathbf{w}_{n+1/2}^h - \varphi^h \left(\mathbf{w}_{n+1/2}^h \right), \mathbf{w}_{n+1/2}^h \right) \geq 0. \tag{4.10}$$

Then applying (4.9), (4.10) to (4.8), combining-like terms and simplifying provides

$$\frac{1}{2\Delta t} \left(\left\| \mathbf{w}_{n+1}^h \right\|^2 - \left\| \mathbf{w}_n^h \right\|^2 \right) + \frac{\mathrm{Re}}{2} \left| \mathbf{w}_{n+1/2}^h \right|_1^2 + \chi \left\| \mathbf{w}_{n+1/2}^h \right\|_{\star h}^2 \leq \frac{\mathrm{Re}}{2} \| \mathbf{f}_{n+1/2} \|_{-1}^2. \tag{4.11}$$

Summing from $n = 0$ to $m - 1$, we get the desired bound.

Next, let $\mathbf{W}_n^h = \mathbf{w}_{n+1}^h + \mathbf{w}_n^h$. Showing that $\mathbf{W}_n^h = F(\mathbf{W}_n^h)$ has a fixed point will ensure existence of solutions to (4.3). Indeed, if we can show that $\mathbf{W}_0^h = F(\mathbf{W}_0^h)$, then since $\mathbf{w}_0^h$ is given initial data, existence of $\mathbf{w}_1^h$ is immediate. Induction can be applied to prove existence of $(\mathbf{w}_n^h)_{1 \leq n \leq M}$. To this end, since F is compact, it is enough to show (via Leray Schauder) that any solution $\mathbf{W}_{n,\lambda}^h$ of the fixed-point problem $\mathbf{W}_{n,\lambda}^h = \lambda F(\mathbf{W}_{n,\lambda}^h)$ is uniformly bounded with respect to $0 \leq \lambda \leq 1$. Hence, we consider

$$a \left(\mathbf{W}_{n,\lambda}^h, \mathbf{v}^h \right) = \lambda \left(N \left(\mathbf{W}_{n,\lambda}^h \right), \mathbf{v}^h \right). \tag{4.12}$$

Test with $\mathbf{v}^h = \mathbf{W}_{n,\lambda}^h$, use skew-symmetry of the trilinear form and properties of D^h given in Assumption 3.5 and (3.13) to get

$$\lambda \left\| \mathbf{W}_{n,\lambda}^h \right\|^2 + \frac{\Delta t}{2\,\mathrm{Re}} \left| \mathbf{W}_{n,\lambda}^h \right|_1^2 + \frac{\chi \Delta t}{2} \left\| \mathbf{W}_{n,\lambda}^h \right\|_{\star h}^2 \leq 2\lambda \left(\mathbf{w}_n^h, \mathbf{W}_{n,\lambda}^h \right) + \lambda \Delta t \left(\mathbf{f}_{n+1/2}, \mathbf{W}_{n,\lambda}^h \right). \tag{4.13}$$

Duality of $X' \times X$ followed by Young's inequality implies

$$\lambda \Delta t \left(\mathbf{f}_{n+1/2}, \mathbf{W}_{n,\lambda}^h \right) \leq \Delta t \, \mathrm{Re} \, \| \mathbf{f}_{n+1/2} \|_{-1}^2 + \frac{\Delta t}{4\,\mathrm{Re}} \left| \mathbf{W}_{n,\lambda}^h \right|_1^2. \tag{4.14}$$

Since $\mathbf{w}_n^h \in L^2(\Omega)$ from the a priori estimate (4.7), we apply Hölder's and Young's inequalities to get

$$2\lambda \left(\mathbf{w}_n^h, \mathbf{W}_{n,\lambda}^h \right) \leq 2 \left\| \mathbf{w}_n^h \right\|^2 + \frac{\lambda}{2} \left\| \mathbf{W}_{n,\lambda}^h \right\|^2. \tag{4.15}$$

Applying estimates (4.14), (4.15) to (4.13) we get that $|\mathbf{W}_{n,\lambda}^h|_1 \leq C < \infty$ independent of λ. By the Leray-Schauder fixed point theorem, given $\mathbf{w}_n^h$, there exists a solution to the fixed-point theorem $\mathbf{W}_n^h = F(\mathbf{W}_n^h)$. By the induction argument noted above, there exists a solution $\mathbf{w}_n^h$ for each $n = 0, 1, 2, \ldots, M - 1$ to (4.3). Existence of an associated discrete pressure follows by a classical argument, since the pair (X^h, Q^h) satisfies the discrete inf-sup condition (2.15). $\quad \square$

4.2. Convergence Analysis

Under usual regularity assumptions, we summarize the main convergence estimate in Theorem 4.6. Suppose that D represents deconvolution with J-updates.

Theorem 4.6. *Suppose that $(\mathbf{u}, p)$ are strong solutions to (2.3), (2.4), (2.5) and that G, G^h, D, D^h satisfy Assumptions 3.4, 3.5. Suppose further that $\mathbf{u} \in l^2(H^2(\Omega) \cap V) \cap l^\infty(X)$, $\mathbf{u}_t \in l^2(X') \cap L^2(X')$, $\Delta^{\beta+1}\mathbf{u} \in L^2(\Omega)$ for some $0 \le \beta \le J+1$, $p \in l^2(Q)$. If*

$$C \operatorname{Re} \Delta t \|\mathbf{u}_n\|_2^2 < 1, \quad \forall n = 0, 1, \ldots, M \tag{4.16}$$

then,

$$\left\| \mathbf{u}_M - \mathbf{w}_M^h \right\|^2 + \operatorname{Re}^{-1}\Delta t \sum_{n=0}^{M-1} \left| \mathbf{u}_{n+1/2} - \mathbf{w}_{n+1/2}^h \right|_1^2$$

$$\le C \Bigg(\left\| \mathbf{u}_0 - \mathbf{w}_0^h \right\|^2 + E + \|p\|_{l^2(L^2(\Omega))}^2 + \left(\|\nabla \mathbf{u}\|_{l^\infty(L^2(\Omega))}^2 + \chi^2 \right) \|\mathbf{u}\|_{l^2(H^1(\Omega))}^2$$

$$+ \|\mathbf{u}\|_{l^\infty(L^2(\Omega))}^2 + \|\mathbf{u}\|_{l^2(L^2(\Omega))}^2 + \left(\chi^2 + \|\mathbf{u}\|_{l^\infty(H^2(\Omega))}^2 \right) c_{1,J}^2 \left\| \Delta^\beta \mathbf{u} \right\|^2 + c_{2,j}^2 \|\nabla \mathbf{u}\|_{l^2(L^2(\Omega))}^2 \Bigg), \tag{4.17}$$

where $E > 0$ is given in (4.52).

Corollary 4.7 (convergence estimate). *Under the assumptions of Theorem 4.6, suppose further that $(\mathbf{u}, p)$ satisfy the assumptions for (2.16) for some $k \ge 1$ and $s \ge 0$, $\mathbf{u} \in l^\infty(H^k(\Omega)) \cap l^2(H^{k+1}(\Omega))$, $\mathbf{u}_t \in L^2(H^{k-1}(\Omega) \cap H^1(\Omega)) \cap l^\infty(H^1(\Omega))$, $\mathbf{u}_{tt} \in L^2(L^2(\Omega))$, $\mathbf{u}_{ttt} \in L^2(X')$, and $p \in l^2(H^{s+1}(\Omega))$. If $\|\mathbf{u}_0 - \mathbf{w}_0^h\| \le C(h^k + (\alpha\delta^2)^\beta)$ then*

$$\left\| \mathbf{u}_M - \mathbf{w}_M^h \right\|^2 + \operatorname{Re}^{-1}\Delta t \sum_{n=0}^{M-1} \left| \mathbf{u}_{n+1/2} - \mathbf{w}_{n+1/2}^h \right|_1^2 \tag{4.18}$$

$$\le C \left(h^{2k} + h^{2s+2} + \Delta t^4 + c_{1,J}(\delta, \alpha)^2 + c_{2,J}(h, \delta, \alpha)^2 \right).$$

Proof of Theorem 4.6, Corollary 4.7. Suppose that $\mathbf{u}$, p satisfying (2.3), (2.4) also satisfy $\mathbf{u} \in C^0(V)$, $\mathbf{u}_t \in C^0(X')$, and $p \in C^0(Q)$ so that, for each $n = n_0, n_0 + 1, \ldots, M-1$,

$$((\mathbf{u}_t)_{n+1/2}, \mathbf{v}) + \frac{1}{2}(\mathbf{u}_{n+1} \cdot \nabla \mathbf{u}_{n+1}, \mathbf{v}) + \frac{1}{2}(\mathbf{u}_n \cdot \nabla \mathbf{u}_n, \mathbf{v}) + \operatorname{Re}^{-1}(\nabla \mathbf{u}_{n+1/2}, \nabla \mathbf{v}) - (p_{n+1/2}, \nabla \cdot \mathbf{v})$$

$$= (\mathbf{f}_{n+1/2}, \mathbf{v}), \quad \forall \mathbf{v} \in X,$$

$$\nabla \cdot \mathbf{u}_{n+1} = 0. \tag{4.19}$$

The consistency error for the time-discretization $\tau_n^{(1)}(\mathbf{u}, p; \mathbf{v}^h)$ and regularization/time-relaxation error $\tau_n^{(2)}(\mathbf{u}, p; \mathbf{v}^h)$ are given by, for $n = 0, 1, \ldots, M - 1$,

$$\tau_n^{(1)}\left(\mathbf{u}, p; \mathbf{v}^h\right) := \left(\frac{\mathbf{u}_{n+1} - \mathbf{u}_n}{\Delta t} - (\mathbf{u}_t)_{n+1/2}, \mathbf{v}^h\right) + \left(\mathbf{u}_{n+1/2} \cdot \nabla \mathbf{u}_{n+1/2}, \mathbf{v}^h\right)$$

$$- \frac{1}{2}\left(\mathbf{u}_{n+1} \cdot \nabla \mathbf{u}_{n+1}, \mathbf{v}^h\right) - \frac{1}{2}\left(\mathbf{u}_n \cdot \nabla \mathbf{u}_n, \mathbf{v}\right), \tag{4.20}$$

$$\tau_n^{(2)}\left(\mathbf{u}, p; \mathbf{v}^h\right) := -b^h\left(\mathbf{u}_{n+1/2} - \varphi^h(\mathbf{u}_{n+1/2}), \mathbf{u}_{n+1/2}, \mathbf{v}^h\right) + \chi\left(\mathbf{u}_{n+1/2} - \varphi^h(\mathbf{u}_{n+1/2}), \mathbf{v}^h\right),$$

where $\mathbf{v}^h \in X^h$. Write $\tau_n := \tau_n^{(1)} + \tau_n^{(2)}$. Using (4.20), rewrite (4.19) in a form conducive to analyzing the error between the continuous and discrete models:

$$\left(\frac{\mathbf{u}_{n+1} - \mathbf{u}_n}{\Delta t}, \mathbf{v}^h\right) + b^h\left(\varphi^h(\mathbf{u}_{n+1/2}), \mathbf{u}_{n+1/2}, \mathbf{v}^h\right) + \mathrm{Re}^{-1}\left(\nabla \mathbf{u}_{n+1/2}, \nabla \mathbf{v}^h\right)$$

$$- \left(p_{n+1/2}, \nabla \cdot \mathbf{v}^h\right) + \chi\left(\mathbf{u}_{n+1/2} - \varphi^h(\mathbf{u}_{n+1/2}), \mathbf{v}^h\right) = \left(\mathbf{f}_{n+1/2}, \mathbf{v}^h\right) + \tau_n\left(\mathbf{u}, p; \mathbf{v}^h\right). \tag{4.21}$$

Let $\tilde{\mathbf{v}}^h = \mathbf{U}_n^h$ be the L^2-projection of $\mathbf{u}(\cdot, t_n)$ so that $(\boldsymbol{\eta}_{n+1} - \boldsymbol{\eta}_n, \boldsymbol{\phi}_{n+1/2}^h) = 0$. Decompose the velocity error

$$\mathbf{e}_n = \mathbf{w}_n^h - \mathbf{u}_n = \boldsymbol{\phi}_n^h - \boldsymbol{\eta}_n, \qquad \boldsymbol{\phi}_n^h = \mathbf{w}_n^h - \mathbf{U}_n^h, \qquad \boldsymbol{\eta}_n = \mathbf{u}_n - \mathbf{U}_n^h, \tag{4.22}$$

where $\mathbf{U}_n^h \in V^h$. Fix $\tilde{q}_{n+1/2}^h \in Q^h$. Note that $(\tilde{q}_{n+1/2}^h, \nabla \cdot \mathbf{v}^h) = 0$ for any $\mathbf{v}^h \in V^h$. Subtract (4.21) from (4.3), apply (3.13)(b), and test with $\mathbf{v}^h = \boldsymbol{\phi}_{n+1/2}^h$ to get

$$\frac{1}{2\Delta t}\left(\left\|\boldsymbol{\phi}_{n+1}^h\right\|^2 - \left\|\boldsymbol{\phi}_n^h\right\|^2\right) + \mathrm{Re}^{-1}\left|\boldsymbol{\phi}_{n+1/2}^h\right|_1^2 + \chi\left\|\boldsymbol{\phi}_{n+1/2}^h\right\|_{\star h}^2$$

$$= -\left(\tilde{q}_{n+1/2}^h - p_{n+1/2}, \nabla \cdot \boldsymbol{\phi}_{n+1/2}^h\right) - b^h\left(\varphi^h\left(\boldsymbol{\phi}_{n+1/2}^h\right), \mathbf{u}_{n+1/2}, \boldsymbol{\phi}_{n+1/2}^h\right)$$

$$+ b^h\left(\varphi^h\left(\boldsymbol{\eta}_{n+1/2}\right), \mathbf{u}_{n+1/2}, \boldsymbol{\phi}_{n+1/2}^h\right) + b^h\left(\varphi^h\left(\mathbf{w}_{n+1/2}^h\right), \boldsymbol{\eta}_{n+1/2}, \boldsymbol{\phi}_{n+1/2}^h\right)$$

$$+ \mathrm{Re}^{-1}\left(\nabla \boldsymbol{\eta}_{n+1/2}, \nabla \boldsymbol{\phi}_{n+1/2}^h\right) + \chi\left(\boldsymbol{\eta}_{n+1/2} - \varphi^h\left(\boldsymbol{\eta}_{n+1/2}\right), \boldsymbol{\phi}_{n+1/2}^h\right) - \tau_n\left(\mathbf{u}, p; \boldsymbol{\phi}_{n+1/2}^h\right). \tag{4.23}$$

(Spatial discretization error): Fix $\varepsilon > 0$. First, apply Hölder's and Young's inequalities to get

$$\left|\mathrm{Re}^{-1}\left(\nabla \boldsymbol{\eta}_{n+1/2}, \nabla \boldsymbol{\phi}_{n+1/2}^h\right) + \left(p_{n+1/2} - \tilde{q}_{n+1/2}^h, \nabla \cdot \boldsymbol{\phi}_{n+1/2}^h\right)\right|$$

$$\leq C\mathrm{Re}^{-1}\left|\boldsymbol{\eta}_{n+1/2}\right|_1^2 + C\,\mathrm{Re}\left\|p_{n+1/2} - \tilde{q}_{n+1/2}^h\right\|^2 + \frac{1}{\varepsilon\,\mathrm{Re}}\left|\boldsymbol{\phi}_{n+1/2}^h\right|_1^2. \tag{4.24}$$

Apply (3.12) and duality estimate on $X \times X'$ to get

$$\left| \chi\left(\boldsymbol{\eta}_{n+1/2} - \varphi^h\left(\boldsymbol{\eta}_{n+1/2} \right), \phi^h_{n+1/2} \right) \right| \leq C\chi^2 \operatorname{Re} \left\| \boldsymbol{\eta}_{n+1/2} \right\|^2_{-1} + \frac{1}{\varepsilon \operatorname{Re}} \left| \phi^h_{n+1/2} \right|_1. \qquad (4.25)$$

We bound the convective terms next. First, $\mathbf{u} \in H^2(\Omega)$ and estimate (2.11)(b) give

$$\left| b^h\left(\varphi^h\left(\phi^h_{n+1/2} \right), \mathbf{u}_{n+1/2}, \phi^h_{n+1/2} \right) \right| \leq C \operatorname{Re} \left\| \varphi^h\left(\phi^h_{n+1/2} \right) \right\|^2 \| \mathbf{u}_{n+1/2} \|^2_2 + \frac{1}{\varepsilon \operatorname{Re}} \left| \phi^h_{n+1/2} \right|^2_1 \qquad (4.26)$$

and $\mathbf{u} \in L^\infty(X)$ with (2.11)(a) give

$$\left| b^h\left(\varphi^h\left(\boldsymbol{\eta}_{n+1/2} \right), \mathbf{u}_{n+1/2}, \phi^h_{n+1/2} \right) \right| \leq C \operatorname{Re} \| \nabla \mathbf{u} \|^2_{L^\infty(L^2(\Omega))} \left| \varphi^h\left(\boldsymbol{\eta}_{n+1/2} \right) \right|^2_1 + \frac{1}{\varepsilon \operatorname{Re}} \left| \phi^h_{n+1/2} \right|^2_1. \qquad (4.27)$$

Next, rewrite the remaining nonlinear term

$$\begin{aligned}
b^h\left(\varphi^h\left(\mathbf{w}^h_{n+1/2} \right), \boldsymbol{\eta}_{n+1/2}, \phi^h_{n+1/2} \right) &= b^h\left(\varphi^h(\mathbf{u}_{n+1/2}), \boldsymbol{\eta}_{n+1/2}, \phi^h_{n+1/2} \right) \\
&\quad - b^h\left(\varphi^h\left(\boldsymbol{\eta}_{n+1/2} \right), \boldsymbol{\eta}_{n+1/2}, \phi^h_{n+1/2} \right) \\
&\quad + b^h\left(\varphi^h\left(\phi^h_{n+1/2} \right), \boldsymbol{\eta}_{n+1/2}, \phi^h_{n+1/2} \right).
\end{aligned} \qquad (4.28)$$

Once again, $\mathbf{u} \in L^\infty(X)$ and (2.11)(a) give

$$\begin{aligned}
\left| b^h\left(\varphi^h(\mathbf{u}_{n+1/2}), \boldsymbol{\eta}_{n+1/2}, \phi^h_{n+1/2} \right) \right. &\left. + b^h\left(\varphi^h\left(\boldsymbol{\eta}_{n+1/2} \right), \boldsymbol{\eta}_{n+1/2}, \phi^h_{n+1/2} \right) \right| \\
&\leq C \operatorname{Re} \left\| \nabla \varphi^h(\mathbf{u}) \right\|_{L^\infty(L^2(\Omega))} \left\| \varphi^h(\mathbf{u}) \right\|_{L^\infty(L^2(\Omega))} \left| \boldsymbol{\eta}_{n+1/2} \right|^2_1 \\
&\quad + C \operatorname{Re} \left| \varphi^h\left(\boldsymbol{\eta}_{n+1/2} \right) \right|_1 \left| \varphi^h\left(\boldsymbol{\eta}_{n+1/2} \right) \right|_1 \left| \boldsymbol{\eta}_{n+1/2} \right|^2_1 + \frac{1}{\varepsilon \operatorname{Re}} \left| \phi^h_{n+1/2} \right|^2_1.
\end{aligned} \qquad (4.29)$$

Lastly, estimate (2.11)(a) and inverse inequality (2.17) give

$$\begin{aligned}
\left| b^h\left(\varphi^h\left(\phi^h_{n+1/2} \right), \boldsymbol{\eta}_{n+1/2}, \phi^h_{n+1/2} \right) \right| \\
\leq C \operatorname{Re} h^{-1} \left\| \varphi^h\left(\phi^h_{n+1/2} \right) \right\|^2 \left| \boldsymbol{\eta}_{n+1/2} \right|^2_1 + \frac{1}{\varepsilon \operatorname{Re}} \left| \phi^h_{n+1/2} \right|^2_1.
\end{aligned} \qquad (4.30)$$

Convergence Analysis of a Fully Discrete Family of Iterated Deconvolution Methods for
Turbulence Modeling with Time Relaxation

21

Apply estimates $\|\varphi^h(\mathbf{v})\| \leq \|\mathbf{v}\|$ and $|\varphi^h(\mathbf{v})|_1 \leq d_1|\mathbf{v}|_1$ from Assumption 3.5 along with estimates (4.26)–(4.30) to get

$$
\begin{aligned}
&\left| b^h\left(\varphi^h\left(\phi^h_{n+1/2}\right), \mathbf{u}_{n+1/2}, \phi^h_{n+1/2}\right) + b^h\left(\varphi^h\left(\boldsymbol{\eta}_{n+1/2}\right), \mathbf{u}_{n+1/2}, \phi^h_{n+1/2}\right) \right. \\
&\left. + b^h\left(\varphi^h\left(\mathbf{w}^h_{n+1/2}\right), \boldsymbol{\eta}_{n+1/2}, \phi^h_{n+1/2}\right) \right| \\
&\quad \leq \frac{4}{\varepsilon \operatorname{Re}}\left|\phi^h_{n+1/2}\right|^2_1 + C\operatorname{Re}\left(\|\mathbf{u}_{n+1/2}\|^2_2 + h^{-1}\left|\boldsymbol{\eta}_{n+1/2}\right|^2_1\right)\left\|\phi^h_{n+1/2}\right\|^2 \\
&\quad + Cd_1\operatorname{Re}\left(\|\nabla\mathbf{u}\|^2_{L^\infty(L^2(\Omega))} + \left|\boldsymbol{\eta}_{n+1/2}\right|^2_1\right)\left|\boldsymbol{\eta}_{n+1/2}\right|^2_1.
\end{aligned}
\tag{4.31}
$$

(Time discretization error): First, apply duality estimate on $X \times X'$ to get

$$
\left|\left(\frac{\mathbf{u}_{n+1}-\mathbf{u}_n}{\Delta t} - (\mathbf{u}_t)_{n+1/2}, \phi^h_{n+1/2}\right)\right| \leq C\operatorname{Re}\left\|\frac{\mathbf{u}_{n+1}-\mathbf{u}_n}{\Delta t} - (\mathbf{u}_t)_{n+1/2}\right\|^2_{-1} + \frac{1}{\varepsilon \operatorname{Re}}\left|\phi^h_{n+1/2}\right|^2_1.
\tag{4.32}
$$

Taylor-expansion about $t_{n+1/2}$ with integral remainder gives

$$
\begin{aligned}
&\frac{1}{2}(\mathbf{u}_{n+1} \cdot \nabla\mathbf{u}_{n+1}, \mathbf{v}) + \frac{1}{2}(\mathbf{u}_n \cdot \nabla\mathbf{u}_n, \mathbf{v}) \\
&\quad = (\mathbf{u}(\cdot, t_{n+1/2}) \cdot \nabla\mathbf{u}(\cdot, t_{n+1/2}), \mathbf{v}) \\
&\qquad + \frac{1}{2}\int_{t_{n+1/2}}^{t^{n+1}} (t_{n+1}-t)\frac{d^2}{dt^2}(\mathbf{u}(\cdot,t) \cdot \nabla\mathbf{u}(\cdot,t), \mathbf{v})dt \\
&\qquad + \frac{1}{2}\int_{t_n}^{t_{n+1/2}} (t-t_n)\frac{d^2}{dt^2}(\mathbf{u}(\cdot,t) \cdot \nabla\mathbf{u}(\cdot,t), \mathbf{v})dt.
\end{aligned}
\tag{4.33}
$$

Add/subtract $(\mathbf{u}_{n+1/2} \cdot \nabla\mathbf{u}(\cdot, t_{n+1/2}), \mathbf{v})$ and apply (4.33) to get

$$
\begin{aligned}
&(\mathbf{u}_{n+1/2} \cdot \nabla\mathbf{u}_{n+1/2}, \mathbf{v}) - \frac{1}{2}(\mathbf{u}_{n+1} \cdot \nabla\mathbf{u}_{n+1}, \mathbf{v}) - \frac{1}{2}(\mathbf{u}_n \cdot \nabla\mathbf{u}_n, \mathbf{v}) \\
&\quad = (\mathbf{u}_{n+1/2} \cdot \nabla(\mathbf{u}_{n+1/2} - \mathbf{u}(\cdot, t_{n+1/2})), \mathbf{v}) + ((\mathbf{u}_{n+1/2} - \mathbf{u}(\cdot, t_{n+1/2})) \cdot \nabla\mathbf{u}(\cdot, t_{n+1/2}), \mathbf{v}) \\
&\qquad - \frac{1}{2}\int_{t_{n+1/2}}^{t_{n+1}} (t_{n+1}-t)\int(\mathbf{u}_{tt} \cdot \nabla\mathbf{u} + \mathbf{u} \cdot \nabla\mathbf{u}_{tt} + 2\mathbf{u}_t \cdot \nabla\mathbf{u}_t) \cdot \mathbf{v}dt \\
&\qquad - \frac{1}{2}\int_{t_n}^{t_{n+1/2}} (t-t_n)\int(\mathbf{u}_{tt} \cdot \nabla\mathbf{u} + \mathbf{u} \cdot \nabla\mathbf{u}_{tt} + 2\mathbf{u}_t \cdot \nabla\mathbf{u}_t) \cdot \mathbf{v}dt.
\end{aligned}
\tag{4.34}
$$

Majorize either directly or with (2.7)(a) to get

$$\left| (\mathbf{u}_{n+1/2} \cdot \nabla \mathbf{u}_{n+1/2}, \mathbf{v}) - \frac{1}{2} (\mathbf{u}_{n+1} \cdot \nabla \mathbf{u}_{n+1}, \mathbf{v}) - \frac{1}{2} (\mathbf{u}_n \cdot \nabla \mathbf{u}_n, \mathbf{v}) \right|$$

$$\leq C \|\nabla \mathbf{u}\|_{l^\infty(L^2(\Omega))} (|\mathbf{u}_{n+1}|_1 + |\mathbf{u}_n|_1) \left| \mathbf{v}^h \right|_1, \tag{4.35}$$

or with (2.7)(b), (2.7)(c) and Hölder's inequality (in time) applied to (4.34) to get

$$\left| (\mathbf{u}_{n+1/2} \cdot \nabla \mathbf{u}_{n+1/2}, \mathbf{v}) - \frac{1}{2} (\mathbf{u}_{n+1} \cdot \nabla \mathbf{u}_{n+1}, \mathbf{v}) - \frac{1}{2} (\mathbf{u}_n \cdot \nabla \mathbf{u}_n, \mathbf{v}) \right|$$

$$\leq C \|\mathbf{u}\|_{l^\infty(H^2(\Omega))} \|\mathbf{u}_{n+1/2} - \mathbf{u}(\cdot, t_{n+1/2})\| \left| \mathbf{v}^h \right|_1$$

$$+ \frac{C \Delta t^{3/2}}{\sqrt{t^{n+1/2}}} \|\mathbf{u}\|_{l^\infty(n,n+1;H^2(\Omega))} \left(\int_{t_n}^{t_{n+1}} t \, \|\mathbf{u}_{tt}(\cdot, t)\|^2 dt \right)^{1/2} \left| \mathbf{v}^h \right|_1 \tag{4.36}$$

$$+ \frac{C \Delta t^{3/2}}{\sqrt{t^{n+1/2}}} \|\mathbf{u}_t\|_{l^\infty(n,n+1;L^2)} \left(\int_{t_n}^{t_{n+1}} t \|\mathbf{u}_t(\cdot, t)\|_2^2 \, dt \right)^{1/2} \left| \mathbf{v}^h \right|_1.$$

Then, to prove Corollary 4.7, apply (4.32), (4.36) with Young's inequality give

$$\left| \tau_n^{(1)} \left(\mathbf{u}, p; \phi_{n+1/2}^h \right) \right| \leq \frac{2}{\varepsilon \, \mathrm{Re}} \left| \phi_{n+1/2}^h \right|_1^2 + C \, \mathrm{Re} \left\| \frac{\mathbf{u}_{n+1} - \mathbf{u}_n}{\Delta t} - (\mathbf{u}_t)_{n+1/2} \right\|_{-1}^2$$

$$+ C \, \mathrm{Re} \, \Delta t^3 \|\mathbf{u}\|_{l^\infty(H^2)}^2 \|\mathbf{u}_{tt}\|_{L^2(t^n,t^{n+1};L^2(\Omega))}^2 + C \, \mathrm{Re} \, \Delta t^3 \|\mathbf{u}_t\|_{l^\infty(L^2(\Omega))}^2 \|\mathbf{u}_t\|_{L^2(t^n,t^{n+1};H^2(\Omega))}^2. \tag{4.37}$$

We apply (4.35) instead of (4.36) to prove Theorem 4.6:

$$\left| \tau_n^{(1)} \left(\mathbf{u}, p; \phi_{n+1/2}^h \right) \right| \leq \frac{2}{\varepsilon \, \mathrm{Re}} \left| \phi_{n+1/2}^h \right|_1^2 + C \, \mathrm{Re} \left\| \frac{\mathbf{u}_{n+1} - \mathbf{u}_n}{\Delta t} - (\mathbf{u}_t)_{n+1/2} \right\|_{-1}^2$$

$$+ C \, \mathrm{Re} \, \|\nabla \mathbf{u}\|_{l^\infty(L^2(\Omega))}^2 \left(|\mathbf{u}_{n+1}|_1^2 + |\mathbf{u}_n|_1^2 \right). \tag{4.38}$$

(Deconvolution error): Next, add/subtract $\varphi(\mathbf{u}_{n+1/2})$ we write

$$\tau_n^{(2)} \left(\mathbf{u}, p; \mathbf{v}^h \right) = -b^h \left(\mathbf{u}_{n+1/2} - \varphi(\mathbf{u}_{n+1/2}), \mathbf{u}_{n+1/2}, \phi_{n+1/2}^h \right)$$

$$- b^h \left(\left(\varphi(\mathbf{u}_{n+1/2}) - \varphi^h(\mathbf{u}_{n+1/2}) \right), \mathbf{u}_{n+1/2}, \phi_{n+1/2}^h \right)$$

$$+ \chi \left(\mathbf{u}_{n+1/2} - \varphi(\mathbf{u}_{n+1/2}), \phi_{n+1/2}^h \right) + \chi \left(\varphi(\mathbf{u}_{n+1/2}) - \varphi^h(\mathbf{u}_{n+1/2}), \phi_{n+1/2}^h \right). \tag{4.39}$$

Then duality on $X \times X'$ and Young's inequalities give

$$
\begin{aligned}
\left| \chi\left(\mathbf{u}_{n+1/2} - \varphi(\mathbf{u}_{n+1/2}), \phi_{n+1/2}^h\right) + \chi\left(\varphi(\mathbf{u}_{n+1/2}) - \varphi^h(\mathbf{u}_{n+1/2}), \phi_{n+1/2}^h\right) \right| & \\
\leq C\chi^2 \operatorname{Re} \left\| \mathbf{u}_{n+1/2} - \varphi(\mathbf{u}_{n+1/2}) \right\|_{-1}^2 & \\
+ C\chi^2 \operatorname{Re} \left\| \varphi(\mathbf{u}_{n+1/2}) - \varphi^h(\mathbf{u}_{n+1/2}) \right\|_{-1}^2 + \frac{1}{\varepsilon \operatorname{Re}} \left| \phi_{n+1/2}^h \right|_1^2,
\end{aligned}
\tag{4.40}
$$

and $\mathbf{u} \in H^2(\Omega)$ along with (2.11)(b) give

$$
\begin{aligned}
\left| b^h\left(\mathbf{u}_{n+1/2} - \varphi(\mathbf{u}_{n+1/2}), \mathbf{u}_{n+1/2}, \phi_{n+1/2}^h\right) + b^h\left(\varphi(\mathbf{u}_{n+1/2}) - \varphi^h(\mathbf{u}_{n+1/2}), \mathbf{u}_{n+1/2}, \phi_{n+1/2}^h\right) \right| & \\
\leq C \operatorname{Re} \left\| \mathbf{u}_{n+1/2} \right\|_2^2 \left(\left\| \mathbf{u}_{n+1/2} - \varphi(\mathbf{u}_{n+1/2}) \right\|^2 + \left\| \varphi(\mathbf{u}_{n+1/2}) - \varphi^h(\mathbf{u}_{n+1/2}) \right\|^2 \right) + \frac{1}{\varepsilon \operatorname{Re}} \left| \phi_{n+1/2}^h \right|_1^2.
\end{aligned}
\tag{4.41}
$$

The estimates (4.37), (4.40), (4.41) with identity (4.39) give

$$
\left| \tau_n^{(1)}\left(\mathbf{u}, p; \phi_{n+1/2}^h\right) + \tau_n^{(2)}\left(\mathbf{u}, p; \phi_{n+1/2}^h\right) \right| \leq \frac{4}{\varepsilon \operatorname{Re}} \left| \phi_{n+1/2}^h \right|_1^2 + C \operatorname{Re} E_n.
\tag{4.42}
$$

Then estimates (3.17) and (3.23) give

$$
\begin{aligned}
E_n := {}& \left(\chi^2 + \|\mathbf{u}\|_{l^\infty(H^2(\Omega))}^2 \right) \left(c_{1,j}(\alpha, \delta)^2 \left\| \Delta^{j+1} \mathbf{u}_{n+1/2} \right\|^2 + c_{2,j}(h, \alpha, \delta)^2 \|\mathbf{u}_{n+1/2}\|_{k+1}^2 \right) \\
& + \left\| \frac{\mathbf{u}_{n+1} - \mathbf{u}_n}{\Delta t} - (\mathbf{u}_t)_{n+1/2} \right\|_{-1}^2 + \|\mathbf{u}\|_{l^\infty(H^2)}^2 \|\mathbf{u}_{n+1/2} - \mathbf{u}(\cdot, t_{n+1/2})\|^2 \\
& + \Delta t^3 \|\mathbf{u}\|_{l^\infty(H^2)}^2 \|\mathbf{u}_{tt}\|_{L^2(t^n, t^{n+1}; L^2(\Omega))}^2 + \Delta t^3 \|\mathbf{u}_t\|_{l^\infty(L^2(\Omega))}^2 \|\mathbf{u}_t\|_{L^2(t^n, t^{n+1}; H^2(\Omega))}^2.
\end{aligned}
\tag{4.43}
$$

Apply estimates from (4.24), (4.25), (4.31), (4.37), (4.42) to (4.23). Set $\varepsilon = 20$ and absorb all terms including $|\phi_{n+1/2}^h|_1$ from the right into left-hand-side of (4.23). Sum the resulting inequality on both sides from $n = 0$ to $n = M - 1$ to get

$$
\begin{aligned}
\left\| \phi_M^h \right\|^2 + \operatorname{Re}^{-1} \Delta t \sum_{n=0}^{M-1} \left| \phi_{n+1/2}^h \right|_1^2 + \chi \Delta t \sum_{n=0}^{M-1} \left\| \phi_{n+1/2}^h \right\|_{*h}^2 & \\
\leq \left\| \phi_0^h \right\|^2 + C \operatorname{Re} \Delta t \sum_{n=0}^{M-1} E_n + C \operatorname{Re} \Delta t \sum_{n=0}^{M-1} \left\| p_{n+1/2} - \tilde{q}_{n+1/2}^h \right\|^2
\end{aligned}
$$

$$+ C \operatorname{Re} d_1^2 \Delta t \sum_{n=0}^{M-1} \left(\|\nabla \mathbf{u}\|_{L^\infty(L^2(\Omega))}^2 + \left| \boldsymbol{\eta}_{n+1/2} \right|_1^2 \right) \left| \boldsymbol{\eta}_{n+1/2} \right|_1^2$$

$$+ C \operatorname{Re} \Delta t \sum_{n=0}^{M-1} \left(\|\mathbf{u}_{n+1/2}\|_2^2 + h^{-1} \left| \boldsymbol{\eta}_{n+1/2} \right|_1^2 \right) \left\| \phi_{n+1/2}^h \right\|^2$$

$$+ C\chi^2 \operatorname{Re} \Delta t \sum_{n=0}^{M-1} \left\| \boldsymbol{\eta}_{n+1/2} \right\|_{-1}^2 + C\operatorname{Re}^{-1} \Delta t \sum_{n=0}^{M-1} \left| \boldsymbol{\eta}_{n+1/2} \right|_1^2.$$

$$(4.44)$$

Estimates (2.23), (2.16)(a) imply

$$\sup_n \left| \boldsymbol{\eta}_n \right|_1 \le C \|\nabla \mathbf{u}\|_{l^\infty(L^2(\Omega))}, \qquad \left| \boldsymbol{\eta}_n \right|_1 \le Ch\|\mathbf{u}_n\|_2. \tag{4.45}$$

These estimates applied to (4.44) give

$$\left\| \phi_M^h \right\|^2 + \operatorname{Re}^{-1} \Delta t \sum_{n=0}^{M-1} \left| \phi_{n+1/2}^h \right|_1^2 + \chi \Delta t \sum_{n=0}^{M-1} \left\| \phi_{n+1/2}^h \right\|_{\star h}^2$$

$$\le \left\| \phi_0^h \right\|^2 + C \operatorname{Re} \Delta t \sum_{n=0}^{M-1} E_n + C \operatorname{Re} \Delta t \sum_{n=0}^{M-1} \left\| p_{n+1/2} - \tilde{q}_{n+1/2}^h \right\|^2$$

$$+ C \operatorname{Re} \Delta t \sum_{n=0}^{M-1} \left(\left(d_1^2 \|\nabla \mathbf{u}\|_{L^\infty(L^2(\Omega))}^2 + \operatorname{Re}^{-2} \right) \left| \boldsymbol{\eta}_{n+1/2} \right|_1^2 + \|\mathbf{u}_{n+1/2}\|_2^2 \left\| \phi_{n+1/2}^h \right\|^2 \right)$$

$$+ C\chi^2 \operatorname{Re} \Delta t \sum_{n=0}^{M-1} \left\| \boldsymbol{\eta}_{n+1/2} \right\|_{-1}^2.$$

$$(4.46)$$

Suppose that the Δt-restriction (4.16) is satisfied. Then the discrete Gronwall Lemma 2.2 applies to (4.46) and gives

$$\left\| \phi_M^h \right\|^2 + \operatorname{Re}^{-1} \Delta t \sum_{n=0}^{M-1} \left| \phi_{n+1/2}^h \right|_1^2 + \chi \Delta t \sum_{n=0}^{M-1} \left\| \phi_{n+1/2}^h \right\|_{\star h}^2$$

$$\le G_M \left\| \phi_0^h \right\|^2 + G_M \operatorname{Re} \Delta t \sum_{n=0}^{M-1} E_n$$

$$+ G_M \operatorname{Re} \Delta t \sum_{n=0}^{M-1} \left(\left\| p_{n+1/2} - \tilde{q}_{n+1/2}^h \right\|^2 + \left(d_1^2 \|\nabla \mathbf{u}\|_{L^\infty(L^2(\Omega))}^2 + \operatorname{Re}^{-2} \right) \left| \boldsymbol{\eta}_{n+1/2} \right|_1^2 \right)$$

$$+ G_M \chi^2 \operatorname{Re} \Delta t \sum_{n=0}^{M-1} \left\| \boldsymbol{\eta}_{n+1/2} \right\|_{-1}^2,$$

$$(4.47)$$

Convergence Analysis of a Fully Discrete Family of Iterated Deconvolution Methods for
Turbulence Modeling with Time Relaxation

25

where

$$G_M = C \exp\left(\text{Re}\, \Delta t \sum_{n=0}^{M} g_n \|\mathbf{u}_n\|_2^2 \right), \qquad g_n = \left(1 - C\,\text{Re}\,\Delta t \|\mathbf{u}_n\|_2^2 \right)^{-1}. \tag{4.48}$$

Lastly, the triangle inequality and approximation theory estimates (2.23), (2.16) along with (3.23) applied to (4.47) give

$$\left\| \mathbf{u}_M - \mathbf{w}_M^h \right\|^2 + \text{Re}^{-1}\Delta t \sum_{n=0}^{M-1} \left| \mathbf{u}_{n+1/2} - \mathbf{w}_{n+1/2}^h \right|_1^2$$

$$\leq G_M \left\| \mathbf{e}_0^h \right\|^2 + Ch^{2k}\left(G_M \|\mathbf{u}_0\|_k + \|\mathbf{u}\|_{l^\infty(H^k(\Omega))}^2 + \text{Re}^{-1}\|\mathbf{u}\|_{l^2(H^{k+1}(\Omega))}^2 \right)$$

$$+ G_M\,\text{Re}\,\Delta t \sum_{n=0}^{M-1} E_n + G_M\,\text{Re}\,h^{2s+2}\|p\|_{l^2(H^{s+1}(\Omega))}^2$$

$$+ G_M\,\text{Re}\left(d_1^2 \|\nabla\mathbf{u}\|_{L^\infty(L^2(\Omega))}^2 + \text{Re}^{-2} + \chi^2 h^4 \right) h^{2k}\|\mathbf{u}\|_{l^2(H^{k+1}(\Omega))}^2. \tag{4.49}$$

It remains to bound $\Delta t \sum_n E_n$.

(Theorem 4.6): Suppose that $\partial_t \mathbf{u} \in L^2(X') \cap l^2(X')$. The triangle inequality and (2.18) gives

$$\Delta t \sum_{n=n_0}^{M-1} \left\| \frac{\mathbf{u}_{n+1} - \mathbf{u}_n}{\Delta t} - (\mathbf{u}_t)_{n+1/2} \right\|_{-1}^2 \leq C\left(\|\mathbf{u}_t\|_{L^2(X')}^2 + \|\mathbf{u}_t\|_{l^2(X')}^2 \right). \tag{4.50}$$

Apply (4.38), instead of (4.37), to derive E_n in (4.43). Then

$$\text{Re}\,\Delta t \sum_{n=0}^{M-1} E_n \leq E, \tag{4.51}$$

where

$$E := C\text{Re}\left(\chi^2 + \|\mathbf{u}\|_{l^\infty(H^2(\Omega))}^2 \right)\left(c_{1,j}(\alpha,\delta)^2 \left\| \Delta^{j+1}\mathbf{u}_{n+1/2} \right\|^2 + c_{2,j}(h,\alpha,\delta)^2 \|\mathbf{u}_{n+1/2}\|_{k+1}^2 \right)$$

$$+ C\text{Re}\left(\|\mathbf{u}_t\|_{L^2(X')}^2 + \|\mathbf{u}_t\|_{l^2(X')}^2 + \|\nabla\mathbf{u}\|_{l^\infty(L^2(\Omega))}^2 \|\nabla\mathbf{u}\|_{l^2(L^2(\Omega))}^2 \right). \tag{4.52}$$

(Corollary 4.7): Suppose that $\mathbf{u} \in l^\infty(H^2(\Omega))$, $\mathbf{u}_t \in l^\infty(H^1(\Omega)) \cap L^2(H^1(\Omega))$, $\mathbf{u}_{tt} \in L^2(L^2(\Omega))$, and $\mathbf{u}_{ttt} \in L^2(X')$. Write

$$E := C \operatorname{Re}\left(\chi^2 + \|\mathbf{u}\|^2_{l^\infty(H^2(\Omega))} \right) \left(c_{1,j}(\alpha,\delta)^2 \left\| \Delta^{j+1} \mathbf{u}_{n+1/2} \right\|^2 + c_{2,j}(h,\alpha,\delta)^2 \|\mathbf{u}_{n+1/2}\|^2_{k+1} \right)$$

$$:= C \operatorname{Re}\left(\|\mathbf{u}_{ttt}\|^2_{L^2(X')} + \|\mathbf{u}\|^2_{l^\infty(H^2(\Omega))} \|\mathbf{u}_{tt}\|^2_{L^2(L^2(\Omega))} + \cdots + \|\mathbf{u}_t\|^2_{l^\infty(L^2(\Omega))} \|\mathbf{u}_t\|^2_{L^2(H^2(\Omega))} \right).$$

(4.53)

Then apply (2.19), (2.20) to bound E_n given in (4.43):

$$\operatorname{Re} \Delta t \sum_{n=0}^{M-1} E_n \leq E \Delta t^4.$$

(4.54)

$\square$

5. Applications

We show that the iterated (modified) Tikhonov regularization operator satisfied Assumption 3.4, 3.5 in Section 5.1 and verify the theoretical convergence rate predicted by Theorem 4.6, Corollary 4.7 in Section 5.2.

5.1. Iterated (Modified) Tikhonov-Lavrentiev Regularization

We will prove that $D_{\alpha,J}$, $D^h_{\alpha,J}$ (Definitions 3.14, 3.15) with the differential filter $G = A^{-1}$ satisfies Assumptions 3.4, 3.5. Proposition 3.10 implies that it is enough to show that $D_{\alpha,0}$ satisfies Assumption 3.4. Additionally, we provide sharpened estimates for $d_{1,j}$, $c_{1,j}$, $c_{2,j}$. The key is that $A^{-1} > 0$ is a continuous function of the Laplace operator $-\Delta \geq 0$ and hence they commute (on X). Moreover, $D_{\alpha,0} > 0$ is a continuous function of A^{-1} so that $D_{\alpha,0}$ commutes with A^{-1} and $-\Delta$ (on X).

We first characterize the spectrum of $D_{\alpha,0}$, $D^h_{\alpha,0}$.

Lemma 5.1. *Fix $0 < \alpha \leq 1$. Define $f : (0,1] \to \mathbb{R}$ and $g : (0,1] \to \mathbb{R}$ by*

$$f(x) := \frac{1}{(1-\alpha)x + \alpha}, \qquad g(x) := \frac{x}{(1-\alpha)x + \alpha}.$$

(5.1)

The maps f and g are continuous and $f((0,1]) = [1, \alpha^{-1})$ and $g((0,1]) = (0,1]$.

Proof. The functions f, g are clearly continuous with f *decreasing* and g *increasing* on $(0,1]$. Hence, the range of f is $[1, \alpha^{-1})$ and range of g is $(0,1]$. $\square$

The next result shows that $D_{\alpha,0}$ and $D^h_{\alpha,0}$ satisfy part of Assumptions 3.4, 3.5.

Proposition 5.2. $D_{\alpha,0}$ and $D^h_{\alpha,0}$ (on X and X^h, resp.) are linear, bounded, spd, and commute with A^{-1}, $(A^h)^{-1}$ (resp.). Moreover,

$$\left\|D_{\alpha,0}A^{-1}\right\| \le 1, \qquad |D_{\alpha,0}\overline{\mathbf{u}}|_1 \le |\mathbf{u}|_1 \quad \forall \mathbf{u} \in X,$$

$$\left\|D^h_{\alpha,0}\left(A^h\right)^{-1}\right\| \le 1, \qquad \left|D^h_{\alpha,0}\overline{\mathbf{u}^h}^h\right|_1 \le \left|\mathbf{u}^h\right|_1 \quad \forall \mathbf{u}^h \in X^h. \tag{5.2}$$

Hence $d_1 = 1$ in Assumptions 3.4, 3.5.

Proof. It is immediately clear that $D_{\alpha,0}$, $D^h_{\alpha,0}$ are linear. As a consequence, since $A^{-1} > 0$ with spectrum in $(0,1]$, then $D_{\alpha,0} = f(A^{-1})$ with spectrum contained in $[1,\alpha^{-1})$ so that $D > 0$. Therefore, $D_{\alpha,0}A^{-1} = g(A^{-1})$ with spectrum contained in $(0,1]$. A similar argument shows that $(A^h)^{-1}$ has spectrum in $(0,1]$, $D^h_{\alpha,0}$ has spectrum in $[1,\alpha^{-1})$, and $D^h_{\alpha,0}(A^h)^{-1}$ has spectrum in $(0,1]$. Thus $D_{\alpha,0} > 0$, $D^h_{\alpha,0} > 0$ and $\|D_{\alpha,0}A^{-1}\| \le 1$ and $\|D^h_{\alpha,0}(A^h)^{-1}\| \le 1$. Therefore, $D_{\alpha,0}$ and $D^h_{\alpha,0}$ are bounded and commute with A^{-1} and $(A^h)^{-1}$, respectively, as discussed above.

The second set of inequalities on each line can be proved with an appropriate choice of $\mathbf{v}$ and $\mathbf{v}^h$ in Definitions 3.2 and 3.3. Starting with Definition 3.2, take $\phi = \mathbf{u}$ and choose $\mathbf{v} = \Delta D_{\alpha,0}\overline{\mathbf{u}}$. Then integration by parts and the Cauchy-Schwartz inequality give the result. The discrete form is proved using Definition 3.3 and choosing $\phi = \mathbf{u}^h$ and $\mathbf{v} = \Delta D^h_{\alpha,0}\overline{\mathbf{u}^h}^h$. $\square$

It remains to provide estimates for c_1 and c_2, and sharpened estimates for $c_{1,j}$ and $c_{2,j}$. Indeed, as a direct consequence of Propositions 5.3, 5.4, we have, for each $j = 0, 1, \ldots, J$,

$$c_{1,j} = \left(\alpha\delta^2\right)^{j+1}\left\|\Delta^{j+1}\mathbf{v}\right\|, \quad \forall \mathbf{v} \in H^{j+1}(\Omega),$$

$$c_{2,j} = C\left(h + \left(2^j\alpha\delta^2\right)^{1/2}\right)h^k\max_{0\le n\le j}\|D_{\alpha,n}\overline{\mathbf{v}}\|_{k+1}, \quad \forall \mathbf{v} \in H^{k+1}(\Omega). \tag{5.3}$$

Proposition 5.3. *Let $j = 0, 1, \ldots, J$. Then, for some $0 \le \beta \le j + 1$,*

$$\left\|\mathbf{v} - D_{\alpha,j}\overline{\mathbf{v}}\right\| \le \left(\alpha\delta^2\right)^{\beta}\left\|\Delta^{\beta}\mathbf{v}\right\|, \quad \forall \mathbf{v} \in H^{\beta}(\Omega). \tag{5.4}$$

Proof. Using (1.9), we have

$$D^{-1}_{\alpha,0}\left(D_{\alpha,J}\overline{\mathbf{v}} - D_{\alpha,J-1}\overline{\mathbf{v}}\right) = \overline{\mathbf{v}} - A^{-1}D_{\alpha,J-1}\overline{\mathbf{v}}. \tag{5.5}$$

Subtracting (5.5) from the identity

$$D^{-1}_{\alpha,0}(\mathbf{v} - \mathbf{v}) = \overline{\mathbf{v}} - \overline{\mathbf{v}}, \tag{5.6}$$

gives us

$$D^{-1}_{\alpha,0}\left(\left(I - D_{\alpha,J}A^{-1}\right)\mathbf{v} - \left(I - D_{\alpha,J-1}A^{-1}\right)\mathbf{v}\right) = -A^{-1}\left(I - D_{\alpha,J-1}A^{-1}\right)\mathbf{v}. \tag{5.7}$$

Multiplying by $D_{\alpha,0}$, rearranging, simplifying, and using $A - I = -\delta^2 \Delta$ (Definition 3.2) gives

$$
\begin{aligned}
\left(I - D_{\alpha,J}A^{-1}\right)\mathbf{v} &= \left[-D_{\alpha,0}A^{-1}\left(I - D_{\alpha,J-1}A^{-1}\right) + \left(I - D_{\alpha,J-1}A^{-1}\right)\right]\mathbf{v} \\
&= \left(I - D_{\alpha,J-1}A^{-1}\right)\left(I - D_{\alpha,0}A^{-1}\right)\mathbf{v} \\
&= \left(I - D_{\alpha,J-1}A^{-1}\right)D_{\alpha,0}A^{-1}\left(D_{\alpha,0}^{-1}A - I\right)\mathbf{v} \\
&= \left(I - D_{\alpha,J-1}A^{-1}\right)D_{\alpha,0}A^{-1}\left(\left((1-\alpha)A^{-1} + \alpha I\right)A - I\right)\mathbf{v} \\
&= \left(I - D_{\alpha,J-1}A^{-1}\right)D_{\alpha,0}A^{-1}\alpha(A - I)\mathbf{v} \\
&= -\alpha\delta^2 \Delta D_{\alpha,0}A^{-1}\left(I - D_{\alpha,J-1}A^{-1}\right)\mathbf{v}.
\end{aligned}
\tag{5.8}
$$

Applying recursion, we obtain, for any $0 \le \beta < J$,

$$
\left(I - D_{\alpha,J}A^{-1}\right)\mathbf{v} = \left(-\alpha\delta^2\right)^{\beta}\left(D_{\alpha,J-\beta}A^{-1}\right)^{\beta}\Delta^{\beta}\mathbf{v}.
\tag{5.9}
$$

Thus, taking norms and applying $\|D_{\alpha,J-\beta}A^{-1}\| \le 1$, we get (5.4).   □

Proposition 5.4. *Let $j = 0, 1, \ldots, J$. Then*

$$
\left\|D_{\alpha,j}\overline{\mathbf{w}} - D_{\alpha,j}^h\overline{\mathbf{w}}^h\right\|^2 \le C\left(h^2 + 2^{j+1}\alpha\delta^2\right)h^{2k}\max_{0\le n\le j}\|D_{\alpha,n}\overline{\mathbf{w}}\|_{k+1}^2, \quad \forall \mathbf{w} \in H^{k+1}(\Omega).
\tag{5.10}
$$

Proof. Let $\tilde{\mathbf{v}}_j^h \in X^h$ be the L^2-projection of $D_{\alpha,j}\overline{\mathbf{w}}$. Take $\mathbf{v} = \mathbf{v}^h$ in (3.34). For $j = 1,\ldots,J$, let $e_j = D_{\alpha,j}\overline{\mathbf{w}} - D_{\alpha,j}^h\overline{\mathbf{w}}^h := \eta_j - \phi_j^h$, where $\eta_j := D_{\alpha,j}\overline{\mathbf{w}} - \tilde{\mathbf{v}}_j^h$, and $\phi_j^h := D_{\alpha,j}^h\overline{\mathbf{w}}^h - \tilde{\mathbf{v}}_j^h$. Subtract (3.34) and (3.35) to get

$$
\alpha\delta^2\left(\nabla\phi_j^h, \nabla\mathbf{v}^h\right) + \left(\phi_j^h, \mathbf{v}^h\right) = \alpha\delta^2\left(\nabla\eta_j, \nabla\mathbf{v}^h\right) + \alpha\delta^2\left(\nabla e_{j-1}, \nabla\mathbf{v}^h\right).
\tag{5.11}
$$

Take $\mathbf{v}^h = \phi_j^h$ in (5.11) to get

$$
\alpha\delta^2\left|\phi_j^h\right|_1^2 + \left\|\phi_j^h\right\|^2 = \alpha\delta^2\left(\nabla\eta_j, \nabla\phi_j^h\right) + \alpha\delta^2\left(\nabla e_{j-1}, \nabla\phi_j^h\right).
\tag{5.12}
$$

Fix $\varepsilon > 0$. Apply Hölder's and Young's inequalities to (5.12) to get

$$
\alpha\delta^2\left|\phi_j^h\right|_1^2 + \left\|\phi_j^h\right\|^2 \le \alpha\delta^2\left|\eta_j\right|_1^2 + \varepsilon\alpha\delta^2\left|e_{j-1}\right|_1^2 + \frac{1}{\varepsilon}\alpha\delta^2\left|\phi_j^h\right|_1^2.
\tag{5.13}
$$

Taking $\varepsilon = 1$ and $\varepsilon = 2$ in (5.13) gives

$$\left\|\phi_j^h\right\|^2 \le \alpha\delta^2\left|\boldsymbol{\eta}_j\right|_1^2 + \alpha\delta^2\left|\mathbf{e}_{j-1}\right|_1^2. \tag{5.14}$$

$$\alpha\delta^2\left|\phi_j^h\right|_1^2 + 2\left\|\phi_j^h\right\|^2 \le 2\alpha\delta^2\left|\boldsymbol{\eta}_j\right|_1^2 + 4\alpha\delta^2\left|\mathbf{e}_{j-1}\right|_1^2. \tag{5.15}$$

The triangle inequality and estimate (5.14) give

$$\left\|\mathbf{e}_j\right\|^2 \le \left\|\boldsymbol{\eta}_j\right\|^2 + \alpha\delta^2\left|\boldsymbol{\eta}_j\right|_1^2 + \alpha\delta^2\left|\mathbf{e}_{j-1}\right|_1^2. \tag{5.16}$$

Backward induction, estimate (5.15), and (2.23) give

$$\left\|\mathbf{e}_j\right\|^2 \le |\mathbf{e}_0|_1^2 + \left(h^2 + \alpha\delta^2\left(1 + \sum_{i=0}^{j}2^i\right)\right)\max_{0\le n\le j}\inf_{v^h\in X^h}\left|D_{\alpha,n}\overline{\mathbf{w}} - \mathbf{v}^h\right|_1^2. \tag{5.17}$$

It has been shown (Estimate (2.36) in the proof of Lemma 2.7 [25]) that

$$|\mathbf{e}_0|_1^2 \le C\left(h^2 + \alpha\delta^2\right)h^{2k}|D_0\overline{\mathbf{w}}|_{k+1}^2. \tag{5.18}$$

Note that $\sum_{i=0}^{j}2^i = 2^{j+1} - 1$. Then, along with application of (2.16), we prove (5.10).   □

Corollary 5.5 (convergence estimate). *Under the assumptions of Corollary 4.7, suppose further that, for some $J = 0,1,\ldots$, that $G = A^{-1}$, $G^h = (A^h)^{-1}$, $D = D_{\alpha,J}$, $D^h = D_{\alpha,J}^h$. If $\Delta^\beta\mathbf{u} \in L^2(\Omega)$ for some $0 \le \beta \le J + 1$, then*

$$\left\|\mathbf{u}_M - \mathbf{w}_M^h\right\|^2 + \mathrm{Re}^{-1}\Delta t\sum_{n=0}^{M-1}\left|\mathbf{u}_{n+1/2} - \mathbf{w}_{n+1/2}^h\right|_1^2$$
$$\le C\left(\left(1 + h^2 + 2^J\alpha\delta^2\right)h^{2k} + h^{2s+2} + \Delta t^4 + \left(\alpha\delta^2\right)^{2\beta}\right). \tag{5.19}$$

Proof. Apply estimates for $c_{1,j}$, $c_{2,j}$ from (5.3), resulting from Propositions 5.3, 5.4.   □

5.2. Numerical Testing

This section presents the calculation of a flow with an exact solution to verify the convergence rates of the algorithm. FreeFEM++ [32] was used to run the simulations. The convergence

Table 1: Error and convergence rates for Leray-deconvolution with $J = 0$ for the Taylor-Green vortex with $Re = 10,000$, $\alpha = \sqrt{h}$, and $\delta = \sqrt[4]{h}$. Note the convergence rate is approaching 1 as predicted by (5.21).

m (=1/h)	$\|u - w^h\|_{\infty,0}$	Rate	$\|\nabla(u - w^h)\|_{2,0}$	Rate
20	0.038975		1.651230	
40	0.024334	0.680	1.468510	0.169
60	0.017751	0.778	1.159840	0.582
80	0.013854	0.862	0.935247	0.748
100	0.011255	0.931	0.774285	0.846

Table 2: Error and convergence rates for Leray-deconvolution with $J = 1$ for the Taylor-Green vortex with $Re = 10,000$, $\alpha = \sqrt{h}$, and $\delta = \sqrt[4]{h}$. Note the convergence rate is approaching 2 as predicted by (5.21).

m (=1/h)	$\|u - w^h\|_{\infty,0}$	Rate	$\|\nabla(u - w^h)\|_{2,0}$	Rate
20	0.023384		1.070400	
40	0.009739	1.264	0.640360	0.741
60	0.004997	1.646	0.357779	1.436
80	0.002899	1.892	0.212560	1.810
100	0.001915	1.858	0.136724	1.977

rates are tested against the Taylor-Green vortex problem [13, 33–35]. We use a domain of $\Omega = (0,1) \times (0,1)$ and take $\mathbf{u} = (u_1, u_2)$, where

$$u_1(x,y,t) = -\cos(n\pi x)\sin(n\pi y)e^{-2n^2\pi^2 t/\tau},$$

$$u_2(x,y,t) = \sin(n\pi x)\cos(n\pi y)e^{-2n^2\pi^2 t/\tau}, \qquad (5.20)$$

$$p(x,y,t) = -\frac{1}{4}(\cos(n\pi x) + \cos(n\pi y))e^{-2n^2\pi^2 t/\tau}.$$

The pair $(\mathbf{u}, p)$ is a solution the two-dimensional NSE when $\tau = Re$ and $\mathbf{f} = 0$.

We used CN discretization in time and P2-P1 elements in space according to Problem 1. That is, we used continuous piecewise quadratic elements for the velocity and continuous piecewise linear elements for the pressure. We chose the spatial discretization elements and parameters $n = 1$, $T = 0.5$, $\chi = 0.1$ and $Re = 10,000$ as a illustrative example. We chose $h = 1/m$, $dt = (1/4)h$, $\delta = \sqrt[4]{h}$ and $\alpha = \sqrt{h}$, where m is the number of mesh divisions per side of $[0,1]$. These were chosen so that the result of Corollary 5.5 reduces to

$$\left\|\mathbf{u}_M - \mathbf{w}_M^h\right\| + \left[Re^{-1}\Delta t \sum_{n=0}^{M-1} \left|\mathbf{u}_{n+1/2} - \mathbf{w}_{n+1/2}^h\right|_1^2\right]^{1/2} \leq C\left(h^2 + h^{J+1}\right). \qquad (5.21)$$

We summarize the results in Tables 1 and 2. Table 1 displays error estimates corresponding to no iterations; that is, $J = 0$ in Definition 3.14. For the particular choice of α and δ, the computed errors $\|u - w^h\|_{\infty,0}$ and $\|\nabla(u - w^h)\|_{2,0}$ tend to the predicted convergence rate $\mathcal{O}(h)$. Table 2 displays error estimates corresponding to one update; that is, when $J = 1$ in Definition 3.14. Again, for the particular choice of α and δ, the computed errors $\|u - w^h\|_{\infty,0}$ and $\|\nabla(u - w^h)\|_{2,0}$ tend to the predicted convergence rate $\mathcal{O}(h^2)$.

Convergence Analysis of a Fully Discrete Family of Iterated Deconvolution Methods for
Turbulence Modeling with Time Relaxation

31

6. Conclusion

It is infeasible to resolve all persistent and energetically significant scales down to the Kolmogorov microscale of $\mathcal{O}(\mathrm{Re}^{-3/4})$ for turbulent flows in complex domains using direct numerical simulations in a given time constraint. Regularization methods are used to find approximations to the solution. The modification of iterated Tikhonov-Lavrentiev to the modified iterated Tikhonov-Lavrentiev deconvolution in Definition 3.14 is a highly accurate method of solving the deconvolution problem in the Leray-deconvolution model, with errors $\mathbf{u} - D_{\alpha,0}\overline{\mathbf{u}} = \mathcal{O}((\alpha\delta^2)^{J+1})$ when applied to the differential filter. We use this result to show that under a regularity assumption, the error between the solutions to the NSE and to the Leray deconvolution model with time relaxation using the modified iterated Tikhonov-Lavrentiev deconvolution and discretized with CN in time and FE's in space are $\mathcal{O}(h^k(h + \sqrt{\alpha\delta^2}) + h^{s+1} + \Delta t^2 + (\alpha\delta^2)^{J+1})$.

We also examined the Taylor-Green vortex problem using Problem 1 with the deconvolution in Definition 3.14. We use this problem because it has an exact analytic solution to the NSE. The regularization parameters α and δ were chosen so that the convergence of the approximate solution to the error would be $\mathcal{O}(h^{J+1})$ for $J = 0$ and $J = 1$. The convergence rates calculated correspond to those predicted, that is $\mathcal{O}(h^1)$ for $J = 0$ and $\mathcal{O}(h^2)$ for $J = 1$.

References

[1] G. P. Galdi, "An introduction to the Navier-Stokes initial-boundary value problem," in *Fundamental Directions in Mathematical Fluid Mechanics*, G. Galdi, J. Heywood, and R. Rannacher, Eds., Advances in Mathematical Fluid Mechanics, pp. 1–70, Birkhäuser, Basle, Switzerland, 2000.

[2] C. Foias, D. D. Holm, and E. S. Titi, "The Navier-Stokes-alpha model of fluid turbulence," *Physica D.*, vol. 152/153, pp. 505–519, 2001.

[3] W. Layton, C. C. Manica, M. Neda, and L. G. Rebholz, "Numerical analysis and computational testing of a high accuracy Leray-deconvolution model of turbulence," *Numerical Methods for Partial Differential Equations*, vol. 24, no. 2, pp. 555–582, 2008.

[4] W. Layton, C. C. Manica, M. Neda, and L. G. Rebholz, "Numerical analysis and computational comparisons of the NS-alpha and NS-omega regularizations," *Computer Methods in Applied Mechanics and Engineering*, vol. 199, no. 13–16, pp. 916–931, 2010.

[5] W. Layton and M. Neda, "Truncation of scales by time relaxation," *Journal of Mathematical Analysis and Applications*, vol. 325, no. 2, pp. 788–807, 2007.

[6] S. Stolz, N. A. Adams, and L. Kleiser, "The approximate deconvolution model for large-eddy simulation of compressible flows and its application to shock-turbulent-boundary-layer interaction," *Physics of Fluids*, vol. 13, no. 10, pp. 2985–3001, 2001.

[7] A. Dunca, *Space averaged Navier-Stokes equations in the presence of walls [Ph.D. thesis]*, University of Pittsburgh, 2004.

[8] J. Leray, "Sur le mouvement d'un liquide visqueux emplissant l'espace," *Acta Mathematica*, vol. 63, no. 1, pp. 193–248, 1934.

[9] M. Germano, "Differential filters of elliptic type," *The Physics of Fluids*, vol. 29, no. 6, pp. 1757–1758, 1986.

[10] M. Bertero and P. Boccacci, *Introduction to Inverse Problems in Imaging*, Institute of Physics Publishing, Bristol, UK, 1998.

[11] J. T. King and D. Chillingworth, "Approximation of generalized inverses by iterated regularization," *Numerical Functional Analysis and Optimization*, vol. 1, no. 5, pp. 499–513, 1979.

[12] W. Layton and L. Rebholz, "Approximate deconvolution models of turbulence approximate," in *Analysis, Phenomenology and Numerical Analysis*, Springer Lecture Notes in Mathematics, Philadelphia, Pa, USA, 2012.

[13] I. Stanculescu and C. C. Manica, "Numerical analysis of Leray-Tikhonov deconvolution models of fluid motion," *Computers and Mathematics with Applications*, vol. 60, no. 5, pp. 1440–1456, 2010.

[14] C. Brezinski, M. Redivo-Zaglia, G. Rodriguez, and S. Seatzu, "Extrapolation techniques for ill-conditioned linear systems," *Numerische Mathematik*, vol. 81, no. 1, pp. 1–29, 1998.

[15] H. Gfrerer, "An a posteriori parameter choice for ordinary and iterated Tikhonov regularization of ill-posed problems leading to optimal convergence rates," *Mathematics of Computation*, vol. 49, no. 180, pp. 507–522, 1987.

[16] U. Hämarik, R. Palm, and T. Raus, "Use of extrapolation in regularization methods," *Journal of Inverse and Ill-Posed Problems*, vol. 15, no. 3, pp. 277–294, 2007.

[17] U. Hämarik and U. Tautenhahn, "On the monotone error rule for parameter choice in iterative and continuous regularization methods," *Bachelor of Information Technology*, vol. 41, no. 5, pp. 1029–1038, 2001.

[18] M. Hanke and C. W. Groetsch, "Nonstationary iterated Tikhonov regularization," *Journal of Optimization Theory and Applications*, vol. 98, no. 1, pp. 37–53, 1998.

[19] A. S. Leonov, "On the accuracy of Tikhonov regularizing algorithms and the quasi-optimal choice of regularization parameter," *Doklady Mathematics*, vol. 44, pp. 711–716, 1992.

[20] H. W. Engl, "On the choice of the regularization parameter for iterated Tikhonov regularization of ill-posed problems," *Journal of Approximation Theory*, vol. 49, no. 1, pp. 55–63, 1987.

[21] H. W. Engl, M. Hanke, and A. Neubauer, *Regularization of Inverse Problems*, vol. 375 of *Mathematics and its Applications*, Kluwer Academic Publishers, Dordrecht, The Netherlands, 1996.

[22] G. Vainikko and A. Y. Veretennikov, *Iterative Procedures in Ill-Posed Problems*, Nauka, Moscow, Russia, 1982.

[23] V. Girault and P.-A. Raviart, *Finite Element Methods for Navier-Stokes Equations*, vol. 5 of *Springer Series in Computational Mathematics*, Springer, Berlin, Germany, 1986.

[24] S. C. Brenner and L. R. Scott, *The Mathematical Theory of Finite Element Methods*, vol. 15 of *Texts in Applied Mathematics*, Springer, New York, NY, USA, 1994.

[25] C. Manica and S. K. Merdan, "Convergence analysis of the finite element method for a fundamental model in turbulence," *Mathematical Models and Methods in Applied Sciences*, vol. 22, no. 11, 24 pages, 2012.

[26] J. G. Heywood and R. Rannacher, "Finite-element approximation of the nonstationary Navier-Stokes problem. IV. Error analysis for second-order time discretization," *SIAM Journal on Numerical Analysis*, vol. 27, no. 2, pp. 353–384, 1990.

[27] J. G. Heywood and R. Rannacher, "Finite element approximation of the nonstationary Navier-Stokes problem. I. Regularity of solutions and second-order error estimates for spatial discretization," *SIAM Journal on Numerical Analysis*, vol. 19, no. 2, pp. 275–311, 1982.

[28] Y. He, "A fully discrete stabilized finite-element method for the time-dependent Navier-Stokes problem," *IMA Journal of Numerical Analysis*, vol. 23, no. 4, pp. 665–691, 2003.

[29] Y. He, "The Euler implicit/explicit scheme for the 2D time-dependent Navier-Stokes equations with smooth or non-smooth initial data," *Mathematics of Computation*, vol. 77, no. 264, pp. 2097–2124, 2008.

[30] Y. He and J. Li, "A penalty finite element method based on the Euler implicit/explicit scheme for the time-dependent Navier-Stokes equations," *Journal of Computational and Applied Mathematics*, vol. 235, no. 3, pp. 708–725, 2010.

[31] Y. He and J. Li, "Numerical implementation of the Crank-Nicolson/Adams-Bashforth scheme for the time-dependent Navier-Stokes equations," *International Journal for Numerical Methods in Fluids*, vol. 62, no. 6, pp. 647–659, 2010.

[32] O. Pironneau, F. Hecht, and J. Morice, "FreeFEM++," http://www.freefem.org.

[33] A. J. Chorin, "Numerical solution of the Navier-Stokes equations," *Mathematics of Computation*, vol. 22, pp. 745–762, 1968.

[34] V. John and W. J. Layton, "Analysis of numerical errors in large eddy simulation," *SIAM Journal on Numerical Analysis*, vol. 40, no. 3, pp. 995–1020, 2002.

[35] D. Tafti, "Comparison of some upwind-biased high-order formulations with a second-order central-difference scheme for time integration of the incompressible Navier-Stokes equations," *Computers & Fluids*, vol. 25, no. 7, pp. 647–665, 1996.

Solution of Nonlinear Volterra-Fredholm Integrodifferential Equations via Hybrid of Block-Pulse Functions and Lagrange Interpolating Polynomials

Hamid Reza Marzban and Sayyed Mohammad Hoseini

Department of Mathematical Sciences, Isfahan University of Technology, P.O. Box 8415683111, Isfahan, Iran

Correspondence should be addressed to Hamid Reza Marzban, hmarzban@cc.iut.ac.ir

Academic Editor: Alfredo Bermudez De Castro

An efficient hybrid method is developed to approximate the solution of the high-order nonlinear Volterra-Fredholm integro-differential equations. The properties of hybrid functions consisting of block-pulse functions and Lagrange interpolating polynomials are first presented. These properties are then used to reduce the solution of the nonlinear Volterra-Fredholm integro-differential equations to the solution of algebraic equations whose solution is much more easier than the original one. The validity and applicability of the proposed method are demonstrated through illustrative examples. The method is simple, easy to implement and yields very accurate results.

1. Introduction

Integral and integrodifferential equations have many applications in various fields of science and engineering such as biological models, industrial mathematics, control theory of financial mathematics, economics, electrostatics, fluid dynamics, heat and mass transfer, oscillation theory, queuing theory, and so forth [1].

It is well known that it is extremely difficult to analytically solve nonlinear integrodifferential equations. Indeed, few of these equations can be solved explicitly. So it is required to devise an efficient approximation scheme for solving these equations. So far, several numerical methods are developed. The solution of the first order integrodifferential equations has been obtained by the numerical integration methods such as Euler-Chebyshev [2] and Runge-Kutta methods [3].

Moreover, a differential transform method for solving integrodifferential equations was introduced in [4]. Shidfar et al. [5] applied the homotopy analysis method for solving the nonlinear Volterra and Fredholm integrodifferential equations. As a concrete example, we can express the mathematical model of cell-to-cell spread of HIV-1 in tissue cultures considered by Mittler et al. [6]. Yalcinbas and Sezer [7] proposed an approximation scheme based on Taylor polynomials for solving the high-order linear Volterra-Fredholm integrodifferential equations of the following form:

$$\sum_{j=0}^{q} f_j(t) y^{(j)}(t) = f(t) + \lambda_1 \int_0^t k_1(t,s) y(s) ds + \lambda_2 \int_0^{t_f} k_2(t,s) y(s) ds, \quad 0 \le t, s \le t_f. \tag{1.1}$$

Maleknejad and Mahmoudi [8] developed a numerical method by using Taylor polynomials to solve the following type of nonlinear Volterra-Fredholm integrodifferential equations:

$$\sum_{j=0}^{q} f_j(t) y^{(j)}(t) = f(t) + \lambda_1 \int_0^t k_1(t,s) [y(s)]^p ds + \lambda_2 \int_0^{t_f} k_2(t,s) y(s) ds, \quad 0 \le t, s \le t_f. \tag{1.2}$$

Darania and Ivaz [9] suggested an efficient analytical and numerical procedure for solving the most general form of nonlinear Volterra-Fredholm integrodifferential equations

$$\sum_{j=0}^{q} f_j(t) y^{(j)}(t) = f(t) + \lambda_1 \int_0^t k_1(t,s) g_1(s,y(s)) ds + \lambda_2 \int_0^{t_f} k_2(t,s) g_2(s,y(s)) ds, \tag{1.3}$$

under the mixed conditions

$$\sum_{j=0}^{q-1} \left(\alpha_{ij} y^{(j)}(0) + \beta_{ij} \, y^{(j)}(t_f) \right) = \mu_i, \quad i = 0,1,\ldots,q-1, \tag{1.4}$$

where λ_1 and λ_2 are constants and $0 \le t,s \le t_f$. Moreover, $f(t)$, $k_1(t,s)$, $k_2(t,s)$, $g_1(s,y(s))$, $g_2(s,y(s))$, and $f_j(t)$, $j = 0,1,\ldots,q$, are functions that have suitable derivatives on the interval $0 \le t,s \le t_f$.

These kinds of equations can be found in numerous applications such as electro-dynamics, electromagnetic, biomechanics, and elasticity [10–12]. Akyüz and Sezer [13] presented a Taylor polynomial approach for solving high-order linear Fredholm integrodifferential equations in the most general form. Streltsov [14] developed an effective numerical method based on the use of Chebyshev and Legendre polynomials for solving Fredholm integral equations. Babolian et al. [15] suggested an effective direct method to determine the numerical solution of the specific nonlinear Volterra-Fredholm integrodifferential equations. Their approach was based on triangular functions. In [16, 17], the variational iteration method (VIM) was employed for solving integral and integrodifferential equations. In addition, iterative and noniterative methods for the solution of nonlinear Volterra integrodifferential equations were presented and their local convergence was proved. The iterative methods provide a sequence solution and make use of fixed-point theory whereas the noniterative ones result in series solutions and also make use of the fixed-point principles

[18]. In recent years, the meshless methods have gained more attention not only by mathematicians but also in the engineering community. In [19], the meshless moving least square method was employed for solving nonlinear Fredholm integrodifferential equations.

In the present paper, we introduce a new numerical method for solving (1.3) with the mixed conditions (1.4). The method is based on a hybrid of block-pulse functions and Lagrange interpolating polynomials. The associated operational matrices of integration and product together with the Kronecker property of hybrid functions are then used to reduce the solution of (1.3) and (1.4) to the solution of nonlinear algebraic equations whose solution is much more easier than the original one.

The paper is organized as follows. In Section 2, we describe the basic properties of hybrid functions. Section 3 is devoted to the solution of the nonlinear Volterra-Fredholm integrodifferential equations. Finally, in Section 4, various types of integrodifferential equations are given to demonstrate the efficiency and the accuracy of the proposed method.

2. Hybrid Functions

2.1. Properties of Hybrid Functions

In order to define hybrid functions, we first divide the interval $[0, t_f)$ into N equidistant subintervals $[((n-1)/N) t_f, (n/N) t_f)$, for $n = 1, 2, \ldots, N$. The hybrid of block-pulse functions and Lagrange interpolating polynomials, $b_{nm}(t), n = 1, 2, \ldots, N, m = 0, 1, \ldots, M-1$, are defined on the interval $[0, t_f)$ as, [20],

$$b_{nm}(t) = \begin{cases} L_m\left(\dfrac{2N}{t_f} t - 2n + 1\right), & t \in \left[\left(\dfrac{n-1}{N}\right) t_f, \dfrac{n}{N} t_f\right), \\ 0, & \text{otherwise,} \end{cases} \tag{2.1}$$

where n and m are the orders of block-pulse functions and Lagrange interpolating polynomials, respectively. Here, $L_m(t)$ are defined in [21] as

$$L_m(t) = \prod_{\substack{i=0 \\ i \neq m}}^{M-1} \left(\frac{t - \tau_i}{\tau_m - \tau_i}\right), \quad m = 0, 1, \ldots, M-1, \tag{2.2}$$

in which $\tau_i, i = 0, 1, \ldots, M-1$ are the zeros of Legendre polynomial of order M, with the Kronecker property

$$L_m(\tau_i) = \delta_{mi} = \begin{cases} 1, & \text{if } i = m, \\ 0, & \text{if } i \neq m. \end{cases} \tag{2.3}$$

where δ_{mi} is the Kronecker delta function. No explicit formulas are known for the points τ_i. However, they can be computed numerically using existing subroutines [21]. Since $b_{nm}(t)$ consists of block-pulse functions and Lagrange interpolating polynomials, which are both complete and orthogonal, the set of hybrid of block-pulse functions and Lagrange interpolating polynomials is a complete orthogonal set in the Hilbert space $\mathcal{L}^2[0, t_f]$. A more detailed error analysis concerning this type of hybrid functions was discussed in [22].

2.2. *Function Approximation*

A function $f(t) \in \mathcal{L}^2[0, t_f)$ may be expanded as

$$f(t) = \sum_{n=1}^{\infty} \sum_{m=0}^{\infty} f_{nm}\, b_{nm}(t). \tag{2.4}$$

If the infinite series in (2.4) is truncated, then (2.4) can be written as

$$f(t) \simeq \sum_{n=1}^{N} \sum_{m=0}^{M-1} f_{nm}\, b_{nm}(t) = F^T B(t), \tag{2.5}$$

where

$$F = \left[f_{10}, \ldots, f_{1(M-1)}, \ldots, f_{N0}, \ldots, f_{N(M-1)} \right]^T,$$
$$B(t) = \left[b_{10}(t), \ldots, b_{1(M-1)}(t), \ldots, b_{N0}(t), \ldots, b_{N(M-1)}(t) \right]^T. \tag{2.6}$$

In (2.6), f_{nm}, $n = 1, 2, \ldots, N$, $m = 0, 1, \ldots, M-1$ are the expansion coefficients of the function $f(t)$ in the nth subinterval $[((n-1)/N)t_f, (n/N)t_f)$, and $b_{nm}(t)$, $n = 1, 2, \ldots, N$, $m = 0, 1, \ldots, M-1$ are defined in (2.1). In order to evaluate the coefficients f_{nm}, we use Gaussian nodes which are the zeros of Legendre polynomial of order M. Let τ_m, $m = 0, 1, \ldots, M-1$ be the Gaussian nodes which are defined in $[-1, 1]$ and let t_{nm}, $n = 1, 2, \ldots, N$, $m = 0, 1, \ldots, M-1$ be the corresponding Gaussian nodes in the nth subinterval $[((n-1)/N)\, t_f, (n/N)\, t_f]$. The relation between τ_m and t_{nm} is given by

$$t_{nm} = \frac{t_f}{2N}(\tau_m + 2\,n - 1). \tag{2.7}$$

With the aid of (2.3) we have

$$b_{nm}(t_{ij}) = \begin{cases} 1, & \text{if } i = n,\ j = m, \\ 0, & \text{otherwise.} \end{cases} \tag{2.8}$$

Using (2.8), the coefficients f_{nm}, can be obtained as

$$f_{nm} = f(t_{nm}). \tag{2.9}$$

Now, let $g(t, s)$ be a function of two independent variables defined for $t \in [0, t_f]$ and $s \in [0, s_f]$. Then $g(t, s)$ can be expanded in terms of hybrid functions as follows:

$$g(t, s) \simeq B^T(t)GB(s), \tag{2.10}$$

Solution of Nonlinear Volterra-Fredholm Integrodifferential Equations via Hybrid of Block-Pulse
Functions and Lagrange Interpolating Polynomials

37

where G is a matrix of order $NM \times NM$ whose elements can be calculated for $i, n = 1, 2, \ldots, N$, and $j, m = 0, 1, \ldots, M - 1$ as follows:

$$G_{(n-1)M+m,(i-1)M+j} = g(t_{nm}, s_{ij}), \qquad (2.11)$$

in which s_{ij} can be determined by the following relation:

$$s_{ij} = \frac{s_f}{2N}(\tau_j + 2i - 1). \qquad (2.12)$$

The integration of the vector $B(t)$ can also be expanded in terms of hybrid functions as

$$\int_0^t B(s)\,ds \simeq PB(t), \qquad (2.13)$$

where P is the $NM \times NM$ operational matrix of integration and given by, [20],

$$P = \begin{bmatrix} E & H & H & \cdots & H \\ 0 & E & H & \cdots & H \\ 0 & 0 & E & \cdots & H \\ \vdots & \vdots & \vdots & & \vdots \\ 0 & 0 & 0 & \cdots & E \end{bmatrix}, \qquad (2.14)$$

in which E and H can be obtained in the following manner.
    Let

$$E = (e_{ij}), \qquad H = (h_{ij}), \qquad (2.15)$$

then, for $i, j = 0, 1, \ldots, M - 1$, we have

$$e_{ij} = \frac{t_f}{2N} \int_{-1}^{\tau_j} L_i(t)dt,$$

$$h_{ij} = \frac{t_f}{2N} \int_{-1}^{1} L_i(t)dt, \qquad (2.16)$$

where $\tau_j, j = 0, 1, \ldots, M - 1$ are the zeros of Legendre polynomial of order M. It is noted that E is the associated operational matrix of integration for Lagrange interpolating polynomials on interval $[((n-1)/N)t_f, (n/N)\ t_f]$.
    Integrating (2.5) from 0 to t and using (2.13) imply

$$\int_0^t f(s)ds \simeq F^T PB(t). \qquad (2.17)$$

From (2.13), by using the interpolating property of hybrid functions we get

$$\int_0^{t_{nm}} f(s)ds \simeq F^T PB(t_{nm}) = \sum_{i=1}^{n-1}\sum_{j=0}^{M-1} f_{ij}h_{jm} + \sum_{j=0}^{M-1} f_{nj}e_{jm}. \tag{2.18}$$

2.3. The Integration of the Cross-Product of Two Hybrid Function Vectors

The following property of the cross-product of two hybrid function vectors will also be used. Integrating the square matrix $B(t)B^T(t)$ from 0 to t_f, and using the orthogonality of hybrid functions, we get

$$W = \int_0^{t_f} B(t)B^T(t)dt, \tag{2.19}$$

where W is a block diagonal matrix of order $NM \times NM$ given by

$$W = \begin{bmatrix} A & 0 & \cdots & 0 \\ 0 & A & \cdots & 0 \\ \vdots & \vdots & \ddots & \vdots \\ 0 & 0 & \cdots & A \end{bmatrix}, \tag{2.20}$$

moreover A is also a diagonal matrix of order $M \times M$ given as

$$A = \mathrm{diag}[a_0, a_1, \ldots, a_{M-1}]. \tag{2.21}$$

According to the orthogonality of hybrid functions, the values of a_m, for $m = 0, 1, \ldots, M-1$, can be computed as follows:

$$a_m = \frac{t_f}{2N}\int_{-1}^1 L_m^2(t)dt. \tag{2.22}$$

2.4. The Operational Matrix of Derivative

Let d/dt denote the derivative operator. By expanding $(d/dt)B(t)$ in terms of $B(t)$ we get

$$\frac{d}{dt}B(t) = D^*B(t). \tag{2.23}$$

In (2.23), D^* is called the operational matrix of derivative for hybrid functions which is a block diagonal matrix of order $NM \times NM$ and is given by

$$D^* = \begin{bmatrix} D & 0 & \cdots & 0 \\ 0 & D & \cdots & 0 \\ \vdots & \vdots & \ddots & \vdots \\ 0 & 0 & \cdots & D \end{bmatrix}, \tag{2.24}$$

where D is also a matrix of order $M \times M$ and the (i, j) th element of this matrix denoted by $[D]_{ij}$ for $i, j = 0, 1, \ldots, M-1$ can be calculated as follows:

$$[D]_{ij} = \frac{dL_i}{dt}(t_{ij}) = \frac{2N}{t_f} L_i'\left(\frac{2N}{t_f} t_{ij} - 2j + 1\right) = \frac{2N}{t_f} L_i'(\tau_j). \tag{2.25}$$

Let $f^{(1)}(t)$ denote the first derivative of $f(t)$ with respect to t. By expanding $f^{(1)}(t)$ in terms of hybrid functions and using (2.5) and (2.23) we obtain

$$f^{(1)}(t) \simeq \sum_{n=1}^{N} \sum_{m=0}^{M-1} f^{(1)}(t_{nm})b_{nm}(t) = F^T D^* B(t). \tag{2.26}$$

According to (2.8) we get

$$f^{(1)}(t_{nm}) = F^T D^* B(t_{nm}) = \sum_{l=0}^{M-1} f_{nl}[D]_{lm}. \tag{2.27}$$

Let $f^{(j)}(t)$ denote the jth derivative of $f(t)$ with respect to t. In order to obtain the relation between the expansion coefficients of $f^{(j)}(t)$ and $f(t)$, we use the following fact:

$$\frac{d^j}{dt^j}B(t) = (D^*)^j B(t), \tag{2.28}$$

where $(D^*)^j$ is the jth power of the matrix D^* given in (2.23). Equation (2.28) can easily be obtained from (2.23) by differentiating $f(t)$, j times with respect to t, $j = 2, 3, \ldots, q$. It should be noted that the matrix $(D^*)^j$ has the following structure:

$$(D^*)^j = \begin{bmatrix} D^j & 0 & \cdots & 0 \\ 0 & D^j & \cdots & 0 \\ \vdots & \vdots & \ddots & \vdots \\ 0 & 0 & \cdots & D^j \end{bmatrix}. \tag{2.29}$$

By expanding $f^{(j)}(t)$ in terms of hybrid functions and using (2.5) and (2.28) we get

$$f^{(j)}(t) \simeq \sum_{n=1}^{N} \sum_{m=0}^{M-1} f^{(j)}(t_{nm}) \, b_{nm}(t) = F^T (D^*)^j B(t). \tag{2.30}$$

According to (2.8) it immediately follows that

$$f^{(j)}(t_{nm}) = F^T (D^*)^j B(t_{nm}) = \sum_{l=0}^{M-1} f_{nl} \left[D^j \right]_{lm}. \tag{2.31}$$

where $[D^j]_{lm}$ denotes the (l, m) th element of the matrix D^j.

3. Solution of the Nonlinear Volterra-Fredholm Integrodifferential Equations

Consider the nonlinear Volterra-Fredholm integrodifferential equation given in (1.3). Since the set of hybrid functions is a complete orthogonal set in the Hilbert space $\mathcal{L}^2[0, t_f)$, we can expand any function in this space in terms of hybrid functions.

Let

$$y(t) = \sum_{n=1}^{N} \sum_{m=0}^{M-1} y_{nm} b_{nm}(t) = Y^T B(t), \tag{3.1}$$

where

$$y_{nm} = y(t_{nm}). \tag{3.2}$$

Using (2.30), for $j = 1, 2, \ldots, q$, we have

$$y^{(j)}(t) = \sum_{n=1}^{N} \sum_{m=0}^{M-1} y^{(j)}(t_{nm}) b_{nm}(t). \tag{3.3}$$

where

$$y^{(j)}(t_{nm}) = \sum_{l=0}^{M-1} y_{nl} \left[D^j \right]_{lm}. \tag{3.4}$$

Moreover, by expanding $k_2(t, s)$ in terms of hybrid functions we get

$$k_2(t, s) = B^T(t) K_2 B(s), \tag{3.5}$$

Solution of Nonlinear Volterra-Fredholm Integrodifferential Equations via Hybrid of Block-Pulse
Functions and Lagrange Interpolating Polynomials

41

where K_2 is a matrix of order $NM \times NM$ and can be calculated similar to matrix G in (2.10). From (2.5) and (2.9) we obtain

$$
\begin{aligned}
g_2(t, y(t)) &= \sum_{n=1}^{N} \sum_{m=0}^{M-1} g_2(t_{nm}, y(t_{nm})) b_{nm}(t) \\
&= \sum_{n=1}^{N} \sum_{m=0}^{M-1} g_2(t_{nm}, y_{nm}) b_{nm}(t) = B^T(t) G_2,
\end{aligned}
\tag{3.6}
$$

where G_2 is a vector of order $NM \times 1$. From (2.19), (3.5) and (3.6), we have

$$
\int_0^1 k_2(t, s) g_2(s, y(s)) ds = \int_0^1 B^T(t) K_2 B(s) B^T(s) G_2 ds = B^T(t) V,
\tag{3.7}
$$

where

$$
V = K_2 W G_2.
\tag{3.8}
$$

It should be noted that V is a vector of order $NM \times 1$. Substituting (3.1)–(3.7) in (1.3) yields

$$
\sum_{j=0}^{q} f_j(t) y^{(j)}(t) = f(t) + \lambda_1 \int_0^t k_1(t, s) g_1(s, y(s)) ds + \lambda_2 B^T(t) V.
\tag{3.9}
$$

We now collocate (1.3) at NM points. For a suitable collocation points, we use Gaussian nodes $t_{nm}, n = 1, 2, \ldots, N, m = 0, 1, \ldots, M - 1$, which are introduced in (2.7). Therefore we have

$$
\sum_{j=0}^{q} f_j(t_{nm}) y^{(j)}(t_{nm}) = f(t_{nm}) + \lambda_1 \int_0^{t_{nm}} k_1(t_{nm}, s) g_1(s, y(s)) ds + \lambda_2 B^T(t_{nm}) V.
\tag{3.10}
$$

Also, using (2.8) we get

$$
B^T(t_{nm}) V = V_{(n-1)M+m},
\tag{3.11}
$$

where $V_{(n-1)M+m}$ denotes the $((n-1)M + m)$th element of vector V. Consequently, with the aid of (2.18) and (3.10) we obtain

$$
\sum_{j=0}^{q} f_j(t_{nm}) y^{(j)}(t_{nm}) = f(t_{nm}) + \lambda_2 V_{(n-1)M+m}
$$

$$
+ \lambda_1 \left(\sum_{i=1}^{n-1} \sum_{j=0}^{M-1} k_1(t_{nm}, s_{ij}) g_1(s_{ij}, y_{ij}) h_{jm} \right.
$$

$$
\left. + \sum_{j=0}^{M-1} k_1(t_{nm}, s_{nj}) g_1(s_{nj}, y_{nj}) e_{jm} \right),
$$

(3.12)

where $y^{(j)}(t_{nm})$ for $n = 1, 2, \ldots, N$, $m = 0, 1, \ldots, M-1$ can be calculated from (3.4). Equation (3.12) gives NM nonlinear equations. Finally, we approximate the mixed conditions given in (1.4). From (1.4), (3.1), (3.3), and the definition of hybrid functions we obtain

$$
\sum_{j=0}^{q-1} \left(\alpha_{ij} \sum_{m=0}^{M-1} y^{(j)}(t_{1m}) b_{1m}(0) + \beta_{ij} \sum_{m=0}^{M-1} y^{(j)}(t_{Nm}) b_{Nm}(t_f) \right) = \mu_i, \quad i = 0, 1, \ldots, q-1.
$$

(3.13)

In order that the approximate solution obtained by the present method be continuous, we impose the continuity condition at the points $(n/N)t_f$, $n = 1, 2, \ldots, N-1$. Equations. (3.12)-(3.13) together with the continuity conditions give $N(M+1)+q-1$, nonlinear equations that can be solved by using the well-known Tau method [23] for the unknown coefficients y_{nm}, $n = 1, 2, \ldots, N$, $m = 0, 1, \ldots M-1$.

4. Illustrative Examples

In this section, four examples are given to demonstrate the efficiency, the accuracy, and the applicability of the proposed method.

Example 4.1. Consider the nonlinear Volterra-Fredholm integrodifferential equation

$$
t^2 y''(t) + 2y'(t) = 2 - \frac{5}{6} t + \frac{1}{2} t e^{-t^2} + \int_0^t t s e^{-y^2(s)} ds + \int_0^1 t y^2(s) ds,
$$

$$
y(0) = 0, \qquad y'(0) = 1,
$$

(4.1)

whose exact solution is given in [9] as $y(t) = t$. We applied the present method to solve this problem.

Let us define

$$
E = \left(\int_0^{t_f} (y_e(t) - y(t))^2 dt \right)^{1/2},
$$

(4.2)

Table 1: The computational results of $\mathcal{L}^2$-norm error for Example 4.1.

Present method	E
$N = 2, M = 8$	$9.28e - 12$
$N = 2, M = 9$	$6.32e - 12$
$N = 2, M = 10$	$1.85e - 13$
$N = 4, M = 8$	$1.11e - 13$
$N = 4, M = 9$	$6.26e - 15$
$N = 4, M = 10$	$9.29e - 17$

Table 2: The computational results of $\mathcal{L}^2$-norm error for Example 4.2.

Present method	E
$N = 2, M = 8$	$8.83e - 12$
$N = 2, M = 9$	$3.37e - 12$
$N = 2, M = 10$	$2.15e - 13$
$N = 4, M = 8$	$7.00e - 14$
$N = 4, M = 9$	$4.63e - 15$
$N = 4, M = 10$	$9.70e - 17$

where $y(t)$ and $y_e(t)$ denote the approximate solution obtained by the present method and the exact solution, respectively. In Table 1, the computational results of the $\mathcal{L}^2$-norm error denoted by E between the approximate solution and the exact solution for different values of N and M are given. The simulation results reported in Table 1, indicate that there is an excellent agreement between the approximate and exact solution.

Example 4.2. As the second example, consider the nonlinear Volterra integrodifferential equation

$$y'(t) = 2t - \frac{1}{2}\sin\left(t^4\right) + \int_0^t t^2 s \cos\left(t^2 y(s)\right) ds,$$

$$y(0) = 0,$$

(4.3)

whose exact solution is given in [19] as $y(t) = t^2$. We applied the present method to solve this problem. In Table 2, the computational results of the $\mathcal{L}^2$-norm error between the approximate solution and the exact solution for different values of N and M are summarized. The simulation results obtained by the present method are superior to those reported in [19].

Example 4.3. Let us consider the nonlinear Volterra integrodifferential equation

$$y^{(4)}(t) = 1 + \int_0^t e^{-s} y^2(s) ds, \quad t \in [0,1],$$

$$y(0) = 1, \qquad y'(0) = 1,$$

$$y(1) = e, \qquad y'(1) = e,$$

(4.4)

Table 3: The computational results of $y(t)$ for Example 4.3.

t	Tau method $n = 10$	Present method $N = 2, M = 10$	Exact
0.0	1.0000000000	1.0000000000	1.0000000000
0.2	1.2214027513	1.2214027582	1.2214027582
0.4	1.4918246782	1.4918246976	1.4918246976
0.6	1.8221187743	1.8221188004	1.8221188004
0.8	2.2255409119	2.2255409285	2.2255409285
1.0	2.7182818285	2.7182818284	2.7182818284

Table 4: The computational results of $\mathcal{L}^2$-norm error for Example 4.3.

Present method	E
$N = 2, M = 8$	$4.64e - 08$
$N = 2, M = 9$	$2.94e - 10$
$N = 2, M = 10$	$3.68e - 11$
$N = 4, M = 8$	$3.01e - 09$
$N = 4, M = 9$	$4.52e - 12$
$N = 4, M = 10$	$5.96e - 13$

whose exact solution is given in [24] as $y(t) = e^t$. We solved this problem by using hybrid functions. In Table 3, the computational results of $y(t)$ obtained by the present method with $N = 2$ and $M = 10$ together with the exact solution and the solution determined by the Tau method [24] are summarized. The computational results obtained by the present method are superior to that reported in [24]. Moreover in Table 4, the $\mathcal{L}^2$-norm error between the approximate solution obtained by the present method and the exact solution for different values of N and M is given. It should be pointed out that only a very small numbers of hybrid basis functions are needed to obtain a quite satisfactory approximation to the exact solution.

Example 4.4. As the last example, consider the nonlinear Volterra-Fredholm integrodifferential equation

$$y'''(t) - ty''(t) = \frac{1}{3}t^3 - 2t + 1 - \int_0^t s\, y(s)ds + \int_0^1 \frac{2te^s}{y(s) - s}ds,$$

$$y(0) = 1, \qquad y'(0) = 1, \qquad y(1) = 1 + e, \tag{4.5}$$

which has the exact solution $y(t) = t + e^t$. To measure the accuracy of proposed approach, the $\mathcal{L}^2$-norm error between the approximate solution obtained by the present method for different values of N and M and the exact solution is given in Table 5. The computational results reveal that the method is convergent.

5. Conclusion

In the present work, a hybrid approximation method was developed for solving the high-order nonlinear Volterra-Fredholm integrodifferential equations. The nice properties of

Solution of Nonlinear Volterra-Fredholm Integrodifferential Equations via Hybrid of Block-Pulse
Functions and Lagrange Interpolating Polynomials

45

Table 5: The computational results of $\mathcal{L}^2$-norm error for Example 4.4.

Present method	E
$N = 2, M = 8$	$2.18e - 08$
$N = 2, M = 9$	$6.15e - 10$
$N = 2, M = 10$	$1.31e - 11$
$N = 4, M = 8$	$6.42e - 10$
$N = 4, M = 9$	$9.28e - 12$
$N = 4, M = 10$	$9.82e - 14$

hybrid functions together with the associated operational matrices of integration, derivative, and cross-product are used to reduce the solution of problem to the solution of nonlinear algebraic equations whose solution is much more easier than the original one. The matrices P, C, and D^* are sparse, hence making the method computationally attractive without sacrificing the accuracy of the solution. Moreover, It was shown that small values for N and M are needed to achieve high accuracy and a satisfactory convergence. The numerical results support this claim. The method is general, easy to implement and yields the desired accuracy in a few terms of hybrid basis functions. The simulation results indicate the convergence and the effectiveness of the proposed approach.

Acknowledgments

The authors are very grateful to the referees for their comments and valuable suggestions which improved the paper.

References

[1] A. D. Polyanin and A. V. Manzhirov, *Handbook of Integral Equations*, Chapman & Hall/CRC, Boca Raton, Fla, USA, 2nd edition, 2008.

[2] P. J. Van der Houwen and B. P. Sommeijer, "Euler-Chebyshev methods for integro-differential equations," *Applied Numerical Mathematics*, vol. 24, no. 2-3, pp. 203–218, 1997.

[3] W. H. Enright and M. Hu, "Continuous Runge-Kutta methods for neutral Volterra integro-differential equations with delay," *Applied Numerical Mathematics*, vol. 24, no. 2-3, pp. 175–190, 1997.

[4] P. Darania and A. Ebadian, "A method for the numerical solution of the integro-differential equations," *Applied Mathematics and Computation*, vol. 188, no. 1, pp. 657–668, 2007.

[5] A. Shidfar, A. Molabahrami, A. Babaei, and A. Yazdanian, "A series solution of the nonlinear Volterra and Fredholm integro-differential equations," *Communications in Nonlinear Science and Numerical Simulation*, vol. 15, no. 2, pp. 205–215, 2010.

[6] J. E. Mittler, B. Sulzer, A. U. Neumann, and A. S. Perelson, "Influence of delayed viral production on viral dynamics in HIV-1 infected patients," *Mathematical Biosciences*, vol. 152, no. 2, pp. 143–163, 1998.

[7] S. Yalcinbas and M. Sezer, "The approximate solution of high-order linear Volterra-Fredholm integro-differential equations in terms of Taylor polynomials," *Applied Mathematics and Computation*, vol. 112, no. 2-3, pp. 291–308, 2000.

[8] K. Maleknejad and Y. Mahmoudi, "Taylor polynomial solution of high-order nonlinear Volterra-Fredholm integro-differential equations," *Applied Mathematics and Computation*, vol. 145, no. 2-3, pp. 641–653, 2003.

[9] P. Darania and K. Ivaz, "Numerical solution of nonlinear Volterra-Fredholm integro-differential equations," *Computers & Mathematics with Applications*, vol. 56, no. 9, pp. 2197–2209, 2008.

[10] F. Bloom, "Asymptotic bounds for solutions to a system of damped integro-differential equations of electromagnetic theory," *Journal of Mathematical Analysis and Applications*, vol. 73, no. 2, pp. 524–542, 1980.

[11] K. Holmåker, "Global asymptotic stability for a stationary solution of a system of integro-differential equations describing the formation of liver zones," *SIAM Journal on Mathematical Analysis*, vol. 24, no. 1, pp. 116–128, 1993.

[12] L. K. Forbes, S. Crozier, and D. M. Doddrell, "Calculating current densities and fields produced by shielded magnetic resonance imaging probes," *SIAM Journal on Applied Mathematics*, vol. 57, no. 2, pp. 401–425, 1997.

[13] A. Akyüz and M. Sezer, "A Taylor polynomial approach for solving high-order linear Fredholm integro-differential equations in the most general form," *International Journal of Computer Mathematics*, vol. 84, no. 4, pp. 527–539, 2007.

[14] I. P. Streltsov, "Application of Chebyshev and Legendre polynomials on discrete point set to function interpolation and solving Fredholm integral equations," *Computer Physics Communications*, vol. 126, no. 1-2, pp. 178–181, 2000.

[15] E. Babolian, Z. Masouri, and S. Hatamzadeh-Varmazyar, "Numerical solution of nonlinear Volterra-Fredholm integro-differential equations via direct method using triangular functions," *Computers & Mathematics with Applications*, vol. 58, no. 2, pp. 239–247, 2009.

[16] N. Bildik, A. Konuralp, and S. Yalçınbaş, "Comparison of Legendre polynomial approximation and variational iteration method for the solutions of general linear Fredholm integro-differential equations," *Computers & Mathematics with Applications*, vol. 59, no. 6, pp. 1909–1917, 2010.

[17] S.-Q. Wang and J.-H. He, "Variational iteration method for solving integro-differential equations," *Physics Letters A*, vol. 367, no. 3, pp. 188–191, 2007.

[18] J. I. Ramos, "Iterative and non-iterative methods for non-linear Volterra integro-differential equations," *Applied Mathematics and Computation*, vol. 214, no. 1, pp. 287–296, 2009.

[19] M. Dehghan and R. Salehi, "The numerical solution of the non-linear integro-differential equations based on the meshless method," *Journal of Computational and Applied Mathematics*, vol. 236, no. 9, pp. 2367–2377, 2012.

[20] H. R. Marzban, S. M. Hoseini, and M. Razzaghi, "Solution of Volterra's population model via block-pulse functions and Lagrange-interpolating polynomials," *Mathematical Methods in the Applied Sciences*, vol. 32, no. 2, pp. 127–134, 2009.

[21] C. Canuto, M. Y. Hussaini, A. Quarteroni, and T. A. Zang, *Spectral Methods in Fluid Dynamics*, Springer, New York, NY, USA, 1987.

[22] H. R. Marzban, H. R. Tabrizidooz, and M. Razzaghi, "A composite collocation method for the nonlinear mixed Volterra-Fredholm-Hammerstein integral equations," *Communications in Nonlinear Science and Numerical Simulation*, vol. 16, no. 3, pp. 1186–1194, 2011.

[23] E. L. Ortiz, "The tau method," *SIAM Journal on Numerical Analysis*, vol. 6, pp. 480–492, 1969.

[24] G. Ebadi, M. Y. Rahimi-Ardabili, and S. Shahmorad, "Numerical solution of the nonlinear Volterra integro-differential equations by the tau method," *Applied Mathematics and Computation*, vol. 188, no. 2, pp. 1580–1586, 2007.

The Optimal L^2 Error Estimate of Stabilized Finite Volume Method for the Stationary Navier-Stokes Problem

Guoliang He,[1] Jian Su,[2] and Wenqiang Dai[3]

[1] *School of Mathematical Science, University of Electronic Science and Technology, Chengdu 610054, China*
[2] *Faculty of Science, Xi'an Jiaotong University, Xi'an 710049, China*
[3] *School of Management, University of Electronic Science and Technology, Chengdu 610054, China*

Correspondence should be addressed to Guoliang He, hegl@uestc.edu.cn

Academic Editor: Raytcho Lazarov

A finite volume method based on stabilized finite element for the two-dimensional stationary Navier-Stokes equations is analyzed. For the P_1–P_0 element, we obtain the optimal L^2 error estimates of the finite volume solution u_h and p_h. We also provide some numerical examples to confirm the efficiency of the FVM. Furthermore, the effect of initial value for iterative method is analyzed carefully.

1. Introduction

In paper [1], G. He and Y. He introduce a finite volume method (FVM) based on the stabilized finite element method for solving the stationary Navier-Stokes problem and obtain the optimal H^1 error estimates for discretization velocity, however, to our dismay, without the optimal L^2 error estimate. It is inspiring that the following further numerical examples tell us that it has nearly second-order convergence rate. So, in this paper, we introduce a new technique to prove the optimal L^2 error of a generalized bilinear form and then gain the optimal L^2 error estimates of the stabilized finite volume method for the stationary Navier-Stokes problem.

For the convenience of analysis, we introduce the following useful notations. Let Ω be a bounded domain in $\mathbb{R}^2$ assumed to have a Lipschitz continuous boundary $\partial\Omega$ and to satisfy a further smooth condition to ensure the weak solution's existence and regularity of

Stokes problem. (For more information, see the A1 assumption stated in [1, 2].) We consider the stationary Navier-Stokes equations

$$-\nu\Delta u + (u \cdot \nabla)u + \nabla p = f, \quad \text{div } u = 0, \ x \in \Omega,$$

$$u|_{\partial\Omega} = 0, \tag{1.1}$$

where $u = (u_1(x), u_2(x))$ represents the velocity vector, $p = p(x)$ the pressure, $f = f(x)$ the prescribed body force, and $\nu > 0$ the viscosity.

For the mathematical setting of problem (1.1), we introduce the following Hilbert spaces:

$$X = \left(H_0^1(\Omega)\right)^2, \quad Y = \left(L^2(\Omega)\right)^2, \quad M = \left\{q \in L^2(\Omega) : \int_\Omega q \, dx = 0\right\},$$

$$H = \left\{v \in L^2(\Omega)^2; \ \text{div } v = 0 \text{ in } \Omega, \ v \cdot \mathbf{n}|_{\partial\Omega} = 0\right\}. \tag{1.2}$$

The spaces $(L^2(\Omega))^m$ $(m = 1, 2, 4)$ are endowed with the usual L^2-scalar product $(\cdot, \cdot)$ and norm $\|\cdot\|_0$, as appropriate. The space X is equipped with the scalar product $(\nabla u, \nabla v)$ and norm $\|\nabla u\|_0$.

Define $Au = -\Delta u$, which is the operator associated with the Navier-Stokes equations. It is positive self-adjoint operator from $D(A) = (H^2(\Omega))^2 \cap X$ onto Y, so, for $\alpha \in \mathbb{R}$, the power A^α of A is well defined. In particular, $D(A^{1/2}) = X$, $D(A^0) = Y$, and

$$\left(A^{1/2}u, \ A^{1/2}v\right) = (\nabla u, \ \nabla v), \quad \left\|A^{1/2}u\right\|_0 = (\nabla u, \ \nabla u)^{1/2}, \tag{1.3}$$

for all $u, v \in X$.

We also introduce the following continuous bilinear forms $a(\cdot, \cdot)$ and $d(\cdot, \cdot)$ on $X \times X$ and $X \times M$, respectively, by

$$a(u, v) = \nu((u, v)) \quad \forall u, v \in X, \quad d(v, q) = -(v, \nabla q) = (q, \text{div } v) \quad \forall v \in X, \ q \in M, \tag{1.4}$$

a generalized bilinear form on $(X, M) \times (X, M)$ by

$$\mathcal{B}((u, p); (v, q)) = a(u, v) - d(v, p) + d(u, q), \tag{1.5}$$

and a trilinear form on $X \times X \times X$ by

$$b(u, v, w) = ((u \cdot \nabla)v, w) + \frac{1}{2}((\text{div } u)v, w). \tag{1.6}$$

Under the above notations, the variational formulation of the problem (1.1) reads as follows: find $(u, p) \in (X, M)$ such that for all $(v, q) \in (X, M)$:

$$\mathcal{B}((u, p); (v, q)) + b(u, u, v) = (f, v). \tag{1.7}$$

The following existence and uniqueness results are classical (see [3]).

Theorem 1.1. *Assume that v and $f \in Y$ satisfy the following uniqueness condition:*

$$1 - \frac{N_1}{v^2}\|f\|_{-1} > 0, \tag{1.8}$$

where

$$N_1 = \sup_{u,v,w \in H_0^1(\Omega)} \frac{b(u,v,w)}{\|A^{1/2}u\|_0 \|A^{1/2}v\|_0 \|A^{1/2}w\|_0}. \tag{1.9}$$

Then the problem (1.7) admits a unique solution $(u,p) \in (D(A) \cap X, H^1(\Omega) \cap M)$ such that

$$\left\|A^{1/2}u\right\|_1 \leq v^{-1}\|f\|_{-1}, \qquad \|u\|_2 + \|p\|_1 \leq c\|f\|_0. \tag{1.10}$$

2. FVM Based on Stabilized Finite Element Approximation

In this section, we consider the FVM for two-dimensional stationary incompressible Navier-Stokes equations (1.1). Let $h > 0$ be a real positive parameter. The finite element subspace (X_h, M_h) of (X, M) is characterized by $T_h = T_h(\Omega)$, a partition of $\overline{\Omega}$ into triangles, assumed to be regular in the usual sense (see [4–7]). The mesh parameter h is given by $h = \max\{h_K\}$, and the set of all interelement boundaries will be denoted by Γ_h. Besides, we also use the configuration based on barycenter of element $K_i \in T_h$ to construct a dual partition T_h^* of T_h, which is shown in Figure 1.

Finite element subspaces of interest in this paper are defined as follows: the continuous piecewise linear velocity subspace

$$X_h = \left\{v \in X : v|_K \in (P_1(K))^2, \ \forall K \in T_h\right\}, \tag{2.1}$$

the piecewise constant pressure subspace

$$M_h = \left\{q \in M : q|_K \in P_0(K), \ \forall K \in T_h\right\}, \tag{2.2}$$

and the dual space of velocity subspace X_h^*

$$X_h^* = \left\{v \in \left(L^2(\Omega)\right)^2 : v|_{K^*} \in (P_0(K^*))^2, \ \forall K^* \in T_h^*\right\}. \tag{2.3}$$

Actually, this choice of X_h^* is the span of the charaicteristic functions of the volume K^*. Note that this mixed finite element method is unstable in the standard Babuška-Brezzi sense [8].

Define the interpolation operator $I_h^* : X_h \rightarrow X_h^*$,

$$I_h^* u_h = \sum_{x_i \in N_h} u_h(x_i)\chi_i(x), \tag{2.4}$$

where $N_h = \{P_i : \text{Vertices of triangles in } T_h\}$.

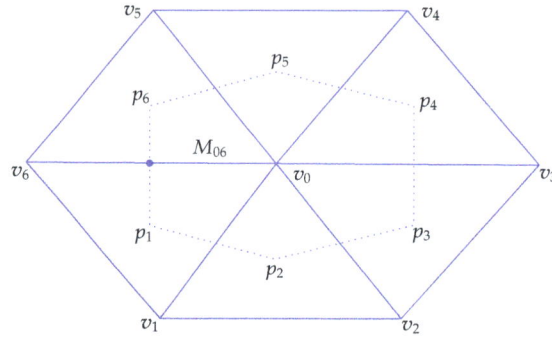

Figure 1: The partition and dual partition of a triangular.

Let us introduce the continuous bilinear forms $\tilde{a}(\cdot,\cdot)$, $\tilde{d}(\cdot,\cdot)$, and $d(\cdot,\cdot)$ on $X_h \times X_h$, $X_h \times M_h$ and $X_h \times M_h$ as follows:

$$\tilde{a}(u_h, I_h^* v_h) = \nu((u_h, I_h^* v_h)) = -\nu \sum_{K_i^* \in T_h^*} \int_{\partial K_i^*} v_h(x_i) \frac{\partial u_h}{\partial \mathbf{n}} ds, \quad \forall u_h, \ v_h \in X_h,$$

$$\tilde{d}(I_h^* v_h, p_h) = (I_h^* v_h, \nabla p_h) = \sum_{K_i^* \in T_h^*} \int_{\partial K_i^*} v_h(x_i) p_h \cdot \mathbf{n} \, ds, \quad \forall u_h \in X_h, \ p_h \in M_h, \qquad (2.5)$$

$$d(u_h, q_h) = -(u_h, \nabla q_h) = (q_h, \text{div } u_h), \quad \forall u_h \in X_h, \ q_h \in M_h,$$

where $\mathbf{n}$ is the out normal vector. We also define the trilinear forms $\tilde{b}(\cdot,\cdot,\cdot)$ on $X_h \times X_h \times X_h$ by

$$\tilde{b}(u_h, v_h, I_h^* w_h) = ((u_h \cdot \nabla)v_h, I_h^* w_h), \qquad (2.6)$$

for all $u_h, v_h, w_h \in X_h$, the right side of term

$$(f, I_h^* v_h) = \sum_{K_i^* \in T_h^*} \int_{K_i^*} v_h(x_i) f \, dx, \quad \forall v_h \in X_h, \qquad (2.7)$$

and a generalized bilinear form on

$$\tilde{B}((u_h, p_h); (I_h^* v_h, q_h)) = \tilde{a}(u_h, I_h^* v_h) - \tilde{d}(I_h^* v_h, p_h) + d(u_h, q_h). \qquad (2.8)$$

Based on the dual partition and bilinear forms defined above, this paper still
introduces the norms and seminorms [1, 9]:

$$\|u_h\|_{0,h} = \left(\sum_{K \in T_h} \|u_h\|_{0,h,K}^2 \right)^{1/2},$$

$$\left\| \tilde{A}_h^{1/2} u_h \right\|_0 = \left(\sum_{K \in T_h} \left\| \tilde{A}_h^{1/2} u_h \right\|_{0,h,K}^2 \right)^{1/2}, \tag{2.9}$$

$$\|u_h\|_{1,h} = \left(\|u_h\|_{0,h}^2 + \left\| \tilde{A}_h^{1/2} u_h \right\|_0^2 \right)^{1/2},$$

where

$$\|u_h\|_{0,h,K} = \left[\frac{S_v}{3} \left(u_{P_i}^2 + u_{P_j}^2 + u_{P_k}^2 \right) \right]^{1/2},$$

$$\left\| \tilde{A}_h^{1/2} u_h \right\|_{0,h,K} = \left\{ \left[\left(\frac{\partial u_h(p)}{\partial x} \right)^2 + \left(\frac{\partial u_h(p)}{\partial y} \right)^2 \right] S_v \right\}^{1/2}, \tag{2.10}$$

with S_v the area of $\Delta v_i v_j v_k$ (see Figure 1).

Formally, there are some differences between $\|u_h\|_{0,h}, \|\tilde{A}_h^{1/2} u_h\|_0$ and $\|u_h\|_0, \|A_h^{1/2} u_h\|_0$,
respectively, but we, actually, have the following results [9–12].

Lemma 2.1. *There exist constants c_1, c_2, independent of h, such that*

$$c_1 \|u_h\|_{0,h} \le \|u_h\|_0 \le c_2 \|u_h\|_{0,h}, \quad \forall u_h \in X_h,$$

$$\left\| \tilde{A}_h^{1/2} u_h \right\|_0 = \left\| A_h^{1/2} u_h \right\|_0, \quad \forall u_h \in X_h. \tag{2.11}$$

So, for simplicity, we also denote $\|u_h\|_{0,h}$ and $\|u_h\|_{1,h}$ by $\|u_h\|_0$ and $\|u_h\|_1$, respectively,
without confusion. Below c (with or without a subscript) is a generic positive constant.

For the above finite element spaces X_h and M_h, it is well known that the following
approximation properties and inverse inequality

$$\left\| A_h^{1/2} v \right\|_0 \le c h^{-1} \|v\|_0, \quad \forall v \in X_h,$$

$$\|v - I_h v\|_0 + h \left\| A_h^{1/2} (v - I_h v) \right\|_0 \le c h^2 \|A_h v\|_0, \quad \forall v \in D(A),$$

$$\|v - I_h^* v\|_0 \le c h \left\| A_h^{1/2} v \right\|_0, \quad \forall v \in X_h, \tag{2.12}$$

$$\|q - J_h q\|_0 \le c h \|q\|_1, \quad \forall q \in H^1(\Omega) \cap M$$

hold (see [4, 13]), where $I_h : D(A) \to X_h$ is the interpolation operator and $J_h : H^1(\Omega) \cap M \to M_h$ is the L^2-orthogonal projection.

In order to define a locally stabilized formulation of the stationary Navier-Stokes problem, we also need *a macroelement partition* Λ_h as follows: Given any subdivision T_h, a macroelement partition Λ_h may be defined such that each macroelement $\mathcal{K}$ is a connected set of adjoining elements from T_h. Every element K must lie in exactly one macroelement. For each $\mathcal{K}$, the set of interelement edges which are strictly in the interior of $\mathcal{K}$ will be denoted by $\Gamma_{\mathcal{K}}$, and the length of an edge $e \in \Gamma_{\mathcal{K}}$ is denoted by h_e. For a macroelement $\mathcal{K}$ the restricted pressure space is given by

$$M_{0,h} = \left\{ q \in L_0^2(\mathcal{K}) : q|_K \in P_0(K), \ \forall K \in \mathcal{K} \right\}. \tag{2.13}$$

With the above choices of the velocity-pressure finite element spaces X_h, X_h^*, M_h and these additional definitions, a locally stabilized formulation of the Navier-Stokes problem (1.1) can be stated as follows.

Definition 2.2 (locally stabilized FVM formulation). Find $(u_h, p_h) \in (X_h, M_h)$, such that for all $(v, q) \in (X_h, M_h)$

$$\widetilde{B}_h\big((u_h, p_h); (I_h^* v, q)\big) + \widetilde{b}(u_h, u_h, I_h^* v_h) = (f, I_h^* v), \tag{2.14}$$

where

$$\widetilde{B}_h\big((u_h, p_h); (I_h^* v, q)\big) = \widetilde{B}\big((u_h, p_h); (I_h^* v, q)\big) + \beta C_h(p_h, q), \quad \forall (u_h, p_h), (v, q) \in (X_h, M_h),$$

$$C_h(p, q) = \sum_{\mathcal{K} \in \Lambda_h} \sum_{e \in \Gamma_{\mathcal{K}}} h_e \int_e [p]_e [q]_e ds, \tag{2.15}$$

for all p, q in the algebraic sum $H^1(\Omega) + M_h$, and $[\cdot]_e$ is the jump operator across $e \in \Gamma_{\mathcal{K}}$ and $\beta > 0$ is the local stabilization parameter.

A general framework for analyzing the locally stabilized formulation (2.14) can be developed using the notion of equivalence class of macroelements. As in Stenberg [14], each equivalence class, denoted by $\mathcal{E}_{\widehat{\mathcal{K}}}$, contains macroelements which are topologically equivalent to a reference macroelement $\widehat{\mathcal{K}}$. See [1, 2] to get some examples.

The following stability results of these mixed methods for the macroelement partition defined above were formally established by Kay and Silvester [6] and Kechkar and Silvester [7]. Throughout the paper we will assume that $\beta \geq \beta_0$.

Theorem 2.3. *Given a stabilization parameter $\beta \geq \beta_0 > 0$, suppose that every macroelement $\mathcal{K} \in \Lambda_h$ belongs to one of the equivalence classes $\mathcal{E}_{\widehat{\mathcal{K}}}$, and that the following macroelement connectivity condition is valid: for any two neighboring macroelements $\mathcal{K}_1$ and $\mathcal{K}_2$ with $\int_{\mathcal{K}_1 \cap \mathcal{K}_2} ds \neq 0$, there exists $v \in X_h$ such that*

$$\text{supp } v \subset \mathcal{K}_1 \cup \mathcal{K}_2, \qquad \int_{\mathcal{K}_1 \cap \mathcal{K}_2} v \cdot nds \neq 0. \tag{2.16}$$

Then,

$$\|C_h(p,q)\|_0 \le c \sum_{K \in T_h} \left(\int_K \left(\|p\|_0^2 + h^2 \|\nabla p\|_0^2 \right) dx \right)^{1/2} \left(\int_K \left(\|q\|_0^2 + h^2 \|\nabla q\|_0^2 \right) dx \right)^{1/2}, \quad (2.17)$$

for all $p, q \in H^1(\Omega) + M_h$, and

$$C_h(p, q_h) = 0, \quad C_h(p_h, q) = 0, \quad C_h(p, q) = 0 \quad \forall p, q \in H^1(\Omega), \ p_h, q_h \in M_h, \quad (2.18)$$

where $c > 0$ is a constant independent of h and β, and β_0 is some fixed positive constant.

3. Error Estimates

In order to derive error estimates of (u_h, p_h) in the FVM, we need the existence and some regularities of the variational problem (2.14) (see [1]).

Lemma 3.1. *Under the assumptions of Theorem 2.3, there exist constants γ and $\alpha > 0$ such that*

$$v \left\| \tilde{A}_h^{1/2} u_h \right\|_0^2 + \beta C_h(p_h, p_h) = \tilde{B}_h((u_h, p_h); (I_h^* u_h, p_h)), \quad (3.1)$$

$$\left| \tilde{B}_h((u_h, p_h); (I_h^* v_h, q_h)) \right| \le \gamma \left(\left\| \tilde{A}_h^{1/2} A u_h \right\|_0 + \|p_h\|_0 \right) \left(\left\| \tilde{A}_h^{1/2} A v_h \right\|_0 + \|q_h\|_0 \right), \quad (3.2)$$

$$\alpha \left(\left\| \tilde{A}_h^{1/2} u_h \right\|_0 + \|p_h\|_0 \right) \le \sup_{(v_h, q_h) \in (X_h, M_h)} \frac{\tilde{B}_h((u_h, p_h); (I_h^* v_h, q_h))}{\left\| \tilde{A}_h^{1/2} v_h \right\|_0 + \|q_h\|_0} \quad (3.3)$$

hold for all (u_h, p_h) and $(v_h, q_h) \in (X_h, M_h)$.

For the trilinear terms $\tilde{b}(u, v_h, I_h^* w_h)$ and $\tilde{b}(v_h, u, I_h^* w_h)$, the following properties are useful [1, 2]. Set

$$N_2 = \sup_{u, v_h, w_h \in H_0^1(\Omega)} \frac{\tilde{b}(u, v_h, I_h^* w)}{\|A^{1/2} u\|_0 \left\| \tilde{A}_h^{1/2} v \right\|_0 \left\| \tilde{A}_h^{1/2} w \right\|_0},$$

$$N_3 = \sup_{u, v_h, w_h \in H_0^1(\Omega)} \frac{\tilde{b}(v_h, u, I_h^* w)}{\left\| \tilde{A}_h^{1/2} v \right\|_0 \|A^{1/2} u\|_0 \left\| \tilde{A}_h^{1/2} w \right\|_0}, \quad (3.4)$$

$$N = \max\{N_1, N_2, N_3\}.$$

Lemma 3.2. *The trilinear form $\tilde{b}$ satisfies*

$$\left| \tilde{b}(u_h, v_h, I_h^* w_h) \right| \le c \left\| \tilde{A}_h u_h \right\|_0^{1/2} \left\| \tilde{A}_h^{1/2} u_h \right\|_0^{1/2} \left\| \tilde{A}_h^{1/2} v_h \right\|_0 \left\| \tilde{A}_h w_h \right\|_0^{1/2} \left\| \tilde{A}_h^{1/2} w_h \right\|_0^{1/2}, \quad (3.5)$$

for any $u_h, v_h, w_h \in X_h$.

Lemma 3.3. *Suppose the assumptions of Theorem 2.3 and (3.4) hold, and the body force f satisfies the following uniqueness condition:*

$$1 - \frac{4N}{v^2} \|f\|_{-1} > 0. \tag{3.6}$$

Then there exists a unique solution (u_h, p_h) of problem (2.14) satisfying the following estimate:

$$v \left\| \tilde{A}_h^{1/2} u_h \right\|_0^2 + \|p_h\|_0^2 \leq \kappa, \tag{3.7}$$

$$\left\| \tilde{A}_h^{1/2} (u - u_h) \right\|_0 + \|p - p_h\|_0 \leq \kappa h. \tag{3.8}$$

In order to derive error estimates of the stabilized finite volume solution (u_h, p_h), we need the following Galerkin projection $(\tilde{R}_h, \tilde{Q}_h) : (X, M) \rightarrow (X_h, M_h)$ defined by

$$\tilde{\mathcal{B}}_h \left(\left(\tilde{R}_h(v, q), \tilde{Q}_h(v, q) \right); (I_h^* v_h, q_h) \right) = \tilde{\mathcal{B}}((v, q); (I_h^* v_h, q_h)) \quad \forall (v_h, q_h) \in (X_h, M_h), \tag{3.9}$$

for each $(v, q) \in (X, M)$.

Note that, due to Lemma 3.1, $(\tilde{R}_h, \tilde{Q}_h)$ is well defined. Now, we derive the following optimal L^2 error estimate of u_h and p_h defined in (3.9). Using an argument similar to ones used by Layton and Tobiska in [15], the following approximate properties can be obtained.

Lemma 3.4. *Under the assumptions of Lemma 3.3, the projection (R_h, Q_h) satisfies*

$$\left\| \tilde{A}_h^{1/2} \left(v - \tilde{R}_h(v, q) \right) \right\|_0 + \left\| q - \tilde{Q}_h(v, q) \right\|_0 \leq c \left(\left\| A^{1/2} v \right\|_0 + \|q\|_0 \right), \tag{3.10}$$

for all $(v, q) \in (X, M)$,

$$\left\| \tilde{A}_h^{1/2} \left(v - \tilde{R}_h(v, q) \right) \right\|_0 + \left\| q - \tilde{Q}_h(v, q) \right\|_0 \leq ch \left(\|Av\|_0 + \left\| A^{1/2} q \right\|_0 \right), \tag{3.11}$$

for all $(v, q) \in (D(A), H^1(\Omega) \cap M)$, and

$$\left\| v - \tilde{R}_h(v, q) \right\|_0 + h \left\| \tilde{A}_h^{1/2} \left(v - \tilde{R}_h(v, q) \right) \right\| + h \left\| q - \tilde{Q}_h(v, q) \right\|_0 \leq ch^2 (|Av|_0 + \|q\|_1), \tag{3.12}$$

for all $(v, q) \in (D(A), H^1(\Omega) \cap M)$.

Proof. Equations (3.10) and (3.11) is the directly from [1]. Next, let $(v, q) \in (D(A), H^1(\Omega) \cap M)$ and introduce the dual Stokes problem: find $(\Phi, \Psi) \in (X, M)$ such that

$$\tilde{\mathcal{B}}((w, r); (\Phi, \Psi)) = \left(w, I_h^* \left(v - \tilde{R}_h(v, q) \right) \right), \quad \forall (w, r) \in (X, M). \tag{3.13}$$

Using the regularity assumption of Stokes problem (See the A1 assumption in [1, 2]), there holds

$$\|\Phi\|_2 + \|\Psi\|_1 \le c\left\|v - \tilde{R}_h(v,q)\right\|_0. \tag{3.14}$$

Now, setting $w = v - \tilde{R}_h(v,q)$, $r = q - \tilde{Q}_h(v,q)$, using (2.18) and (3.9), we obtain that for $(\Phi_h, \Psi_h) = (I_h^* \Phi, J_h \Psi) \in (X_h, M_h)$,

$$
\begin{aligned}
\left\|v - \tilde{R}_h(v,q)\right\|_0^2 &= \tilde{B}\Big(\big(v - \tilde{R}_h(v,q), q - \tilde{Q}_h(v,q)\big); (\Phi, \Psi)\Big) \\
&= \tilde{B}_h\Big(\big(v - \tilde{R}_h(v,q), q - \tilde{Q}_h(v,q)\big); (\Phi, \Psi)\Big) \\
&= \tilde{B}_h\Big(\big(v - \tilde{R}_h(v,q), q - \tilde{Q}_h(v,q)\big); (\Phi - \Phi_h, \Psi - \Psi_h)\Big) \\
&= \tilde{a}\Big(v - \tilde{R}_h(v,q), I_h^*(\Phi - \Phi_h)\Big) \\
&\quad - \tilde{d}\Big(I_h^*(\Phi - \Phi_h), q - \tilde{Q}_h(v,q)\Big) \\
&\quad + d\Big(v - \tilde{R}_h(v,q), \Psi - \Psi_h\Big).
\end{aligned}
\tag{3.15}
$$

For any $v_h \in X_h$, we have [9–11]

$$
\begin{aligned}
\tilde{a}\Big(v - \tilde{R}_h(v,q), I_h^*(\Phi - \Phi_h)\Big) &= \Big(\Delta\big((v - \tilde{R}_h(v,q)) - I_h(v - \tilde{R}_h(v,q))\big), I_h^*(\Phi - \Phi_h)\Big), \\
\left\|\tilde{d}\Big(I_h^*(\Phi - \Phi_h), q - \tilde{Q}_h(v,q)\Big)\right\|_0 &\le c\left\|d\big((\Phi - \Phi_h), q - \tilde{Q}_h(v,q)\big)\right\|_0.
\end{aligned}
\tag{3.16}
$$

Since the dual partition is formed by the barycenter, similar calculation in [10, 11, 16] allows us to have

$$
\begin{aligned}
\left\|\tilde{a}\Big(v - \tilde{R}_h(v,q), I_h^*(\Phi - \Phi_h)\Big)\right\|_0 &\le \kappa h^2 \left\|v - \tilde{R}_h(v,q)\right\|_0 \\
&\quad \times \left(\left\|\tilde{A}_h^{1/2} u_h\right\|_0 + \|u\|_2 + \|f\|_1 + \left\|\tilde{A}_h^{1/2}(v - \tilde{R}_h(v,q))\right\|_0\right),
\end{aligned}
\tag{3.17}
$$

$$
\begin{aligned}
\left\|\tilde{d}\Big(I_h^*(\Phi - \Phi_h), q - \tilde{Q}_h(v,q)\Big)\right\|_0 &\le c\left\|d\big((\Phi - \Phi_h), q - \tilde{Q}_h(v,q)\big)\right\|_0 \\
&\le c\|\nabla(\Phi - \Phi_h)\|_0 \left\|q - \tilde{Q}_h(v,q)\right\|_0.
\end{aligned}
\tag{3.18}
$$

By (3.10), (3.11), and (3.14), we have

$$
\begin{aligned}
\left\|\tilde{d}\Big(I_h^*(\Phi - \Phi_h), q - \tilde{Q}_h(v,q)\Big)\right\|_0 &\le ch^2\|A\Phi\|_0\big(\|Av\|_0 + \|A^{1/2}q\|_0\big) \\
&\le ch^2\left\|v - \tilde{R}_h(v,q)\right\|_0 \big(\|Av\|_0 + \|A^{1/2}q\|_0\big).
\end{aligned}
\tag{3.19}
$$

Since $C_h(p, q_h) = 0$, for all $p \in (H^1(\Omega) \cap M)$, for all $q_h \in M_h$, similarly in [7], we have

$$
\left\| \left(\operatorname{div} \left(v - \tilde{R}_h(v,q) \right), \Psi - \Psi_h \right) \right\|_0 \leq ch \left\| \nabla \left(v - \tilde{R}_h(v,q) \right) \right\|_0 \|\Psi\|_1
$$
$$
\leq \kappa h^2 (\|u\|_2 + \|p\|_1) \left\| \left(v - \tilde{R}_h(v,q) \right) \right\|_0. \tag{3.20}
$$

Finally, combining (3.10) with (3.17), (3.19), and (3.20) yields (3.12).   □

Next, we will derive the following error estimates of the finite element solution (u_h, p_h) defined in Section 2.

Theorem 3.5. *Assume that the assumptions of Lemma 3.3 hold. Then the stabilized finite element solution (u_h, p_h) satisfies the error estimates*

$$
\|u - u_h\|_0 + h \left(\left\| \tilde{A}_h^{1/2}(u - u_h) \right\|_0 + \|p - p_h\|_0 \right) \leq ch^2. \tag{3.21}
$$

Proof. It is well known that the weak solutions $(u, p) \in (D(A) \cap V, H^1(\Omega) \cap M)$. Hence, we derive from (2.14) and (3.9) that for all $(v_h, q_h) \in (X_h, M_h)$

$$
\mathcal{B}_h((e_h, \eta_h); (I_h^* v_h, q_h)) + \tilde{b}(u - u_h, u, I_h^* v_h) + \tilde{b}(u_h, u - u_h, I_h^* v_h) = 0, \tag{3.22}
$$

where $e_h = \tilde{R}_h(u, p) - u_h$ and $\eta_h = \tilde{Q}_h(u, p) - p_h$. Taking $(v, q) = (e_h, \eta_h)$ in (3.22), we obtain

$$
\nu \left\| \tilde{A}_h^{1/2} e_h \right\|^2 + \beta_0 C_h(\eta_h, \eta_h) + \tilde{b}(e_h, u, I_h^* e_h) + \tilde{b}(u, e_h, I_h^* e_h)
$$
$$
\leq \left| b \left(u - \tilde{R}_h(u, p), u, e_h \right) \right| + \left| b \left(u_h, u - \tilde{R}_h(u, p), e_h \right) \right|. \tag{3.23}
$$

We find from (1.10), (3.1), (3.4), and (3.6) that

$$
\nu \left\| \tilde{A}_h^{1/2} e_h \right\|_0^2 - \left| \tilde{b}(e_h, u, I_h^* e_h) \right| - \left| \tilde{b}(u, e_h, I_h^* e_h) \right| \geq \nu \left\| \tilde{A}_h^{1/2} e_h \right\|_0^2 - 2N \left\| A_h^{1/2} u \right\|_0 \left\| \tilde{A}_h^{1/2} e_h \right\|_0^2
$$
$$
\geq \nu \left(1 - 2\|f\|_0 \nu^{-2} \right) \left\| \tilde{A}_h^{1/2} e_h \right\|_0^2. \tag{3.24}
$$

Moreover, By (3.5), (3.11), and Poincaré's estimate, we have

$$
\begin{aligned}
&\left\|\tilde{b}\left(u_h, u - \tilde{R}_h(u,p), e_h\right)\right\|_0 + \left\|\tilde{b}\left(u - \tilde{R}_h(u,p), u, e_h\right)\right\|_0 \\
&\leq \left\|\tilde{b}\left(u, u - \tilde{R}_h(u,p), e_h\right)\right\|_0 + \left\|\tilde{b}\left(u - \tilde{R}_h(u,p), u, e_h\right)\right\|_0 \\
&\quad + \left\|\tilde{b}\left(u - \tilde{R}_h(u,p), u - \tilde{R}_h(u,p), e_h\right)\right\|_0 + \left\|\tilde{b}\left(e_h, u - \tilde{R}_h(u,p), e_h\right)\right\|_0 \\
&\leq c_1 \|Au\|_0 \left\|\tilde{A}_h^{1/2}\left(u - \tilde{R}_h(u,p)\right)\right\|_0 \left\|\tilde{A}_h^{1/2} e_h\right\|_0 \\
&\quad + c_2 \left(\left\|\tilde{A}_h^{1/2}\left(u - \tilde{R}_h(u,p)\right)\right\|_0 + \left\|\tilde{A}_h^{1/2} e_h\right\|_0\right)\left\|\tilde{A}_h^{1/2}\left(u - \tilde{R}_h(u,p)\right)\right\|_0 \left\|\tilde{A}_h^{1/2} e_h\right\|_0 \\
&\leq ch^2 \left\|\tilde{A}_h^{1/2} e_h\right\|_0.
\end{aligned}
\tag{3.25}
$$

Combining the above estimates with (3.24) and using the uniqueness condition (3.4) yield

$$
\left\|\tilde{A}_h^{1/2} e_h\right\|_0 \leq ch^2. \tag{3.26}
$$

Finally, one finds from (3.12) and (3.26) that

$$
\|u - u_h\|_0 \leq \|e_h\|_0 + \|u - R_h(u,p)\|_0 \leq c_3 \left\|\tilde{A}_h^{1/2} e_h\right\|_0 + ch^2 (\|Au\|_0 + \|p\|_1) \leq ch^2. \tag{3.27}
$$

Hence, combining the above estimates with (3.8) gives (3.21). □

4. Numerical Example

For stationary Navier-Stokes problem, the iteration scheme, in general, is

$$
\begin{aligned}
\nu Av + N(v)v + Bp &= f \\
-B^T v + \beta C &= 0.
\end{aligned}
\tag{4.1}
$$

The submatrices occurring in (4.1) correspond to differential operators as $A \sim -\text{diag}(\nu\Delta)$, $N(v) \sim v \cdot \nabla$, $B \sim \nabla$, $-B^T \sim \text{div}$, and $C \sim C_h(\cdot, \cdot)$. The right-hand side f contains information from the source information.

In general, this problem can be solved by the following Newton method:

(1) $R = f - \left(\nu Av^{\text{old}} + N\left(v^{\text{old}}\right)\right)v^{\text{old}} - Bp^{\text{old}}, \qquad r = -B^T v^{\text{old}};$

(2) $\left(\nu Av^{\text{mid}} + N\left(v^{\text{old}}\right)\right)v^{\text{mid}} + N\left(v^{\text{mid}}\right)v^{\text{old}} + Bp^{\text{mid}} = R, \qquad B^T v^{\text{mid}} = r;$ (4.2)

(3) $v^{\text{new}} = v^{\text{old}} + v^{\text{mid}}, \qquad p^{\text{new}} = p^{\text{old}} + p^{\text{mid}},$

where R, r are the so-called nonlinear residual. Actually the difference between (4.2) and (4.1) is that, in computing the corrections v^{mid} and p^{mid} from R, r, the quadratic term $N(v^{\text{mid}})v^{\text{mid}}$ deduced from (4.1) is dropped and gives the linear problem (4.2).

4.1. Numeric Example I

Consider a unit square domain with an exact solution given by

$$u(x,y) = (u_1(x,y), u_2(x,y)), \qquad p(x,y) = 10(2x-1)(2y-1),$$

$$u_1(x,y) = 10x^2(x-1)^2 y(y-1)(2y-1), \qquad u_2(x,y) = -10x(x-1)(2x-1)y^2(y-1)^2.$$
(4.3)

f is determined by (1.1). After some computation using stretched gird, we have the following results.

Figure 2 is the relative error and convergence rate of velocity and pressure when $v = 0.005$, $\beta = 10$. Table 1 lists the different errors and convergence rates of numerical velocity and pressure for the same v and β. From the figure and table, we can see that the almost second-order L^2 convergence is obtained, which confirms our theoretical prediction.

Figure 3 is the L^2 relative error of numerical velocity versus the number of iterate steps for different v. The figure tells us the numerical velocities, in general, converge very fast. Moreover, the figure also tells the bigger the v, the faster the convergence speed, which is consistent with the really case. If v is not too small, for example, $v \geq 0.01$, only several Newton iterations are needed.

4.2. Numeric Example II

The second example is the classical lid-driven flow governed by stationary Navier-Stokes equations in a square cavity. We impose watertight boundary conditions, that is, $u_x(0,1) = u_x(1,1) = 0$ and $u_x = 1$, for $0 < x < 1$. From the streamlines in Figures 4–7, we can see there are some different performances for different v for the problem based on the stretched grid [16, 17] (with 128×128 grid).

The left subplot in Figure 4 is the velocity solution of Stokes problem which serves as the initial guess of Newton method; the right subplot in Figure 4 is the nonconvergence numeric velocity with that initial guess and 9 times Newton iterations. From Figure 4, we can see that there is a different performance from Numerical Example I for the lid-driven flow; if the v is smaller, the initial value needed in Newton iteration has to be nearer the exact solution. The nonconvergence indicates that we must have a good initiate value for Newton iteration. For the $v = 0.001$, the usual Stokes initial value is not sufficient and a better initial value is needed, which can be computed by the following Picard method:

$$(1) \; R = f - \left(vAv^{\text{old}} + N\left(v^{\text{old}}\right)\right)v^{\text{old}} - Bp^{\text{old}}, \qquad r = -B^T v^{\text{old}};$$

$$(2) \; \left(vAv^{\text{mid}} + N\left(v^{\text{old}}\right)\right)v^{\text{mid}} + Bp^{\text{mid}} = R, \qquad B^T v^{\text{mid}} = r; \tag{4.4}$$

$$(3) \; v^{\text{new}} = v^{\text{old}} + v^{\text{mid}}, \qquad p^{\text{new}} = p^{\text{old}} + p^{\text{mid}}.$$

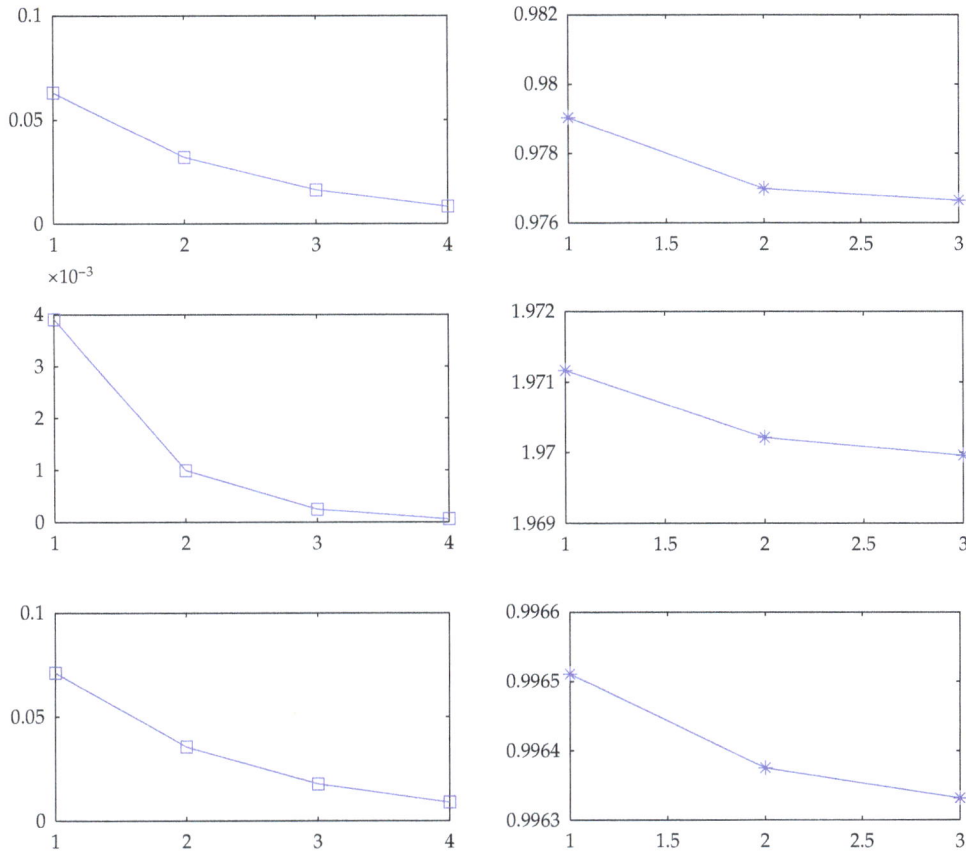

Figure 2: The relative errors and convergence rates of velocity and pressure.

Table 1: Numerical results of the FVM ($\nu = 0.005$, $\beta = 10$).

h	$\|\tilde{A}_h^{1/2}(u - u_h)\|_0 / \|\tilde{A}_h^{1/2}u\|_0$	Con. rate	$\|u - u_h\|_0 / \|u\|_0$	Con. rate	$\|p - p_h\|_0 / \|p\|_0$	Con. rate
1/16	0.06316279	—	0.00391173	—	0.07127911	—
1/32	0.03204341	0.97904713	0.00099767	1.97116812	0.03572585	0.99651095
1/64	0.01627933	0.97698618	0.00025462	1.97021278	0.01790786	0.99637536
1/128	0.00827253	0.97664010	0.00006499	1.96996167	0.00897672	0.99633198

The difference between Picard method and Newton method is that the linear term $N(v^{\text{mid}})v^{\text{old}}$ is also dropped from (4.2), and thus the Picard method commonly referred to as the Oseen system.

The left subplot in Figure 5 is the initial velocity for the Newton iteration based on two Picard iterations without Newton iteration, and the right subplot in Figure 5 is the streamlines of the convergence numeric velocity evaluated at the 2 Picard iterations, using 4 times Newton iterations.

Figures 6 and 7 give the behavior of different iteration results for $\nu = 0.001$ and $\nu = 0.00033$. From these figures, we can see that if ν is smaller, the initial value needed in Newton method should be better. The initial value computed by one or two steps Picard method is already insufficient and thus more Picard iterations are needed. In addition, we

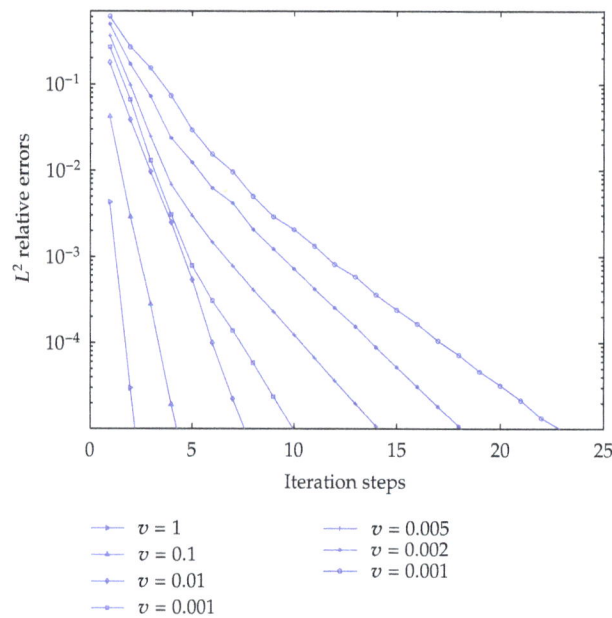

Figure 3: L^2 relative error of numerical velocity versus iteration steps for different ν.

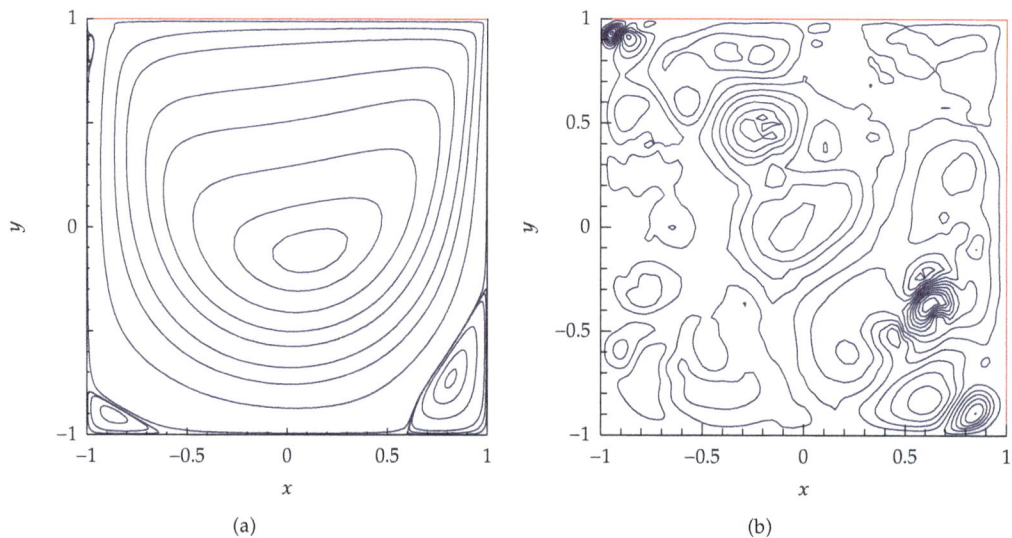

Figure 4: Nonconvergence streamlines for $\nu = 0.001$, $\beta = 10$ ((a) Stokes, (b) N9).

can also see that the convergence speed of Picard is not as fast as Newton: if the initial value is sufficient for Newton iteration, the convergence speed of Newton's method is faster than Picard's method.

Actually, the Picard method corresponds to a simple fixed point iteration strategy for solving (2.14) whose convection coefficient is evaluated at the current velocity. Thus, the rate of convergence of Picard method is only linear in general; whereas, for the added more linear

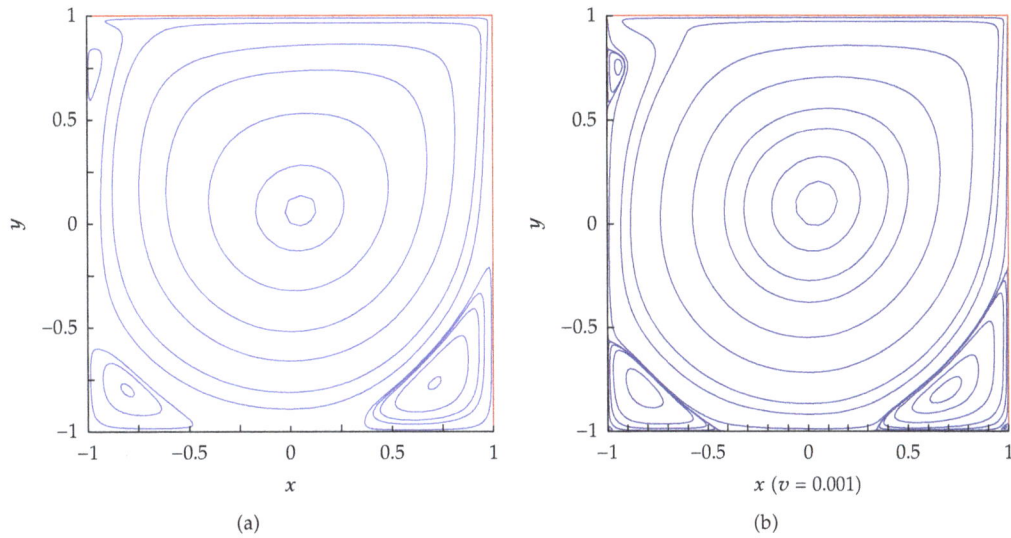

Figure 5: Streamlines for $v = 0.001$, $\beta = 10$ ((a) P2N0, (b) P2N4).

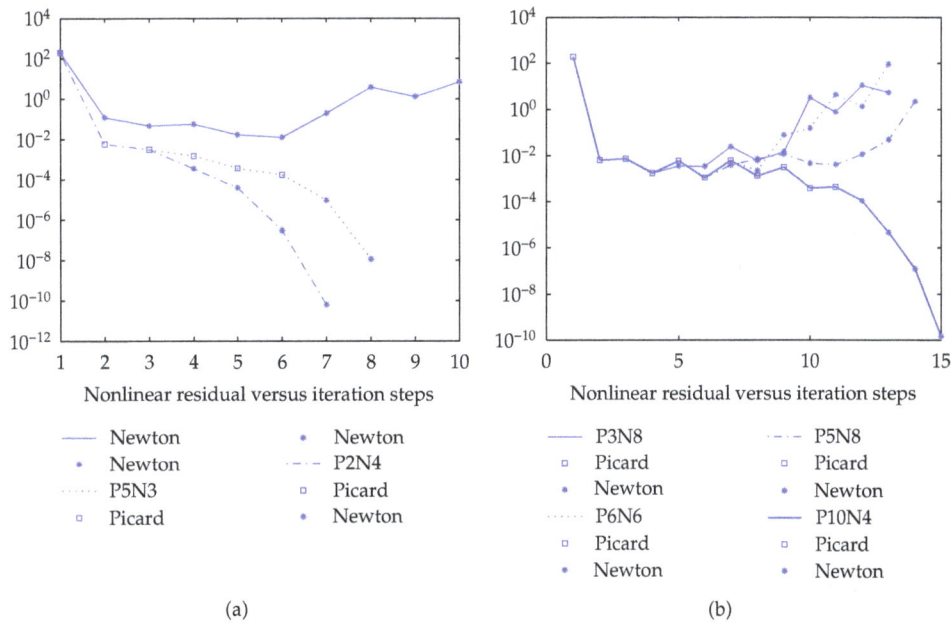

Figure 6: Nonlinear residual versus iteration steps ((a) $v = 0.001$, $\beta = 10$ and (b) $v = 0.00033$, $\beta = 20$).

term, if the initial value is sufficient close to a nonsingular solution, the Newton method is locally convergence quadratic (For more information see [18]).

It is necessary to pay attention to the "finest" number of Picard iteration in the computation of initial value for Newton iteration. Since the convergence radius of the Newton method is proportional to Reynolds number (namely, $1/v$) in general, in these computations, we roughly choose the times of Picard iteration to increase proportionately with Reynolds

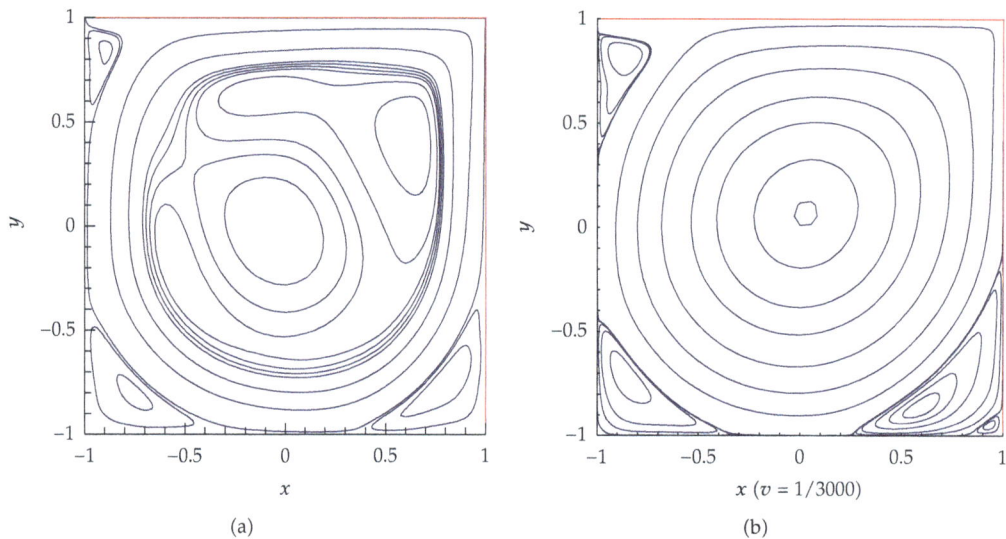

Figure 7: Streamlines for $v = 0.000333, \beta = 30$ ((a) P5N10 (wrong result), (b) P10N5 (right result)).

number. Many numerical tests show that this strategy is good enough for the success of the ensuing Newton iteration.

5. Conclusions

The main work in this paper is the demonstration of the optimal order in the L^2 error of the velocity and emphasis on some aspects of its associated numerical computation. Both the theoretical analysis and numerical results indicate the efficiency of the FVM for stationary Navier-Stokes equations.

Further, numerical computations show the convergence of Newton method is closely related to the viscosity v. Thus, as it is decreased, better and better initial values are needed, whereas the advantage of Picard method is that, relative to Newton method, it has a much large region of trust of convergence. As a result, a good choice is to combine the Newton method with Picard method in computing, and thus more complicated problems can be solved efficiently.

Acknowledgments

The paper was subsidized by the NSF of China 10926133, 60973015, 70901012, and FRFCU ZYGX2009J098 and YF of UESTC 7X0775.

References

[1] G. He and Y. He, "The finite volume method based on stabilized finite element for the stationary Navier-Stokes problem," *Journal of Computational and Applied Mathematics*, vol. 205, no. 1, pp. 651–665, 2007.

[2] Y. He, A. Wang, and L. Mei, "Stabilized finite-element method for the stationary Navier-Stokes equations," *Journal of Engineering Mathematics*, vol. 51, no. 4, pp. 367–380, 2005.

[3] R. Temam, *Navier-Stokes Equations*, vol. 2 of *Studies in Mathematics and its Applications*, North-Holland, Amsterdam, The Nertherlands, 3rd edition, 1984.

[4] P. G. Ciarlet, *The Finite Element Method for Elliptic Problems*, North-Holland, Amsterdam, The Nertherlands, 1978.

[5] V. Girault and P.-A. Raviart, *Finite Element Methods for Navier-Stokes Equations: Theory and Algorithms*, vol. 5 of *Springer Series in Computational Mathematics*, Springer, Berlin, Germany, 1986.

[6] D. Kay and D. Silvester, "A posteriori error estimation for stabilized mixed approximations of the Stokes equations," *SIAM Journal on Scientific Computing*, vol. 21, no. 4, pp. 1321–1336, 1999/00.

[7] N. Kechkar and D. Silvester, "Analysis of locally stabilized mixed finite element methods for the Stokes problem," *Mathematics of Computation*, vol. 58, no. 197, pp. 1–10, 1992.

[8] R. L. Sani, P. M. Gresho, R. L. Lee, and D. F. Griffiths, "The cause and cure (?) of the spurious pressures generated by certain FEM solutions of the incompressible Navier-Stokes equations. I," *International Journal for Numerical Methods in Fluids*, vol. 1, no. 1, pp. 17–43, 1981.

[9] R. Li, Z. Chen, and W. Wu, *Generalized Difference Methods for Differential Equations*, vol. 226 of *Monographs and Textbooks in Pure and Applied Mathematics*, Marcel Dekker, New York, NY, USA, 2000.

[10] R. E. Ewing, T. Lin, and Y. Lin, "On the accuracy of the finite volume element method based on piecewise linear polynomials," *SIAM Journal on Numerical Analysis*, vol. 39, no. 6, pp. 1865–1888, 2002.

[11] S. H. Chou and D. Y. Kwak, "Analysis and convergence of a MAC-like scheme for the generalized Stokes problem," *Numerical Methods for Partial Differential Equations*, vol. 13, no. 2, pp. 147–162, 1997.

[12] X. Ye, "A discontinuous finite volume method for the Stokes problems," *SIAM Journal on Numerical Analysis*, vol. 44, no. 1, pp. 183–198, 2006.

[13] R. A. Adams, *Sobolev Spaces*, Academic Press, New York, NY, USA, 1975.

[14] R. Stenberg, "Analysis of mixed finite elements methods for the Stokes problem: a unified approach," *Mathematics of Computation*, vol. 42, no. 165, pp. 9–23, 1984.

[15] W. Layton and L. Tobiska, "A two-level method with backtracking for the Navier-Stokes equations," *SIAM Journal on Numerical Analysis*, vol. 35, no. 5, pp. 2035–2054, 1998.

[16] G. He, Y. Zhang, and H. Xiaoqin, "The locally stabilized finite volume method for incompressible flow," *Applied Mathematics and Computation*, vol. 187, no. 2, pp. 1399–1409, 2007.

[17] F. H. Harlow and J. E. Welch, "Numerical calculation of time-dependent viscous incompressible flow of fluid with free surface," *Physics of Fluids*, vol. 8, no. 12, pp. 2182–2189, 1965.

[18] H. C. Elman, D. J. Silvester, and A. J. Wathen, *Finite Elements and Fast Iterative Solvers: with Applications in Incompressible Fluid Dynamics*, Numerical Mathematics and Scientific Computation, Oxford University Press, New York, NY, USA, 2005.

4

An Efficient Family of Root-Finding Methods with Optimal Eighth-Order Convergence

Rajni Sharma[1] and Janak Raj Sharma[2]

[1] *Department of Applied Sciences, DAV Institute of Engineering and Technology, Kabirnagar 144008, India*
[2] *Department of Mathematics, Sant Longowal Institute of Engineering and Technology, Longowal 148106, India*

Correspondence should be addressed to Rajni Sharma, rajni_gandher@yahoo.co.in

Academic Editor: Nils Henrik Risebro

We derive a family of eighth-order multipoint methods for the solution of nonlinear equations. In terms of computational cost, the family requires evaluations of only three functions and one first derivative per iteration. This implies that the efficiency index of the present methods is 1.682. Kung and Traub (1974) conjectured that multipoint iteration methods without memory based on n evaluations have optimal order 2^{n-1}. Thus, the family agrees with Kung-Traub conjecture for the case $n = 4$. Computational results demonstrate that the developed methods are efficient and robust as compared with many well-known methods.

1. Introduction

Solving nonlinear equations is one of the most important problems in science and engineering [1, 2]. The boundary value problems arising in kinetic theory of gases, vibration analysis, design of electric circuits, and many applied fields are reduced to solving such equations. In the present era of advance computers, this problem has gained much importance than ever before.

In this paper, we consider iterative methods to find a simple root r of the nonlinear equation $f(x) = 0$, where $f : R \rightarrow R$ be the continuously differentiable real function. Newton's method [1] is probably the most widely used algorithm for solving such equations, which starts with an initial approximation x_0 closer to the root r and generates a sequence of successive iterates $\{x_i\}_0^\infty$ converging quadratically to the root. It is given by the following:

$$x_{i+1} = x_i - \frac{f(x_i)}{f'(x_i)}, \quad i = 0, 1, 2, 3, \dots. \tag{1.1}$$

In order to improve the local order of convergence, a number of ways are considered by many researchers, see [3–26] and references therein. In particular, King [3] developed a one-parameter family of fourth-order methods defined by

$$
w_i = x_i - \frac{f(x_i)}{f'(x_i)},
$$

$$
x_{i+1} = w_i - \frac{f(x_i) + \beta f(w_i)}{f(x_i) + (\beta - 2) f(w_i)} \frac{f(w_i)}{f'(x_i)},
$$

$$(1.2)$$

where w_i is the Newton point and β is a constant.

This family requires two evaluations of the function f and one evaluation of first derivative f' per iteration. The famous Ostrowski's method [4, 5] is a member of this family for the case $\beta = 0$. From practical point of view, the methods (1.2) are important because of higher efficiency than Newton's method (1.1).

Traub [5] has divided iterative methods into two classes, namely, one-point methods and multipoint methods. Each class is further divided into two subclasses, namely, one-point methods with and without memory, and multipoint methods with and without memory. The important aspects related to these classes of methods are order of convergence and computational efficiency. Order of convergence shows the speed with which a given sequence of iterates converges to the root while the computational efficiency concerns with the economy of the entire process. Investigation of one-point methods with and without memory, has demonstrated theoretical restrictions on the order and efficiency of these two categories (see [5]). However, Kung and Traub [6] have conjectured that multipoint iteration methods without memory based on n evaluations have optimal order 2^{n-1}. In particular, with three evaluations a method of fourth-order can be constructed. The King's method (1.2) is a well-known example of fourth-order multipoint methods without memory.

Recently, based on Ostrowski's or King's methods some higher-order multipoint methods have been proposed and analyzed for solving nonlinear equations. For example, Grau and Díaz-Barrero [10], Sharma and Guha [11], and Chun and Ham [12] have developed sixth-order modified Ostrowski's methods each requires three f and one f' evaluations per iteration. Kou et al. [15] presented a family of variants of Ostrowski's method with seventh-order convergence requiring three f and one f' evaluations. With same number of evaluations, Bi et al. [18] developed a seventh-order family of modified King's methods. Bi et al. [19] also presented an eighth-order family of modified King's methods requiring four evaluations which agrees with the Kung-Traub conjecture.

In this paper, we present a new family of eighth-order methods without using second and higher derivatives. In terms of computational cost, it requires the evaluations of three functions and one first derivative per iteration. Thus the present methods provide a new example of multipoint methods without memory that with four evaluations a method of optimum order eight can be achieved as conjectured by Kung and Traub. The performance and effectiveness of the developed family of methods is tested and compared through some test functions.

Contents of the paper are summarized as follows. Some basic definitions relevant to the present work are presented in Section 2. In Section 3, we obtain new methods. Convergence analysis, for establishing eighth-order convergence, is carried out in Section 4. In Section 5, we provide some particular cases of the family. In Section 6, the method is tested and compared with other well-known methods on a number of problems. Concluding remarks are given in Section 7.

2. Basic Definitions

Definition 2.1. Let $f(x)$ be a real function with a simple root r and let $\{x_i\}_{i \in N}$ be a sequence of real numbers that converges towards r. Then, we say that the order of convergence of the sequence is p, if there exits a number $p \in \mathbb{R}^+$ such that

$$\lim_{i \to \infty} \frac{x_{i+1} - r}{(x_i - r)^p} = C, \tag{2.1}$$

for some $C \neq 0$, C is known as the asymptotic error constant.

If $p = 1, 2$ or 3, the sequence is said to have linear convergence, quadratic convergence or cubic convergence, respectively.

Definition 2.2. Let $e_i = x_i - r$ be the error in the ith iteration, we call the relation

$$e_{i+1} = C e_i^p + O\left(e_i^{p+1}\right), \tag{2.2}$$

the error equation.

Definition 2.3. Let n be the number of new pieces of information required by a method. A "piece of information" typically is any evaluation of a function or one of its derivatives. The efficiency of the method is measured by the concept of efficiency index [27] and is defined by the following:

$$E = p^{1/n}, \tag{2.3}$$

where p is the order of the method.

Definition 2.4. Suppose that x_{i+1}, x_i and x_{i-1} are three successive iterations closer to the root r. Then, the computational order of convergence ρ (see [24, 25, 28]) is approximated by using (2.2) as follows:

$$\rho \cong \frac{\ln|(x_{i+1} - r)/x_i - r|}{\ln|x_i - r/x_{i-1} - r|}. \tag{2.4}$$

3. The Method

We consider the iteration scheme of the form

$$w_i = x_i - \frac{f(x_i)}{f'(x_i)},$$

$$z_i = w_i - \omega(\lambda_i)\frac{f(w_i)}{f'(x_i)}, \tag{3.1}$$

$$x_{i+1} = z_i - W(\mu_i)\frac{f(z_i)}{f'(z_i)},$$

where $\lambda_i = f(w_i)/f(x_i)$, $\mu_i = f(z_i)/f(x_i)$, and $\omega(t)$ and $W(t)$ represent the real-valued functions (here onwards called weight functions). This scheme consists of three steps in which the first step represents Newton's method and last two are weighted-Newton steps. It is quite obvious that formula (3.1) requires five evaluations per iteration. However, we can reduce the number of evaluations to four by using some suitable approximation of the derivative $f'(z_i)$. We obtain this approximation by considering the approximation of $f(x)$ by a rational linear function of the form

$$y(x) - y(x_i) = \frac{(x - x_i) + a}{b(x - x_i) + c}, \tag{3.2}$$

where the parameters a, b, and c are determined by the condition that f and y coincide at x_i, w_i and z_i. That means $y(x)$ satisfies the conditions

$$y(x_i) = f(x_i), \qquad y(w_i) = f(w_i), \qquad y(z_i) = f(z_i). \tag{3.3}$$

From (3.2) and first condition of (3.3), it is easy to show that

$$a = 0. \tag{3.4}$$

Substituting the value of a into (3.2) then using the last two conditions of (3.3), after some simple calculations we obtain

$$b(w_i - x_i) + c = \frac{1}{f[x_i, w_i]},$$

$$b(z_i - x_i) + c = \frac{1}{f[x_i, z_i]}, \tag{3.5}$$

where $f[x_i, w_i] = (f(w_i) - f(x_i))/(w_i - x_i)$ and $f[x_i, z_i] = (f(z_i) - f(x_i))/(z_i - x_i)$ are first-divided differences.

Solving these equations, we can obtain b and c as follows:

$$b = \frac{1}{w_i - z_i} \left(\frac{1}{f[x_i, w_i]} - \frac{1}{f[x_i, z_i]} \right),$$

$$c = \frac{1}{w_i - z_i} \left(\frac{x_i - z_i}{f[x_i, w_i]} - \frac{x_i - w_i}{f[x_i, z_i]} \right). \tag{3.6}$$

Differentiation of (3.2) gives

$$y'(x) = \frac{c}{[b(x - x_i) + c]^2}. \tag{3.7}$$

We can now approximate the derivative $f'(x)$ with the derivative $y'(x)$ of rational function (3.2) and obtain

$$f'(z_i) \approx y'(z_i). \tag{3.8}$$

Substituting the values of b and c obtained in (3.6) into (3.7) then using (3.8), we get after simplifications

$$f'(z_i) = \frac{f[x_i, z_i] f[w_i, z_i]}{f[x_i, w_i]}. \tag{3.9}$$

Then the iteration scheme (3.1) in its final form is given by the following:

$$w_i = x_i - \frac{f(x_i)}{f'(x_i)},$$

$$z_i = w_i - \omega(\lambda_i) \frac{f(w_i)}{f'(x_i)},$$

$$x_{i+1} = z_i - W(\mu_i) \frac{f[x_i, w_i] f(z_i)}{f[x_i, z_i] f[w_i, z_i]}, \tag{3.10}$$

where $\lambda_i = f(w_i)/f(x_i)$, $\mu_i = f(z_i)/f(x_i)$, and $\omega(t)$ and $W(t)$ are the weight functions.

Thus the scheme (3.10) defines a new family of multipoint methods with two weight functions $\omega(t)$ and $W(t)$. In the next section, we will see that both of these functions play an important role in establishing eighth-order convergence of the methods.

4. Convergence of the Method

In order to examine the convergence property of the family (3.10), we prove the following theorem.

Theorem 4.1. *Let the function $f : R \to R$ be sufficiently smooth in R. If $f(x)$ has a simple root r in R and x_0 is sufficiently close to r, then the sequence $\{x_i\}$ generated by any method of the family (3.10)* *converges to r with convergence order eight, provided the weight functions $w(t)$ and $W(t)$ satisfy the conditions $w(0) = 1, w'(0) = 2, w''(0) = 8, W(0) = 1, W'(0) = 1,$ and $|w'''(0)| < \infty$.*

Proof. Let $e_i = x_i - r$ be the error in the iterate x_i. Using Taylor's series expansion, we get

$$f(x_i) = f'(r)\left[e_i + A_2 e_i^2 + A_3 e_i^3 + A_4 e_i^4 + A_5 e_i^5 + A_6 e_i^6 + A_7 e_i^7 + A_8 e_i^8 + O\left(e_i^9\right)\right],$$
$$f'(x_i) = f'(r)\left[1 + 2A_2 e_i + 3A_3 e_i^2 + 4A_4 e_i^3 + 5A_5 e_i^4 + 6A_6 e_i^5 + 7A_7 e_i^6 + 8A_8 e_i^7 + O\left(e_i^8\right)\right],$$

(4.1)

where $A_k = f^{(k)}(r)/k! f'(r)$ for $k \in N$, N is the set of natural numbers.

Now,

$$\frac{f(x_i)}{f'(x_i)} = e_i - A_2 e_i^2 - 2\left(-A_2^2 + A_3\right)e_i^3 - \left(4A_2^3 - 7A_2 A_3 + 3A_4\right)e_i^4 - K_1 e_i^5 - K_2 e_i^6 - K_3 e_i^7$$
$$- K_4 e_i^8 + O\left(e_i^9\right),$$

(4.2)

Following are the expressions of K_n $(n = 1, 2, 3, 4)$

$$K_1 = -8A_2^4 + 20A_2^2 A_3 - 6A_3^2 - 10A_2 A_4 + 4A_5,$$

$$K_2 = 16A_2^5 - 52A_2^3 A_3 + 33A_2 A_3^2 + 28A_2^2 A_4 - 17A_3 A_4 - 13A_2 A_5 + 5A_6,$$

$$K_3 = -32A_2^6 + 128A_2^4 A_3 - 126A_2^2 A_3^2 + 18A_3^3 - 72A_2^3 A_4 + 92A_2 A_3 A_4 - 12A_4^2 + 36A_2^2 A_5$$
$$- 22A_3 A_5 - 16A_2 A_6 + 6A_7,$$

$$K_4 = 64A_2^7 - 304A_2^5 A_3 + 408A_2^3 A_3^2 - 135A_2 A_3^3 + 176A_2^4 A_4 - 348A_2^2 A_3 A_4 + 75A_3^2 A_4 + 64A_2 A_4^2$$
$$- 92A_2^3 A_5 + 118A_2 A_3 A_5 - 31A_4 A_5 + 44A_2^2 A_6 - 27A_3 A_6 - 19A_2 A_7 + 7A_8.$$

(4.3)

For the sake of brevity, we omit their specific forms. We will use the same means in the following.

For

$$w_i = x_i - \frac{f(x_i)}{f'(x_i)}$$
$$= r + A_2 e_i^2 + 2\left(-A_2^2 + A_3\right)e_i^3 + \left(4A_2^3 - 7A_2 A_3 + 3A_4\right)e_i^4 + K_1 e_i^5 + K_2 e_i^6 + K_3 e_i^7$$
$$+ K_4 e_i^8 + O\left(e_i^9\right).$$

(4.4)

$$\tilde{e}_i = A_2 e_i^2 + 2\left(-A_2^2 + A_3\right)e_i^3 + \left(4A_2^3 - 7A_2 A_3 + 3A_4\right)e_i^4 + K_1 e_i^5 + K_2 e_i^6 + K_3 e_i^7$$
$$+ K_4 e_i^8 + O\left(e_i^9\right).$$

Using Taylor's series expansion, we get

$$f(w_i) = f'(r)\left[\tilde{e}_i + A_2\tilde{e}_i^2 + A_3\tilde{e}_i^3 + A_4\tilde{e}_i^4 + O\left(e_i^9\right)\right], \tag{4.5}$$

therefore,

$$\frac{f(w_i)}{f'(x_i)} = \tilde{e}_i - 2A_2\tilde{e}_ie_i + \left[\left(4A_2^2 - 3A_3\right)\tilde{e}_ie_i^2 + A_2\tilde{e}_i^2\right]$$

$$+ \left[-2A_2^2\tilde{e}_i^2e_i + \left(-8A_2^3 + 12A_2A_3 - 4A_4\right)\tilde{e}_ie_i^3\right] + M_1e_i^6 + M_2e_i^7 + M_3e_i^8 + O\left(e_i^9\right),$$

$$M_1 = \left(4A_2^3 - 3A_2A_3\right)\frac{\tilde{e}_i^2}{e_i^4} + \left(16A_2^4 - 36A_2^2A_3 + 9A_3^2 + 16A_2A_4 - 5A_5\right)\frac{\tilde{e}_i}{e_i^2} + A_3\frac{\tilde{e}_i^3}{e_i^6},$$

$$M_2 = -2A_2A_3\frac{\tilde{e}_i^3}{e_i^6} + \left(-8A_2^4 + 12A_2^2A_3 - 4A_2A_4\right)\frac{\tilde{e}_i^2}{e_i^4}$$

$$+ \left(-32A_2^5 + 96A_2^3A_3 - 54A_2A_3^2 - 48A_2^2A_4 + 24A_3A_4 + 20A_2A_5 - 6A_6\right)\frac{\tilde{e}_i}{e_i^2},$$

$$M_3 = \left(4A_2^2A_3 - 3A_3^2\right)\frac{\tilde{e}_i^3}{e_i^6} + \left(16A_2^5 - 36A_2^3A_3 + 9A_2A_3^2 + 16A_2^2A_4 - 5A_2A_5\right)\frac{\tilde{e}_i^2}{e_i^4}$$

$$+ \left(64A_2^6 - 240A_2^4A_3 + 216A_2^2A_3^2 - 27A_3^3 + 128A_2^3A_4 - 144A_2A_3A_4\right.$$

$$\left. +16A_4^2 - 60A_2^2A_5 + 30A_3A_5 + 24A_2A_6 - 7A_7\right)\frac{\tilde{e}_i}{e_i^2} + A_3\frac{\tilde{e}_i^4}{e_i^8}. \tag{4.6}$$

Also

$$\lambda_i = \frac{f(w_i)}{f(x_i)} = \frac{\tilde{e}_i}{e_i} - A_2\tilde{e}_i + \left[A_2\frac{\tilde{e}_i^2}{e_i} + \left(A_2^2 - A_3\right)\tilde{e}_ie_i\right] + \left[\left(-A_2^3 + 2A_2A_3 - A_4\right)\tilde{e}_ie_i^2 - A_2^2\tilde{e}_i^2\right]$$

$$+ \left[\left(A_2^4 - 3A_2^2A_3 + A_3^2 + 2A_2A_4 - A_5\right)\tilde{e}_ie_i^3 + \left(A_2^3 - A_2A_3\right)\tilde{e}_i^2e_i + A_3\frac{\tilde{e}_i^3}{e_i}\right]$$

$$+ \left[\left(-A_2^5 + 4A_2^3A_3 - 3A_2A_3^2 - 3A_2^2A_4 + 2A_3A_4 + 2A_2A_5 - A_6\right)\tilde{e}_ie_i^4\right.$$

$$\left. + \left(-A_2^4 + 2A_2^2A_3 - A_2A_4\right)\tilde{e}_i^2e_i^2 - A_2A_3\tilde{e}_i^3\right] + O\left(e_i^7\right). \tag{4.7}$$

Thus, using the Taylor expansion, we get

$$
\begin{aligned}
\omega(\lambda_i) &= \omega(0) + \omega'(0)\lambda_i + \frac{1}{2!}\omega''(0)\lambda_i^2 + \frac{1}{3!}\omega'''(0)\lambda_i^3 + O\left(\lambda_i^4\right) \\
&= \omega(0) + \omega'(0)\frac{\tilde{e}_i}{e_i} + \left[\frac{1}{2}\omega''(0)\frac{\tilde{e}_i^2}{e_i^2} - A_2\omega'(0)\tilde{e}_i\right] \\
&\quad + \left[\frac{1}{6}\omega'''(0)\frac{\tilde{e}_i^3}{e_i^3} + A_2(\omega'(0) - \omega''(0))\frac{\tilde{e}_i^2}{e_i} + \left(A_2^2 - A_3\right)\omega'(0)\tilde{e}_i e_i\right] \\
&\quad + L_1 e_i^4 + L_2 e_i^5 + L_3 e_i^6 + O\left(e_i^7\right),
\end{aligned}
\tag{4.8}
$$

where

$$
\begin{aligned}
L_1 &= \frac{1}{24}\omega^{iv}(0)\frac{\tilde{e}_i^4}{e_i^8} + \frac{1}{2}A_2\left(2\omega''(0) - \omega'''(0)\right)\frac{\tilde{e}_i^3}{e_i^6} \\
&\quad + \left(-A_3\omega''(0) + A_2^2\left(-\omega'(0) + \frac{3}{2}\omega''(0)\right)\right)\frac{\tilde{e}_i^2}{e_i^4} - \left(A_2^3 + 2A_2A_3 - A_4\right)\omega'(0)\frac{\tilde{e}_i}{e_i^2}, \\[2mm]
L_2 &= \frac{1}{120}\omega^{v}(0)\frac{\tilde{e}_i^5}{e_i^{10}} + \frac{1}{6}A_2\left(3\omega'''(0) - \omega^{(iv)}(0)\right) \\
&\quad \times \left(A_3\left(\omega'(0) - \frac{1}{2}\omega'''(0)\right) + A_2^2(-2\omega''(0) + \omega'''(0))\right)\frac{\tilde{e}_i^3}{e_i^6} \\
&\quad + \left(-A_2A_3(\omega'(0) - 3\omega''(0)) + A_2^3(\omega'(0) - 2\omega''(0)) - A_4\,\omega''(0)\right)\frac{\tilde{e}_i^2}{e_i^4} \\
&\quad + \left(A_2^4 - 3A_2^2A_3 + A_3^2 + 2A_2A_4 - A_5\right)\omega'(0)\frac{\tilde{e}_i}{e_i^2}, \\[2mm]
L_3 &= \frac{1}{720}\omega^{vi}(0)\frac{\tilde{e}_i^6}{e_i^{12}} + \frac{1}{24}A_2\left(4\omega^{iv}(0) - \omega^{v}(0)\right)\frac{\tilde{e}_i^5}{e_i^{10}} \\
&\quad + \frac{1}{12}\left(2A_3\left(6\omega''(0) - \omega^{iv}(0)\right) + A_2^2\left(6\omega''(0) - 18\omega'''(0) + 5\omega^{iv}(0)\right)\right)\frac{\tilde{e}_i^4}{e_i^8} \\
&\quad + \left(-A_2A_3(\omega'(0) + 2\omega''(0) - 2\omega'''(0)) + A_2^3\left(3\omega''(0) - \frac{5}{3}\omega'''(0)\right) - \frac{1}{2}A_4\,\omega'''(0)\right)\tilde{e}_i^3 \\
&\quad + \left(2A_2^2A_3(\omega'(0) - 3\omega''(0)) - A_2A_4(\omega'(0) - 3\omega''(0)) + \frac{1}{2}\left(3A_3^2 - 2A_5\right)\omega''(0)\right. \\
&\quad \left. + A_2^4\left(-\omega'(0) + \frac{5}{2}\omega''(0)\right)\right)\frac{\tilde{e}_i^2}{e_i^4} \\
&\quad + \left(-A_2^5 + 4A_2^3A_3 - 3A_2A_3^2 - 3A_2^2A_4 + 2A_3A_4 + 2A_2A_5 - A_6\right)\omega'(0)\frac{\tilde{e}_i}{e_i^2}.
\end{aligned}
\tag{4.9}
$$

Using (4.6) and (4.8), we have

$$
\begin{aligned}
z_i &= w_i - \omega(\lambda_i)\,\frac{f(w_i)}{f'(x_i)} \\[2mm]
&= r + [1 - \omega(0)]\tilde{e}_i - \left[\omega'(0)\frac{\tilde{e}_i^2}{e_i} - 2\omega(0)A_2\tilde{e}_i e_i\right] \\[2mm]
&\quad - \left[\frac{1}{2}\omega''(0)\frac{\tilde{e}_i^3}{e_i^2} + A_2(\omega(0) - 3\omega'(0))\tilde{e}_i^2 + \omega(0)\left(4A_2^2 - 3A_3\right)\tilde{e}_i e_i^2\right] \\[2mm]
&\quad - \left[\frac{1}{2}\omega''(0)\frac{\tilde{e}_i^3}{e_i^2} + A_2(\omega(0) - 3\omega'(0))\tilde{e}_i^2 + \omega(0)\left(4A_2^2 - 3A_3\right)\tilde{e}_i e_i^2\right]e_i^5 \\[2mm]
&\quad - M_6 e_i^6 - M_7 e_i^7 - M_8 e_i^8 + O\left(e_i^9\right),
\end{aligned}
\tag{4.10}
$$

where M_n $(n = 6, 7, 8)$ are the expression about A_n $(n = 2, 3, \ldots, 8)$.

If $(0) = 1$, $\omega'(0) = 2$ and substituting the value of $\tilde{e}_i$ from (4.4), we get

$$
\begin{aligned}
\hat{e}_i &= z_i - r \\[2mm]
&= \left[\left(5 - \frac{1}{2}\omega''(0)\right)A_2^3 - A_2 A_3\right]e_i^4 + \left(-36 + 5\omega''(0) - \frac{1}{6}\omega'''(0)\right)A_2^4 + (32 - 3\omega''(0))A_2^2 A_3 \\[2mm]
&\quad -2A_3^2 - 2A_2 A_4\Big]e_i^5 + M_6 e_i^6 + M_7 e_i^7 + M_8 e_i^8 + O\left(e_i^9\right).
\end{aligned}
\tag{4.11}
$$

Using Taylor's series expansion, we get

$$
f(z_i) = f'(r)\left[\hat{e}_i + A_2\hat{e}_i^2 + O\left(e_i^9\right)\right],
\tag{4.12}
$$

furthermore,

$$
\begin{aligned}
f[x_i, w_i] &= \frac{f(w_i) - f(x_i)}{w_i - x_i} \\[2mm]
&= f'(r)\Big[1 + A_2 e_i + \left(A_2^2 + A_3\right)e_i^2 + \left(-2A_2^3 + 3A_2 A_3 + A_4\right)e_i^3 \\[2mm]
&\quad + \left(4A_2^4 - 8A_2^2 A_3 + 2A_3^2 + 4A_2 A_4 + A_5\right)e_i^4 + O\left(e_i^5\right)\Big].
\end{aligned}
$$

$$f[x_i, z_i] = \frac{f(z_i) - f(x_i)}{z_i - x_i}$$

$$= f'(r)\Big[1 + A_2 e_i + A_3 e_i^2 + A_4 e_i^3$$

$$+ \Big(\Big(5 - \frac{1}{2}\omega''(0)\Big)A_2^4 - A_2^2 A_3 + A_5\Big)e_i^4 + O\big(e_i^5\big)\Big].$$

$$f[w_i, z_i] = \frac{f(z_i) - f(w_i)}{z_i - w_i}$$

$$= f'(r)\Big[1 + A_2^2 e_i^2 - 2\big(A_2^3 - A_2 A_3\big)e_i^3$$

$$+ \Big(\Big(9 - \frac{1}{2}\omega''(0)\Big)A_2^4 - 7A_2^2 A_3 + 3A_2 A_4\Big)e_i^4 + O\big(e_i^5\big)\Big].$$

$$(4.13)$$

Using the above results, we obtain

$$\frac{f[x_i, w_i]}{f[w_i, z_i]f[x_i, z_i]} = \frac{1}{f'(r)}\Big[1 + \big(-A_2^3 + A_2 A_3\big)e_i^3$$

$$+ \big((-7 + \omega''(0))A_2^4 - 4A_2^2 A_3 + 2A_3^2 + A_2 A_4\big)e_i^4 + O\big(e_i^5\big)\Big].$$

$$(4.14)$$

Also

$$\mu_i = \frac{f(z_i)}{f(x_i)} = \frac{\widehat{e}_i}{e_i} - A_2 \widehat{e}_i + O\big(e_i^5\big). \tag{4.15}$$

Thus, using the Taylor expansion and $|W''(0)| < \infty$, we get

$$W(\mu_i) = W(0) + W'(0)\mu_i + O\big(\mu_i^2\big) = W(0) + W'(0)\Big(\frac{\widehat{e}_i}{e_i} - A_2 \widehat{e}_i\Big) + O\big(e_i^5\big). \tag{4.16}$$

Using these results in

$$x_{i+1} = z_i - W(\mu_i)\frac{f[x_i, w_i]f(z_i)}{f[w_i, z_i]\ f[x_i, z_i]}, \tag{4.17}$$

we obtain

$$
\begin{aligned}
e_{i+1} = \widehat{e}_i &- \left[W(0) + W'(0)\frac{\widehat{e}_i}{e_i} - W'(0)A_2\widehat{e}_i + O\left(e_i^5\right) \right] \left[\widehat{e}_i + A_2\widehat{e}_i^2 + O\left(\widehat{e}_i^3\right) \right] \\
&\times \left[1 + \left(-A_2^3 + A_2 A_3\right)e_i^3 + \left(\left(-7 + \omega''(0)\right)A_2^4 - 4A_2^2 A_3 + 2A_3^2 + A_2 A_4\right)e_i^4 + O\left(e_i^5\right) \right] \\
= \widehat{e}_i &- \left[W(0) + W'(0)\frac{\widehat{e}_i}{e_i} - W'(0)A_2\widehat{e}_i \right] \left(\widehat{e}_i + A_2\widehat{e}_i^2\right) \\
&\times \left[1 + \left(-A_2^3 + A_2 A_3\right)e_i^3 + \left(\left(-7 + \omega''(0)\right)A_2^4 - 4A_2^2 A_3 + 2A_3^2 + A_2 A_4\right)e_i^4 \right] + O\left(e_i^9\right) \\
= [1 &- W(0)]\widehat{e}_i - \left[W(0)\left(-A_2^3 + A_2 A_3\right)e_i^3 + W'(0)\frac{\widehat{e}_i}{e_i} \right]\widehat{e}_i - [W(0) - W'(0)]A_2\widehat{e}_i^2 \\
&- \left[\left(-7 + \omega''(0)\right)A_2^4 - 4A_2^2 A_3 + 2A_3^2 + A_2 A_4 \right]W(0)e_i^4\widehat{e}_i + O\left(e_i^9\right) \\
= [1 &- W(0)]\widehat{e}_i - \left[W(0)\left(-A_2^3 + A_2 A_3\right) + W'(0)\left(5 - \frac{1}{2}\omega''(0)A_2^3 - A_2 A_3\right) \right]e_i^3\widehat{e}_i \\
&+ [W'(0) - W(0)]A_2\widehat{e}_i^2 + \left[W(0)\left(\left(7 - \omega''(0)\right)A_2^4 + 4A_2^2 A_3 - 2A_3^2 - A_2 A_4\right) \right. \\
&\left. - W'(0)\left(\left(-36 + 5\omega''(0) - \frac{1}{6}\omega'''(0)\right)A_2^4 + \left(32 - 3\omega''(0)\right)A_2^2 A_3 - 2A_3^2 - 2A_2 A_4\right) \right] \\
&\times e_i^4\widehat{e}_i + O\left(e_i^9\right).
\end{aligned}
$$

$$\tag{4.18}$$

This means that convergence order of the family (3.10) is seventh-order with $W(0) = 1$ and the error equation is

$$
\begin{aligned}
e_{i+1} = &\left[\left(A_2^3 - A_2 A_3\right)\left(\left(5 - \frac{1}{2}\omega''(0)A_2^3 - A_2 A_3\right)\right) \right. \\
&\left. - W'(0)\left(\left(5 - \frac{1}{2}\omega''(0)A_2^3 - A_2 A_3\right)^2\right) \right]e_i^7 + O\left(e_i^8\right),
\end{aligned}
$$

$$\tag{4.19}$$

and if W is any function with $W(0) = 1$, $W'(0) = 1$, and $\omega''(0) = 8$, then the convergence order of any method of the family (3.10) arrives to eight, and the error equation is

$$
e_{i+1} = A_2^2\left(A_2^2 - A_3\right)\left[\left(-5 + \frac{1}{6}\omega'''(0)\right)A_2^3 - 4A_2 A_3 + A_4 \right]e_i^8 + O\left(e_i^9\right).
$$

$$\tag{4.20}$$

Thus if ω and W are any functions with $(0) = 1$, $\omega'(0) = 2$, $\omega''(0) = 8$, $W(0) = 1$, and $W'(0) = 1$, then the eighth-order convergence is established. This completes the proof of the theorem. $\square$

Note that per iteration every method of the family (3.10) uses four pieces of information, namely, $f(x_i), f'(x_i), f(w_i), f(z_i)$ and has eighth-order convergence with the conditions $\omega(0) = 1$, $\omega'(0) = 2$, $\omega''(0) = 8$, $W(0) = 1$, and $W'(0) = 1$, which is in accordance with Kung-Traub conjecture for 4 evaluations.

5. Some Particular Forms

Here, we consider some forms of the functions $\omega(t)$ and $W(t)$ satisfying the conditions of the Theorem 4.1. Based on these forms some methods of the family (3.10) are also presented.

5.1. Forms of $\omega(t)$

Form 1. For the function ω given by the following:

$$\omega_1(t) = 1 + 2t + 4t^2 + \alpha t^3, \tag{5.1}$$

where $\alpha \in R$ is a constant, it is clear that the conditions of Theorem 4.1 are satisfied.

Form 2. For the function ω defined by the following:

$$\omega_2(t) = 1 + \frac{2t}{1 - 2t + \alpha t^2}, \tag{5.2}$$

where $\alpha \in R$, it can be easily seen that this function satisfies the conditions of Theorem 4.1.

Form 3. For the function ω defined by the following:

$$\omega_3(t) = \left(1 + 2\alpha t + \beta t^2\right)^{1/\alpha}, \quad \beta = 2\alpha(\alpha + 1), \tag{5.3}$$

where $\alpha \in R - \{0\}$, it can be seen that this function also satisfies the conditions of Theorem 4.1.

5.2. Forms of $W(t)$

Form 1. For the function W given by the following:

$$W_1(t) = 1 + t + \gamma t^2, \tag{5.4}$$

where $\gamma \in R$, it can be seen the function $W_1(t)$ satisfies the conditions of Theorem 4.1.

Form 2. For the function W defined by the following:

$$W_2(t) = 1 + \frac{t}{1 + \gamma t}, \tag{5.5}$$

where $\gamma \in R$, it is simple to see that $W_2(t)$ satisfies the conditions of Theorem 4.1

Form 3. For the function *W* defined by the following:

$$W_3(t) = (1 + \gamma t)^{1/\gamma}, \tag{5.6}$$

where $\gamma \in R - \{0\}$, again it can be seen that $W_3(t)$ satisfies the conditions of Theorem 4.1.

5.3. Forms of Methods

To form a concrete method we can take any combination of the above defined $w(t)$ and $W(t)$. For simplicity, we consider only three such combinations. For example, by taking $w_2(t)$ with $W_i(t)$, $i = 1, 2, 3$ the following methods can be formed

Method 1. Taking $w_2(t)$ and $W_1(t)$, we get a new two-parameter family of eighth-order methods

$$w_i = x_i - \frac{f(x_i)}{f'(x_i)},$$

$$z_i = w_i - \frac{f^2(x_i) + \alpha f^2(w_i)}{f^2(x_i) - 2f(x_i)f(w_i) + \alpha f^2(w_i)} \frac{f(w_i)}{f'(x_i)}, \tag{5.7}$$

$$x_{i+1} = z_i - \left[1 + \frac{f(z_i)}{f(x_i)} + \gamma \frac{f^2(z_i)}{f^2(x_i)}\right] \frac{f[x_i, w_i] \, f(z_i)}{f[w_i, z_i] f[x_i, z_i]}.$$

Method 2. Considering $w_2(t)$ and $W_2(t)$, we get another new two-parameter family of eighth-order methods

$$w_i = x_i - \frac{f(x_i)}{f'(x_i)},$$

$$z_i = w_i - \frac{f^2(x_i) + \alpha f^2(w_i)}{f^2(x_i) - 2f(x_i)f(w_i) + \alpha f^2(w_i)} \frac{f(w_i)}{f'(x_i)}, \tag{5.8}$$

$$x_{i+1} = z_i - \left[\frac{f(x_i) + (1 + \gamma)f(z_i)}{f(x_i) + \gamma f(z_i)}\right] \frac{f[x_i, w_i] \, f(z_i)}{f[w_i, z_i] \, f[x_i, z_i]}.$$

Table 1: Test functions.

$f(x)$	r
$f_1(x) = x^5 + x^4 + 4x^2 - 15$	1.3474280989683050
$f_2(x) = \sin(x) - x/3$	2.2788626600758283
$f_3(x) = 10xe^{-x^2} - 1$	1.6796306104284499
$f_4(x) = \cos(x) - x$	0.73908513321516065
$f_5(x) = e^{-x^2 + x + 2} - 1$	-1.0000000000000000
$f_6(x) = e^{-x} + \cos(x)$	1.7461395304080124
$f_7(x) = \ln(x^2 + x + 2) - x + 1$	4.1525907367571583
$f_8(x) = \arcsin(x^2 - 1) - x/2 + 1$	0.5948109683983692
$f_9(x) = xe^{x^2} - \sin^2 x + 3\cos x + 5$	-1.2076478271309189

Method 3. Considering now $w_2(t)$ and $W_3(t)$, we get another new two-parameter family of eighth-order methods

$$w_i = x_i - \frac{f(x_i)}{f'(x_i)},$$

$$z_i = w_i - \frac{f^2(x_i) + \alpha f^2(w_i)}{f^2(x_i) - 2f(x_i)f(w_i) + \alpha f^2(w_i)} \frac{f(w_i)}{f'(x_i)}, \tag{5.9}$$

$$x_{i+1} = z_i - \left[1 + \gamma \frac{f(z_i)}{f(x_i)}\right]^{1/\gamma} \frac{f[x_i, w_i] \; f(z_i)}{f[w_i, z_i] \; f[x_i, z_i]}.$$

The proposed families require three evaluations of the function f and one evaluation of first derivative f' per iteration, and achieve eighth-order convergence. Thus the efficiency index (E) defined by (2.3) of the present methods (3.10) is $E = \sqrt[4]{8} \approx 1.682$ which is better than $E = \sqrt{2} \approx 1.414$ of Newton's method, $E = \sqrt[3]{4} \approx 1.587$ of King's [3] and Ostrowski's [4] methods, $E = \sqrt[4]{6} \approx 1.565$ of sixth-order methods [10–12] and $E = \sqrt[4]{7} \approx 1.627$ of seventh-order methods [15, 18].

6. Numerical Examples

We employ the present methods (4.1), and (4.4) denoted by M81, M82 and M83, respectively to solve some nonlinear equations and compare with Newton's method (NM) defined by (1.1), the eighth-order method developed by Cordero et al. [23] denoted by C8 and defined as follows:

$$w_i = x_i - \frac{f(x_i)}{f'(x_i)},$$

$$z_i = x_i - \frac{f(x_i) - f(w_i)}{f(x_i) - 2f(w_i)} \frac{f(x_i)}{f'(x_i)},$$

$$u_i = z_i - \left(\frac{f(x_i) - f(w_i)}{f(x_i) - 2f(w_i)} + \frac{1}{2}\frac{f(z_i)}{f(w_i) - 2f(z_i)}\right)\frac{f(z_i)}{f'(x_i)}, \tag{6.1}$$

$$x_{i+1} = u_i - \frac{3(\beta_2 + \beta_3)(u_i - z_i)}{\beta_1(u_i - z_i) + \beta_2(w_i - x_i) + \beta_3(z_i - x_i)} \frac{f(z_i)}{f'(x_i)},$$

Table 2: Comparison of methods using same total number of function evaluations for all methods (TFE = 12).

	NM	C8 $\beta_1=\beta_3=0, \beta_2=1$	L8 $\alpha_1=\alpha_2=1$	P8	T8 $\alpha=0$	B81 $\alpha=1$	B82 $\alpha=\gamma=1$	B83 $\alpha=1$	M81 $\alpha=\gamma=1$	M82 $\alpha=\gamma=1$	M83 $\alpha=\gamma=1$
$f_1,\ x_0=1.6$											
$\|x_i - r\|$	$9.75e-41$	$4.35e-338$	$4.51e-300$	$0.00e+00$	$3.67e-305$	$6.55e-304$	$8.24e-304$	$9.60e-304$	$0.00e+00$	$1.40e-349$	$1.70e-350$
$\|f(x_i)\|$	$3.61e-39$	$1.61e-336$	$1.67e-298$	$0.00e+00$	$1.36e-303$	$2.43e-302$	$3.05e-302$	$3.55e-302$	$6.96e-350$	$5.19e-348$	$6.28e-349$
ρ	2.0000000	7.9999999	7.9999999	8.0000000	7.9999999	7.9999989	7.9999989	7.9999989	7.9999997	7.9999997	7.9999997
$f_2,\ x_0=2.0$											
$\|x_i - r\|$	$4.27e-57$	$0.00e+00$	$4.76e-334$	$0.00e+00$	$2.88e-333$	$0.00e+00$	$0.00e+00$	$0.00e+00$	$0.00e+00$	$0.00e+00$	$0.00e+00$
$\|f(x_i)\|$	$4.20e-57$	$0.00e+00$	$4.68e-334$	$0.00e+00$	$2.83e-333$	$0.00e+00$	$0.00e+00$	$0.00e+00$	$0.00e+00$	$0.00e+00$	$0.00e+00$
ρ	2.0000000	8.0000000	7.9999993	8.0000000	7.9999994	8.0000000	8.0000000	8.0000000	8.0000000	8.0000000	8.0000000
$f_3,\ x_0=1.8$											
$\|x_i - r\|$	$4.41e-58$	$0.00e+00$	$0.00e+00$	$0.00e+00$	$0.00e+00$	$0.00e+00$	$0.00e+00$	$0.00e+00$	$0.00e+00$	$0.00e+00$	$0.00e+00$
$\|f(x_i)\|$	$1.22e-57$	$0.00e+00$	$0.00e+00$	$0.00e+00$	$0.00e+00$	$0.00e+00$	$0.00e+00$	$0.00e+00$	$0.00e+00$	$0.00e+00$	$0.00e+00$
ρ	2.0000000	8.0000000	8.0000000	8.0000000	8.0000000	8.0000000	8.0000000	8.0000000	7.9999997	7.9999997	7.9999997
$f_4,\ x_0=1.0$											
$\|x_i - r\|$	$1.80e-83$	$0.00e+00$	$0.00e+00$	$0.00e+00$	$0.00e+00$	$0.00e+00$	$0.00e+00$	$0.00e+00$	$0.00e+00$	$0.00e+00$	$0.00e+00$
$\|f(x_i)\|$	$3.00e-83$	$0.00e+00$	$0.00e+00$	$0.00e+00$	$0.00e+00$	$0.00e+00$	$0.00e+00$	$0.00e+00$	$0.00e+00$	$0.00e+00$	$0.00e+00$
ρ	2.0000000	8.0000000	8.0000000	8.0000000	8.0000000	8.0000000	8.0000000	8.0000000	8.0000000	8.0000000	8.0000000
$f_5,\ x_0=-0.5$											
$\|x_i - r\|$	$3.46e-27$	$7.71e-202$	$8.70e-163$	$5.99e-246$	$3.75e-162$	$3.12e-222$	$1.20e-221$	$2.98e-221$	$4.57e-195$	$7.97e-194$	$1.98e-194$
$\|f(x_i)\|$	$1.04e-26$	$2.31e-201$	$2.61e-162$	$1.80e-245$	$1.13e-161$	$9.36e-222$	$3.61e-221$	$8.93e-221$	$1.37e-194$	$2.39e-193$	$5.95e-194$
ρ	2.0000000	7.9999339	7.9996776	7.9999847	7.9996400	8.0000580	8.0000594	8.0000603	7.9995556	7.9995321	7.9995437
$f_6,\ x_0=2.0$											
$\|x_i - r\|$	$7.97e-85$	$0.00e+00$	$0.00e+00$	$0.00e+00$	$0.00e+00$	$0.00e+00$	$0.00e+00$	$0.00e+00$	$0.00e+00$	$0.00e+00$	$0.00e+00$
$\|f(x_i)\|$	$9.24e-85$	$0.00e+00$	$0.00e+00$	$0.00e+00$	$0.00e+00$	$0.00e+00$	$0.00e+00$	$0.00e+00$	$0.00e+00$	$0.00e+00$	$0.00e+00$
ρ	2.0000000	8.0000000	8.0000000	8.0000001	8.0000000	8.0000000	8.0000000	8.0000000	8.0000000	8.0000000	8.0000000
$f_7,\ x_0=3.2$											
$\|x_i - r\|$	$4.66e-74$	$0.00e+00$	$0.00e+00$	$0.00e+00$	$0.00e+00$	$0.00e+00$	$0.00e+00$	$0.00e+00$	$0.00e+00$	$0.00e+00$	$0.00e+00$
$\|f(x_i)\|$	$2.81e-74$	$0.00e+00$	$0.00e+00$	$0.00e+00$	$0.00e+00$	$0.00e+00$	$0.00e+00$	$0.00e+00$	$0.00e+00$	$0.00e+00$	$0.00e+00$
ρ	2.0000000	8.0000000	8.0000000	8.0000000	8.0000000	8.0000000	8.0000000	8.0000000	8.0000000	8.0000000	8.0000000
$f_8,\ x_0=1.0$											
$\|x_i - r\|$	$1.68e-54$	$4.42e-342$	$4.21e-309$	$0.00e+00$	$1.20e-309$	$0.00e+00$	$5.83e-346$	$1.68e-341$	$0.00e+00$	$0.00e+00$	$0.00e+00$
$\|f(x_i)\|$	$1.78e-54$	$4.68e-342$	$4.46e-309$	$0.00e+00$	$1.27e-309$	$0.00e+00$	$6.17e-346$	$1.78e-341$	$0.00e+00$	$0.00e+00$	$0.00e+00$
ρ	2.0000000	8.0000004	8.0000018	8.0000001	8.0000021	8.0000012	8.0000009	8.0000014	8.0000001	8.0000001	8.0000001
$f_9,\ x_0=-1$											
$\|x_i - r\|$	$8.63e-33$	$0.00e+00$	$1.76e-339$	$0.00e+00$	$0.00e+00$	$4.78e-218$	$3.12e-214$	$8.95e-212$	$1.83e-216$	$3.85e-218$	$2.63e-217$
$\|f(x_i)\|$	$1.75e-31$	$0.00e+00$	$3.58e-338$	$0.00e+00$	$0.00e+00$	$9.71e-217$	$6.34e-213$	$1.82e-210$	$3.72e-215$	$7.81e-217$	$5.34e-216$
ρ	2.0000000	8.0000001	8.0000002	8.0000000	8.0000000	7.9999987	7.9999985	7.9999984	8.0001475	8.0001379	8.0001426

where $\beta_1, \beta_2, \beta_3 \in R$ and $\beta_2 + \beta_3 \neq 0$, eighth-order method developed by Liu and Wang [22] denoted by L8 and defined as follows:

$$w_i = x_i - \frac{f(x_i)}{f'(x_i)},$$

$$z_i = w_i - \frac{f(x_i)}{f(x_i) - 2f(w_i)} \frac{f(w_i)}{f'(x_i)}, \tag{6.2}$$

$$x_{i+1} = z_i - \left[\left(\frac{f(x_i) - f(w_i)}{f(x_i) - 2f(w_i)} \right)^2 + \frac{f(z_i)}{f(w_i) - \alpha_1 f(z_i)} + \frac{4f(z_i)}{f(x_i) + \alpha_2 f(z_i)} \right] \frac{f(z_i)}{f'(x_i)},$$

where $\alpha_1, \alpha_2 \in R$, eighth-order method developed by Petković et al. [20] denoted by P8 and defined as follows:

$$w_i = x_i - \frac{f(x_i)}{f'(x_i)},$$

$$z_i = w_i - \frac{f(x_i)}{f(x_i) - 2f(w_i)} \frac{f(w_i)}{f'(x_i)}, \tag{6.3}$$

$$x_{i+1} = z_i - \frac{(1 + a_4(z_i - x_i))^2 f(z_i)}{a_2 - a_1 a_4 + a_3(z_i - x_i)(2 + a_4(z_i - x_i))},$$

where $a_1 = f(x_i)$, $a_3 = (f'(x_i)f[w_i, z_i] - f[x_i, w_i]f[x_i, z_i])/((x_i f[w_i, z_i] + (w_i f(z_i) - z_i f(w_i))/(w_i - z_i)) - f(x_i))$,

$$a_4 = \frac{a_3}{f[x_i, w_i]} + \frac{f'(x_i) - f[x_i, w_i]}{(w_i - x_i)f[x_i, w_i]}, \qquad a_2 = f'(x_i) + a_4 f(x_i), \tag{6.4}$$

eighth-order method developed by Thukral and Petković [21] denoted by T8 and defined as follows:

$$w_i = x_i - \frac{f(x_i)}{f'(x_i)},$$

$$z_i = w_i - \frac{f(x_i)}{f(x_i) - 2f(w_i)} \frac{f(w_i)}{f'(x_i)}, \tag{6.5}$$

$$x_{i+1} = z_i - \left[\frac{f^2(x_i)}{f^2(x_i) - 2f(x_i)f(w_i) - f^2(w_i)} + \frac{f(z_i)}{f(w_i) - \alpha f(z_i)} + \frac{4f(z_i)}{f(x_i)} \right] \frac{f(z_i)}{f'(x_i)},$$

where $\alpha \in R$, eighth-order methods presented in Section 3 of [19] by Bi et al. denoted by B81, B82 and B83.

The test functions and root r correct up to 16 decimal places are displayed in Table 1. The first eight functions we have selected are same as in [19]. The last function is selected from [18]. In Table 2, we exhibit the absolute values of the difference of root r and its approximation x_i, where r is computed with 350 significant digits and x_i is calculated by costing the same total number of function evaluations (TFE) for each method. The TFE is counted as sum of the number of evaluations of the function itself plus the number of evaluations of the derivatives. In the calculations, 12 TFE are used by each method. That means 6 iterations are used for NM and 3 iterations for the remaining methods. The absolute values of the function $|f(x_i)|$ and the computational order of convergence (ρ) are also displayed in Table 2. It can be observed that the computed results, displayed in Table 2, overwhelmingly support the theory of convergence and efficiency analyses discussed in the previous sections. From the results, it can be concluded that the proposed methods are competitive with existing methods and possess quick convergence for good initial approximations. Among the eighth-order methods, we are not able to select one as the best. For some initial guess one is better while for other initial guess the another one would be appropriate. Thus the present methods can be of practical interest.

7. Conclusions

In this work, we have obtained a new simple and elegant family of eighth-order multipoint methods for solving nonlinear equations. Thus, one requires three evaluations of the function f and one of its first-derivative f' per full step and therefore, the efficiency index of the present methods is 1.682 which is better than the efficiency index of Newton method, fourth-order methods, sixth-order methods, and seventh-order methods.

Many numerical applications use higher precision in their computations. In these types of applications, numerical methods of higher-order are important. The numerical results show that the methods associated with a multiprecision arithmetic floating point are very useful, because these methods yield a clear reduction in number of iterations. Finally, we conclude that the methods presented in this paper are preferable to other recognized efficient methods, namely, Newton's method, King's methods, sixth-order methods [10–12], seventh-order methods [15, 18], etc.

References

[1] J. M. Ortega and W. C. Rheinboldt, *Iterative Solution of Nonlinear Equations in Several Variables*, Academic Press, New York, NY, USA, 1970.
[2] S. C. Chapra and R. P. Canale, *Numerical Methods for Engineers*, McGraw-Hill Book Company, New York, NY, USA, 1988.
[3] R. F. King, "A family of fourth order methods for nonlinear equations," *SIAM Journal on Numerical Analysis*, vol. 10, pp. 876–879, 1973.
[4] A. M. Ostrowski, *Solution of Equations in Euclidean and Banach Spaces*, Academic Press, New York, NY, USA, 1960.
[5] J. F. Traub, *Iterative Methods for the Solution of Equations*, Prentice-Hall, Englewood Cliffs, NJ, USA, 1964.
[6] H. T. Kung and J. F. Traub, "Optimal order of one-point and multipoint iteration," *Journal of the Association for Computing Machinery*, vol. 21, pp. 643–651, 1974.
[7] P. Jarratt, "Some efficient fourth order multipoint methods for solving equations," *BIT*, vol. 9, pp. 119–124, 1969.
[8] J. A. Ezquerro, M. A. Hernández, and M. A. Salanova, "Construction of iterative processes with high order of convergence," *International Journal of Computer Mathematics*, vol. 69, no. 1-2, pp. 191–201, 1998.
[9] J. M. Gutiérrez and M. A. Hernández, "An acceleration of Newton's method: super-Halley method," *Applied Mathematics and Computation*, vol. 117, no. 2-3, pp. 223–239, 2001.
[10] M. Grau and J. L. Díaz-Barrero, "An improvement to Ostrowski root-finding method," *Applied Mathematics and Computation*, vol. 173, no. 1, pp. 450–456, 2006.

[11] J. R. Sharma and R. K. Guha, "A family of modified Ostrowski methods with accelerated sixth order convergence," *Applied Mathematics and Computation*, vol. 190, no. 1, pp. 111–115, 2007.

[12] C. Chun and Y. Ham, "Some sixth-order variants of Ostrowski root-finding methods," *Applied Mathematics and Computation*, vol. 193, no. 2, pp. 389–394, 2007.

[13] J. Kou, "The improvements of modified Newton's method," *Applied Mathematics and Computation*, vol. 189, no. 1, pp. 602–609, 2007.

[14] J. Kou and Y. Li, "An improvement of the Jarratt method," *Applied Mathematics and Computation*, vol. 189, no. 2, pp. 1816–1821, 2007.

[15] J. Kou, Y. Li, and X. Wang, "Some variants of Ostrowski's method with seventh-order convergence," *Journal of Computational and Applied Mathematics*, vol. 209, no. 2, pp. 153–159, 2007.

[16] C. Chun, "Some improvements of Jarratt's method with sixth-order convergence," *Applied Mathematics and Computation*, vol. 190, no. 2, pp. 1432–1437, 2007.

[17] S. K. Parhi and D. K. Gupta, "A sixth order method for nonlinear equations," *Applied Mathematics and Computation*, vol. 203, no. 1, pp. 50–55, 2008.

[18] W. Bi, H. Ren, and Q. Wu, "New family of seventh-order methods for nonlinear equations," *Applied Mathematics and Computation*, vol. 203, no. 1, pp. 408–412, 2008.

[19] W. Bi, H. Ren, and Q. Wu, "Three-step iterative methods with eighth-order convergence for solving nonlinear equations," *Journal of Computational and Applied Mathematics*, vol. 225, no. 1, pp. 105–112, 2009.

[20] L. D. Petković, M. S. Petković, and J. Džunić, "A class of three-point root-solvers of optimal order of convergence," *Applied Mathematics and Computation*, vol. 216, no. 2, pp. 671–676, 2010.

[21] R. Thukral and M. S. Petković, "A family of three-point methods of optimal order for solving nonlinear equations," *Journal of Computational and Applied Mathematics*, vol. 233, no. 9, pp. 2278–2284, 2010.

[22] L. Liu and X. Wang, "Eighth-order methods with high efficiency index for solving nonlinear equations," *Applied Mathematics and Computation*, vol. 215, no. 9, pp. 3449–3454, 2010.

[23] A. Cordero, J. R. Torregrosa, and M. P. Vassileva, "Three-step iterative methods with optimal eighth-order convergence," *Journal of Computational and Applied Mathematics*, vol. 235, no. 10, pp. 3189–3194, 2011.

[24] S. K. Khattri and T. Log, "Derivative free algorithm for solving nonlinear equations," *Computing*, vol. 92, no. 2, pp. 169–179, 2011.

[25] S. K. Khattri and I. K. Argyros, "Sixth order derivative free family of iterative methods," *Applied Mathematics and Computation*, vol. 217, no. 12, pp. 5500–5507, 2011.

[26] Y. H. Geum and Y. I. Kim, "A biparametric family of optimally convergent sixteenth-order multipoint methods with their fourth-step weighting function as a sum of a rational and a generic two-variable function," *Journal of Computational and Applied Mathematics*, vol. 235, no. 10, pp. 3178–3188, 2011.

[27] W. Gautschi, *Numerical Analysis*, Birkhäuser, Boston, Mass, USA, 1997.

[28] S. Weerakoon and T. G. I. Fernando, "A variant of Newton's method with accelerated third-order convergence," *Applied Mathematics Letters*, vol. 13, no. 8, pp. 87–93, 2000.

5

Two-Level Stabilized Finite Volume Methods for Stationary Navier-Stokes Equations

Anas Rachid,[1] Mohamed Bahaj,[2] and Noureddine Ayoub[2]

[1] *École Nationale Supérieure des Arts et Métiers-Casablanca, Université Hassan II, B.P. 150, Mohammedia, Morocco*
[2] *Department of Mathematics and Computing Sciences, Faculty of Sciences and Technology, University Hassan 1st, B.P. 577, Settat, Morocco*

Correspondence should be addressed to Anas Rachid, rachid.anas@gmail.com

Academic Editor: Weimin Han

We propose two algorithms of two-level methods for resolving the nonlinearity in the stabilized finite volume approximation of the Navier-Stokes equations describing the equilibrium flow of a viscous, incompressible fluid. A macroelement condition is introduced for constructing the local stabilized finite volume element formulation. Moreover the two-level methods consist of solving a small nonlinear system on the coarse mesh and then solving a linear system on the fine mesh. The error analysis shows that the two-level stabilized finite volume element method provides an approximate solution with the convergence rate of the same order as the usual stabilized finite volume element solution solving the Navier-Stokes equations on a fine mesh for a related choice of mesh widths.

1. Introduction

We consider a two-level method for the resolution of the nonlinear system arising from finite volume discretizations of the equilibrium, incompressible Navier-Stokes equations:

$$-\nu\Delta u + (u \cdot \nabla)u + \nabla p = f \quad \text{in } \Omega, \tag{1.1}$$

$$\nabla \cdot u = 0 \quad \text{in } \Omega, \tag{1.2}$$

$$u = 0 \quad \text{in } \partial\Omega, \tag{1.3}$$

where $u = (u_1(x), u_2(x))$ is the velocity vector, $p = p(x)$ is the pressure, $f = f(x)$ is the body force, $\nu > 0$ is the viscosity of the fluid, and $\Omega \subset \mathbb{R}^2$, the flow domain, is assumed to be bounded, to have a Lipschitz-continuous boundary $\partial\Omega$, and to satisfy a further condition stated in (H1).

Finite volume method is an important numerical tool for solving partial differential equations. It has been widely used in several engineering fields, such as fluid mechanics, heat and mass transfer, and petroleum engineering. The method can be formulated in the finite difference framework or in the Petrov-Galerkin framework. Usually, the former one is called finite volume method [1, 2], MAC (marker and cell) method [3], or cell-centered method [4], and the latter one is called finite volume element method (FVE) [5–7], covolume method [8], or vertex-centered method [9, 10]. We refer to the monographs [11, 12] for general presentations of these methods. The most important property of FVE is that it can preserve the conservation laws (mass, momentum, and heat flux) on each control volume. This important property, combined with adequate accuracy and ease of implementation, has attracted more people to do research in this field.

On the other hand, the two-level finite element strategy based on two finite element spaces on one coarse and one fine mesh has been widely studied for steady semilinear elliptic equations [13, 14] and the Navier-Stokes equations [15–22]. For the finite volume element method, Bi and Ginting [23] have studied two-grid finite volume element method for linear and nonlinear elliptic problems; Chen et al. [24] have applied two-grid methods for solving a two-dimensional nonlinear parabolic equation using finite volume element method. Chen and Liu [25] have also studied this method for semilinear parabolic problems. However, to the best of our knowledge, there is no two-level finite volume convergence analysis for the Navier-Stokes equations in the literature.

In this paper we aim to combine FVE method based on $P_1 - P_0$ macroelement with two-level strategy to solve the two-dimensional Navier-Stokes (1.1)–(1.3). The heart of the analysis is the use of a transfer operator to connect finite volume and finite element estimations which will lead to more difficult term to estimate. We choose the two-grid spaces as two conforming finite element spaces X_H and X_h on one coarse grid with mesh size H and one fine grid with mesh size $h \ll H$. We propose two algorithms of two-level method for resolving the nonlinearity in the stabilized finite volume approximation of the problem (1.1)–(1.3): the simple and Newton algorithms. First we prove that the simple two-level stabilized finite volume solution (u^h, p^h) is the following error estimate:

$$\left\| u - u^h \right\|_1 + \left\| p - p^h \right\|_0 \le C\left(h + H^2 \right). \tag{1.4}$$

Second we prove that the Newton two-level stabilized finite volume solution (u^h, p^h) is the following error estimate:

$$\left\| u - u^h \right\|_1 + \left\| p - p^h \right\|_0 \le C\left(h + H^3 |\log h|^{1/2} \right), \tag{1.5}$$

where C denotes some generic constant which may stand for different values at its different occurrences.

Hence, the two-level algorithms achieve asymptotically optimal approximation as long as the mesh sizes satisfy $h = O(H^2)$ for the simple two-level stabilized finite volume solution and $h = O(H^3|\log h|^{1/2})$ for the Newton two-level stabilized finite volume solution. As a result, solving the nonlinear Navier-Stokes equations will not be much more difficult than solving one single linearized equation.

The rest of this paper is organized as follows. In the next section, we introduce some notations and construct a FVE scheme. In Section 3 we recall same preliminary estimates of the stabilized finite volume approximations. Finally the two-level FVE algorithms and the improved error estimates are presented and established in Section 4.

2. Finite Volume Scheme

2.1. Notations

We will use $\| \cdot \|_m$ and $| \cdot |_m$ to denote the norm and seminorm of the Sobolev space $(H^m(\Omega))^d$, $d = 1, 2$. Let $H_0^1(\Omega)$ be the standard Sobolev subspace of $H^1(\Omega)$ of functions vanishing on $\partial\Omega$. We introduce the following notations:

$$X = \left(H_0^1(\Omega)\right)^2, \qquad Y = L_0^2(\Omega) = \left\{ q : q \in L^2(\Omega), \int_\Omega q = 0 \right\}. \tag{2.1}$$

The scalar product and norm in Y are denoted by the usual $L^2(\Omega)$ inner product $(\cdot, \cdot)$ and $\| \cdot \|_0$, respectively. As mentioned above, we need a further assumption on Ω.

H1

Assume that Ω is regular so that the unique solution $(v, q) \in X \times Y$ of the steady Stokes problem

$$-\Delta v + \nabla q = g, \qquad \nabla \cdot v = 0; \quad v|_{\partial\Omega} = 0 \tag{2.2}$$

for a prescribed $g \in (L^2(\Omega))^2$ exists and satisfies

$$\|v\|_2 + \|q\|_1 \le C\|g\|_0, \tag{2.3}$$

where $C > 0$ is a constant depending on Ω.

The weak formulation of the problem (1.1)–(1.3) is to find $(u, p) \in X \times Y$ such that

$$a(u, v) - d(v, p) + b(u, u, v) = (f, v), \quad \forall v \in X,$$
$$d(u, q) = 0, \quad \forall q \in Y, \tag{2.4}$$

where the bilinear forms $a(\cdot,\cdot)$, $d(\cdot,\cdot)$ and the trilinear form $b(\cdot,\cdot,\cdot)$ are given by

$$a(u,v) = v(\nabla u, \nabla v) = v\int_\Omega \nabla u : \nabla v\, dx, \quad \forall u,v \in X,$$

$$d(v,q) = (\nabla\cdot v, q) = \int_\Omega q\nabla\cdot v\, dx, \quad \forall u, q \in X \times Y,$$

$$b(u,w,v) = ((u\cdot\nabla w),v) + \frac{1}{2}((\nabla\cdot u)w,v)$$

$$= \frac{1}{2}((u\cdot\nabla w),v) - \frac{1}{2}((u\cdot\nabla v),w) \quad \forall u,v,w \in X.$$

(2.5)

Introducing the generalized bilinear form on $(X \times Y)^2$ by

$$B((u,p);(v,q)) = a(u,v) - d(v,p) + d(u,q), \tag{2.6}$$

we can rewrite (2.4) in a compact form: find $(u,p) \in X \times Y$ such that

$$B((u,p);(v,q)) + b(u,u,v) = (f,v), \quad \forall(v,q) \in X \times Y. \tag{2.7}$$

Let $\mathcal{T}_h$ be a quasi-uniform triangulation of Ω with $h = \max h_K$, where h_K is the diameter of the triangle $K \in \mathcal{T}_h$. We assume that the partition $\mathcal{T}_h$ has been obtained from a macrotriangular partition Λ_h by joining the sides of each element of Λ_h. Every element $K \in \mathcal{T}_h$ must lie in exactly one macroelement $\mathcal{K}$, which implies that macroelements do not overlap. For each $\mathcal{K}$, the set of interelement edges which are strictly in the interior of $\mathcal{K}$ will be denoted by $\Gamma_\mathcal{K}$, and the length of an edge $e \in \Gamma_\mathcal{K}$ is denoted by h_e.

Based on this triangulation, let X_h be the standard conforming finite element subspace of piecewise linear velocity,

$$X_h = \{v \in C(\Omega)\cap X : v|_K \text{ is linear}, \forall K \in \mathcal{T}_h; v|_{\partial\Omega} = 0\}, \tag{2.8}$$

and let Y_h be the piecewise constant pressure subspace

$$Y_h = \{q \in Y : q|_K \text{ is constant}, \forall K \in \mathcal{T}_h\}. \tag{2.9}$$

It is well known that the standard $P_1 - P_0$ element does not satisfy the inf-sup condition and cannot be applied to problem (1.1)–(1.3) directly. But a locally stabilized method based on the macroelement can be used to yield adequate approximations [6].

In order to describe the FVE method for solving problem (1.1)–(1.3), we will introduce a dual partition $\mathcal{T}_h^*$ based upon the original partition $\mathcal{T}_h$ whose elements are called control volumes. We construct the control volumes in the same way as in [5, 26]. Let z_K be the barycenter of $K \in \mathcal{T}_h$. We connect z_K with line segments to the midpoints of the edges of K, thus partitioning K into three quadrilaterals K_z, $z \in Z_h(K)$, where $Z_h(K)$ are the vertices of K. Then with each vertex $z \in Z_h = \cup_{K\in\mathcal{T}_h} Z_h(K)$, we associate a control volume V_z, which consists of the union of the subregions K_z, sharing the vertex z. Thus we finally obtain

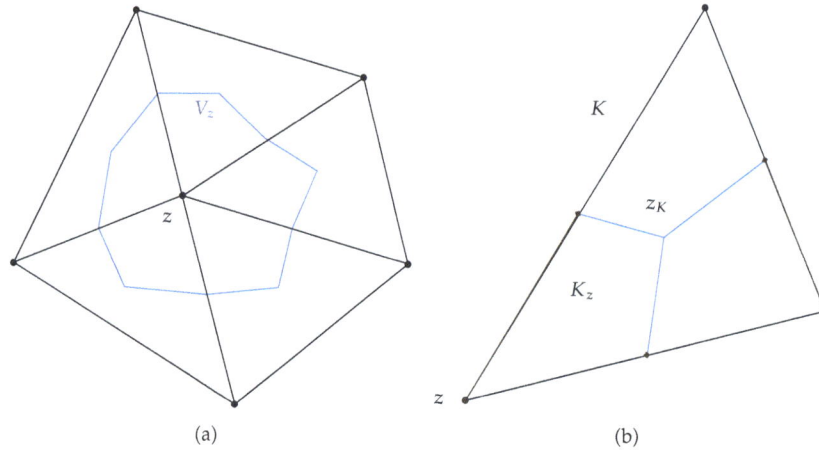

Figure 1: Left-hand side: a sample region with blue lines indicating the corresponding control volume V_z. Right-hand side: a triangle K partitioned into three subregions K_z.

a group of control volumes covering the domain Ω, which is called the dual partition $\mathcal{T}_h^*$ of the triangulation $\mathcal{T}_h$. We denote by Z_h^0 the set of interior vertices.

We call the partition $\mathcal{T}_h^*$ regular or quasi-uniform if there exists a positive constant $C > 0$ such that

$$C^{-1}h^2 \leq \text{meas}(V_z) \leq Ch^2, \quad V_z \in \mathcal{T}_h^*. \tag{2.10}$$

If the finite element triangulation $\mathcal{T}_h$ is quasi-uniform, then the dual partition $\mathcal{T}_h^*$ is also quasi-uniform [23].

2.2. Construction of the FVE Scheme

We formulate the FVE method for the problem (1.1)–(1.3) as follows: given a $z \in Z_h^0$ and $K \in \mathcal{T}_h$, integrating (1.1) over the associated control volume V_z and (1.1) over the element K and applying Green's formula, we obtain an integral conservation form

$$-\nu \int_{\partial V_z} \nabla u n \, ds + \int_{\partial V_z} p n \, ds + \int_{V_z} u \cdot \nabla u \, dx = \int_{V_z} f \, dx, \quad \forall z \in Z_h^0, \tag{2.11}$$

$$\int_K \nabla \cdot u \, dx = 0, \quad \forall K \in \mathcal{T}_h, \tag{2.12}$$

where n denotes the unit outer normal vector to ∂V_z (Figure 1).

Let $I_h^* : X_h \to X_h^*$ be the transfer operator defined by

$$I_h^* v = \sum_{z \in Z_h^0} v(z) \chi_z, \quad \forall v \in X_h, \tag{2.13}$$

where

$$X_h^* = \left\{ v = (v_1, v_2) \in \left(L^2(\Omega) \right)^2 : v_i|_{V_z} \text{ is constant}, i = 1, 2 \ \forall z \in Z_h^0 \right\}, \tag{2.14}$$

and χ_z is the characteristic function of the control volume V_z. The operator I_h^* satisfies [26]

$$\|I_h^* v\|_0 \leq C\|v\|_0, \quad \forall v \in X_h. \tag{2.15}$$

Now for an arbitrary $I_h^* v$, we multiply (2.11) by $v(z)$ and sum over all $z \in Z_h^0$ to get

$$a_h(u, I_h^* v) - d_h(I_h^* v, p) + b_h(u, u, I_h^* v) = (f, I_h^* v), \quad \forall v \in X_h. \tag{2.16}$$

Here $a_h : X \times X_h \to \mathbb{R}$, $d_h : X_h \times Y \to \mathbb{R}$ and $b_h : X \times X \times X_h \to \mathbb{R}$ are defined by

$$a_h(u, I_h^* v) = -v \sum_{z \in Z_h^0} v(z) \int_{\partial V_z} \nabla u n \, ds,$$

$$d_h(I_h^* v, p) = \sum_{z \in Z_h^0} v(z) \int_{\partial V_z} p n \, ds, \tag{2.17}$$

$$b_h(u, u, I_h^* v) = \sum_{z \in Z_h^0} v(z) \int_{V_z} (u \cdot \nabla) u \, dx.$$

We also define the trilinear forms $\tilde{b}(\cdot, \cdot, \cdot)$ and $\overline{b}(\cdot, \cdot, \cdot)$ on $X \times X \times X_h$ by

$$\tilde{b}(u, v, I_h^* w) = ((u \cdot \nabla)v, I_h^* w) + \frac{1}{2}((\nabla \cdot u)v, I_h^* w),$$

$$\overline{b}(u, v, w - I_h^* w) = ((u \cdot \nabla)v, w - I_h^* w) + \frac{1}{2}((\nabla \cdot u)v, w - I_h^* w). \tag{2.18}$$

To formulate the discrete problem so as to eliminate any such potential difficulties, we rewrite (2.16) as follows:

$$a_h(u, I_h^* v) - d_h(I_h^* v, p) + \tilde{b}(u, u, I_h^* v) = (f, I_h^* v), \quad \forall v \in X_h. \tag{2.19}$$

We multiply (2.12) by $q \in Y_h$ and sum over all $K \in \mathcal{T}_h$: then, we obtain

$$b(u, q) = 0, \quad \forall q \in Y_h. \tag{2.20}$$

Now we rewrite (2.19) and (2.20) to a variational form similar to finite element problems. The locally stabilized FVE scheme is to find $(u_h, p_h) \in X_h \times Y_h$ such that

$$a_h(u_h, I_h^* v_h) - d_h(I_h^* v_h, p_h) + \tilde{b}(u_h, u_h, I_h^* v_h) = (f, I_h^* v_h), \quad \forall v_h \in X_h,$$

$$d(u_h, q_h) - \beta c_h(p_h, q_h) = 0, \quad \forall q_h \in Y_h, \tag{2.21}$$

where

$$c_h(p, q) = \sum_{\mathcal{K} \in \Lambda_h} \sum_{e \in \Gamma_\mathcal{K}} h_e \int_e [p]_e [q]_e ds \tag{2.22}$$

is a stabilized form defined on $(H^1(\Omega) + Y_h)^2$, $[\cdot]_e$ is the jump operator across the edge e, and $\beta > 0$ is the local stabilization parameter. It is trivial that $c_h(p, q_h) = c_h(p_h, q) = c_h(p, q) = 0$, for all $p, q \in H^1(\Omega)$, for all $p_h, q_h \in Y_h$.

A general framework for analyzing the locally stabilized formulation (2.21) can be developed using the notion of equivalence class of macroelements. As in Stenberg [27] each equivalence class, denoted by $\mathcal{E}_{\widehat{\mathcal{K}}}$, contains macroelements which are topologically equivalent to a reference macroelement $\widehat{\mathcal{K}}$.

Let

$$\overline{B}_h((u_h, p_h); (I_h^* v_h, q_h)) = a_h(u_h, I_h^* v_h) - d_h(I_h^* v_h, p_h) + d(u_h, q_h), \tag{2.23}$$

$$B_h((u_h, p_h); (I_h^* v_h, q_h)) = \overline{B}_h((u_h, p_h); (I_h^* v_h, q_h)) + \beta c_h(p_h, q_h). \tag{2.24}$$

We can rewrite (2.21) in a compact form: find $(u_h, p_h) \in X_h \times Y_h$ such that

$$B_h((u_h, p_h); (I_h^* v_h, q_h)) + \tilde{b}(u_h, u_h, I_h^* v_h) = (f, I_h^* v_h), \quad \forall (v_h, q_h) \in X_h \times Y_h. \tag{2.25}$$

3. Technical Preliminaries

This section considers preliminary estimates which will be very useful in the error estimates of two-level finite volume solution (u^h, p^h).

The following lemma gives the boundedness of the trilinear form $b(\cdot, \cdot, \cdot)$.

Lemma 3.1 (see [21]). *The following estimates hold:*

$$b(u, w, v) = -b(u, v, w),$$

$$|b(u, w, v)| \le \frac{1}{2} C_0 \|u\|_0^{1/2} |u|_1^{1/2} \left(|w|_1 \|v\|_0^{1/2} |v|_1^{1/2} + \|w\|_0^{1/2} \|w\|_1^{1/2} \|v\|_1^{1/2} \right), \quad \forall u, v, w \in X,$$

$$|b(u, v, w)| + |b(v, u, w)| + |b(w, u, v)| \le C_1 |u|_1 \|v\|_2 \|w\|_0,$$

$$\forall u \in X, \; v \in \left(H^2(\Omega) \right)^2 \cap X, \; w \in Y,$$

$$|b(u, v, w)| \le C |\log h|^{1/2} |u|_1 |v|_1 \|w\|_0, \quad \forall u, v, w \in X_h. \tag{3.1}$$

Here and after C_i, $i = 1, 2$, and C are positive constants depending only on the data (ν, f, Ω).

The existence and uniqueness results of (2.7) can be found in [28, 29].

Theorem 3.2. *Assume that $\nu > 0$ and $f \in Y$ satisfy the following uniqueness condition:*

$$1 - \frac{N_1}{\nu^2} \|f\|_{-1} > 0, \tag{3.2}$$

where

$$N_1 = \sup_{u,v,w \in X} \frac{b(u, v, w)}{|u|_1 |v|_1 |w|_1}. \tag{3.3}$$

Then the problem (2.7) admits a unique solution $(u, p) \in (H_0^1(\Omega)^2 \cap X, H^1(\Omega) \cap Y)$ such that

$$|u|_1 \le \frac{1}{\nu} \|f\|_{-1}, \qquad \|u\|_2 + \|p\|_1 \le C \|f\|_0. \tag{3.4}$$

In [30] the following lemma was proved, which shows that the finite volume element bilinear forms $a_h(\cdot, I_h^* \cdot)$ and $d_h(I_h^* \cdot, \cdot)$ are equal to the finite element ones, respectively.

Lemma 3.3. *For any $u_h, v_h \in X_h$, and $q_h \in Y_h$, one has*

$$a_h(u_h, I_h^* v_h) = a(u_h, v_h),$$
$$d_h(I_h^* v_h, q_h) = d(v_h, q_h). \tag{3.5}$$

The following theorem establishes the weak coercivity of (2.24) [6, 31].

Theorem 3.4. *Given a a stabilization parameter $\beta \ge \beta_0$, suppose that every macroelement $\mathcal{K} \in \Lambda_h$ belongs to one of the equivalence classes $\mathcal{E}_{\widehat{\mathcal{K}}}$ and that the following macroelement connectivity* *condition is valid: for any two neighboring macroelements $\mathcal{K}_1$ and $\mathcal{K}_2$ with $\int_{\mathcal{K}_1 \cap \mathcal{K}_2} \neq 0$, there exists $v \in X_h$ such that*

$$\text{supp } v \subset \mathcal{K}_1 \cap \mathcal{K}_2 \int_{\mathcal{K}_1 \cap \mathcal{K}_2} v \cdot nds \neq 0. \tag{3.6}$$

Then

$$\alpha_1 \left(|u_h|_1 + \|p_h\|_0 \right) \leq \sup_{(v_h, q_h) \in X_h \times Y_h} \frac{B_h \left((u,p); (I_h^* v_h, q_h) \right)}{|v_h|_1 + \|q_h\|_0}, \tag{3.7}$$

for all $(u_h, p_h) \in X_h \times Y_h$, and

$$\left| B_h \left((u,p); (I_h^* v_h, q_h) \right) \right| \leq \alpha_2 \left(|u_h|_1 + \|p_h\|_0 \right) \left(|v_h|_1 + \|q_h\|_0 \right) \quad (v_h, q_h) \in X_h \times Y_h, \tag{3.8}$$

where $\alpha_1, \alpha_2 > 0$ are constants independent of h and β, β_0 is some fixed positive constant, and n is the out-normal vector.

Next, we establish the existence and the uniqueness of FVE scheme (2.25), by the fixed-point theorem, in the following.

Theorem 3.5 (see [6]). *Suppose the assumptions of Theorems 3.2 and 3.4 hold, and the body force f satisfies the following uniqueness condition*

$$1 - \frac{4N}{\nu^2} \|f\|_{-1} > 0. \tag{3.9}$$

Then the variation problem (2.25) admits a unique solution $(u_h, p_h) \in (X_h \times Y_h)$ such that

$$|u_h|_1 \leq \frac{1}{\nu} \|f\|_{-1}, \qquad \|p_h\|_0 \leq \frac{\alpha_2}{\alpha_1} \|f\|_{-1} + \frac{4\alpha_2 N}{\alpha_1 \nu^2} \|f\|_{-1}, \tag{3.10}$$

where

$$N = \max\{CN_1, N_2\}, \qquad N_2 = \sup_{u,v,w \in X} \frac{\tilde{b}(u, v, I_h^* w)}{|u|_1 |v|_1 |w|_1}. \tag{3.11}$$

For the error estimate, we introduce the Galerkin projection $(R_h, Q_h) : X \times Y \to X_h \times Y_h$ defined by

$$B_h \left((R_h(v,q), Q_h(v,q)); (I_h^* v_h, q_h) \right) = \overline{B}_h \left((v,q); (I_h^* v_h, q_h) \right) \tag{3.12}$$

for each $(v, q) \in X \times Y$ and all $(v_h, q_h) \in X_h \times Y_h$. We obtain the following results by using the standard Galerkin finite element [6, 17].

Theorem 3.6. *Under the assumptions of Theorems 3.2 and 3.4, the projection (R_h, Q_h) satisfies*

$$|v - R_h(v,q)|_1 + \|q - Q_h(v,q)\|_0 \le C(|v|_1 + \|q\|_0), \tag{3.13}$$

for all $(v,q) \in X \times Y$ and

$$\|v - R_h(v,q)\|_0 + h(|v - R_h(v,q)|_1 + \|q - Q_h(v,q)\|_0) \le Ch^2(\|v\|_2 + \|q\|_1), \tag{3.14}$$

for all $(v,q) \in (D(A) \cap X) \times (H^1(\Omega) \cap Y)$.

Then the optimal error estimates can be obtained as follows.

Theorem 3.7 (see [6, 32]). *Under the assumptions of Theorems 3.2, 3.4, 3.5, and 3.6, the solution (u_h, p_h) of (2.25) satisfies*

$$\|u - u_h\|_0 + h(|u - u_h|_1 + \|p - p_h\|_0) \le Ch^2. \tag{3.15}$$

4. Two-Level FVE Algorithms and Its Error Analysis

In this section, we will present two-level stabilized finite volume element algorithm for (1.1)–(1.3) and derive some optimal bounds for errors. The idea of the two-level method is to reduce the nonlinear problem on a fine mesh into a linear system on a fine mesh by solving a nonlinear problem on a coarse mesh. The basic mechanisms are two quasi-uniform triangulations of Ω, $\mathcal{T}_H$, and $\mathcal{T}_h$, with two different mesh sizes H and $h(h \ll H)$, and the corresponding solutions spaces (X_H, Y_H) and (X_h, Y_h), which satisfy $(X_H, Y_H) \subset (X_h, Y_h)$ and will be called the coarse and the fine spaces, respectively. Now find (u^h, p^h) as follows.

Algorithm 4.1 (Simple two-level stabilized FVE approximation). We have the following steps:

Step 1. On the coarse mesh $\mathcal{T}_H$, solve the stabilized Navier-Stokes problem.
  Find $(u_H, p_H) \in X_H \times Y_H$ such that, for all $(v_H, q_H) \in X_H \times Y_H$,

$$B_H((u_H, p_H); (I_H^* v_H, q_H)) + \tilde{b}(u_H, u_H, I_H^* v_H) = (f, I_H^* v_H). \tag{4.1}$$

Step 2. On the fine mesh $\mathcal{T}_h$, solve the stabilized linear Stokes problem.
  Find $(u^h, p^h) \in X_h \times Y_h$ such that, for all $(v_h, q_h) \in X_h \times Y_h$,

$$B_h\left(\left(u^h, p^h\right); (I_h^* v_h, q_h)\right) + \tilde{b}(u_H, u_H, I_h^* v_h) = (f, I_h^* v_h). \tag{4.2}$$

Next, we study the convergence of (u^h, p^h) to (u, p) in some norms. For convenience, we set $e = R_h(u,p) - u^h$ and $\eta = Q_h(u,p) - p^h$.

Theorem 4.2. *Under the assumptions of Theorems 3.2, 3.4, 3.5, and 3.6 for H and h, the simple two-level stabilized FVE solution (u^h, p^h) satisfies the following error estimates:*

$$\left|u - u^h\right|_1 + \left\|p - p^h\right\|_0 \le C\left(h + H^2\right). \tag{4.3}$$

Proof. Subtracting (4.2) from (2.25) and using the Galerkin projection (3.12), it is easy to see that

$$B_h((e, \eta); (I_h^* v_h, q_h)) + b(u, u - u_H, v_h) + b(u - u_H, u, v_h) - b(u - u_H, u - u_H, v_h)$$

$$+ \overline{b}(u - u_H, u - u_H, v_h - I_h^* v_h) - \overline{b}(u - u_H, u, v_h - I_h^* v_h) \tag{4.4}$$

$$- \overline{b}(u, u - u_H, v_h - I_h^* v_h) + \overline{b}(u, u, v_h - I_h^* v_h) = (f, v_h - I_h^* v_h),$$

for all $(v_h, q_h) \in X_h \times Y_h$. Due to (3.7), Lemma 3.1, (3.15), and (4.4), we have

$$\alpha_1(|e|_1 + \|\eta\|_0) \leq \sup_{(v_h, q_h) \in X_h \times Y_h} \frac{B_h((u, p); (I_h^* v_h, q_h))}{|v_h|_1 + \|q_h\|_0}$$

$$\leq C\|u\|_2 \|u - u_H\|_0 + C|u - u_H|_1^2 + Ch\|u\|_2 |u - u_H|_1 + Ch \tag{4.5}$$

$$\leq C\left(H^2 + h\right),$$

which, along with (3.14), yields

$$|u - u^h|_1 + \|p - p^h\|_0 \leq |u - R_h(u, p)|_1 + \|p - Q_h(u, p)\|_0 + |e|_1 + \|\eta\|_0$$

$$\leq C\left(h + H^2\right). \tag{4.6}$$
$\square$

Algorithm 4.3 (The Newton two-level stabilized FVE approximation). We have the following steps:

Step 1. On the coarse mesh $\mathcal{T}_H$, solve the stabilized Navier-Stokes problem.
Find $(u_H, p_H) \in X_H \times Y_H$ by (4.1).

Step 2. On the fine mesh $\mathcal{T}_h$, solve the stabilized linear Stokes problem.
Find $(u^h, p^h) \in X_h \times Y_h$ such that, for all $(v_h, q_h) \in X_h \times Y_h$,

$$B_h\left((u^h, p^h); (I_h^* v_h, q_h)\right) + b_h\left(u^h, u_H, I_h^* v_h\right) + b_h\left(u_H, u^h, I_h^* v_h\right) = (f, I_h^* v_h) + b_h(u_H, u_H, I_h^* v_h). \tag{4.7}$$

Now, we will study the convergence of the Newton two-level stabilized finite element solution (u^h, p^h) to (u, p) in some norms. To do this, let us set $e = R_h(u, p) - u^h$, $E = u - R_h(u, p)$, and $\eta = Q_h(u, p) - p^h$.

Theorem 4.4. *Under the assumptions of Theorems 3.2, 3.4, 3.5, and 3.6 for H and h, the Newton two-level stabilized FVE solution (u^h, p^h) satisfies the following error estimates:*

$$|u - u^h|_1 + \|p - p^h\|_0 \leq C\left(h + |\log h|^{1/2} H^3\right). \tag{4.8}$$

Proof. Subtracting (4.7) from (2.25), using the Galerkin projection (3.12) and taking $(v_h, q_h) = (e, \eta)$, we get

$$
\begin{aligned}
B_h\big((e,\eta);(I_h^* e,\eta)\big) &+ b(E,u,e) + b(R_h - u_H, u - u_H, e) + b(u_H, E, e) \\
&- \overline{b}(R_h - u_H, R_h - u_H, e - I_h^* e) - \overline{b}(R_h, R_h - u, e - I_h^* e) - \overline{b}(R_h, u, e - I_h^* e) \\
&- b(e, u_H, e) + \overline{b}(e, u_H, e - I_h^* e) + \overline{b}(u_H, e, e - I_h^* e) = (f, e - I_h^* e).
\end{aligned}
\tag{4.9}
$$

Using Lemma 3.1 and Theorems 3.2, 3.5, and 3.7, we obtain

$$
\begin{aligned}
\big| b(E,u,e) + b(u_H, E, e) + (f, e - I_h^* e) \big| &\leq C(|u|_1 + |u_H|_1)|E|_1 |e|_1 + \|f\|_0 \|e - I_h^* e\|_0 \\
&\leq Ch |e|_1,
\end{aligned}
\tag{4.10}
$$

$$
\begin{aligned}
|b(R_h - u_H, u - u_H, e)| &= |b(R_h - u_H, e, u - u_H)| \\
&\leq C|\log h|^{1/2} |R_h - u_H|_1 |e|_1 |u - u_H|_0 \\
&\leq C|\log h|^{1/2} |R_h - u|_1 + |u - u_H|_1 |e|_1 |u - u_H|_0 \\
&\leq C|\log h|^{1/2} H^3 |e|_1.
\end{aligned}
\tag{4.11}
$$

$$
\begin{aligned}
\big| \overline{b}(R_h - u_H, R_h - u_H, e - I_h^* e) \big| &\leq C|\log h|^{1/2} |R_h - u_H|_1 |R_h - u_H|_1 \|e - I_h^* e\|_0 \\
&\leq C|\log h|^{1/2} H^3 |e|_1.
\end{aligned}
\tag{4.12}
$$

$$
\big| \overline{b}(R_h, R_h - u, e - I_h^* e) \big| \leq N |R_h|_1 |R_h - u|_1 |e|_1 \leq Ch |e|_1,
\tag{4.13}
$$

$$
\big| \overline{b}(R_h - u, e - I_h^* e) \big| \leq C|R_h|_1 \|u\|_2 \|e - I_h^* e\|_0 \leq Ch |e|_1 \leq Ch |e|_1,
\tag{4.14}
$$

$$
\begin{aligned}
\nu |e|_1^2 &- |b(e, u_H, e)| - \big| \overline{b}(e, u_H, e - I_h^* e) \big| - \big| \overline{b}(u_H, e, e - I_h^* e) \big| \\
&\geq \nu |e|_1^2 - 3N |u_H|_1 |e|_1^2 \\
&\geq \nu \left(1 - \frac{3N}{\nu^2} \|f\|_{-1} \right) |e|_1^2.
\end{aligned}
\tag{4.15}
$$

Combining (4.10)–(4.15) with (4.9) yields

$$
|e|_1 \leq C \left(h + |\log h|^{1/2} H^3 \right).
\tag{4.16}
$$

Thanks to (3.7), (4.9), Theorems 3.2 and 3.5, and estimates (4.10)–(4.14) and (4.16), we have

$$
\begin{aligned}
\|\eta\|_0 &\leq (\alpha_1)^{-1} \sup_{(v_h,q_h)\in X_h\times Y_h} \frac{B_h((e,\eta);(I_h^*v_h,q_h))}{|v_h|_1 + \|q_h\|_0} \\
&\leq C\left(h + \log|h|^{1/2}H^3 + |e|_1\right) \\
&\leq C\left(h + \log|h|^{1/2}H^3\right).
\end{aligned}
\tag{4.17}
$$

Combining (4.16) and (4.17) with (3.14) and Theorem 3.2 yields (2.13). □

References

[1] Z. Q. Cai, "On the finite volume element method," *Numerische Mathematik*, vol. 58, no. 7, pp. 713–735, 1991.
[2] R. E. Ewing, T. Lin, and Y. Lin, "On the accuracy of the finite volume element method based on piecewise linear polynomials," *SIAM Journal on Numerical Analysis*, vol. 39, no. 6, pp. 1865–1888, 2002.
[3] S. H. Chou and D. Y. Kwak, "Analysis and convergence of a MAC-like scheme for the generalized Stokes problem," *Numerical Methods for Partial Differential Equations*, vol. 13, no. 2, pp. 147–162, 1997.
[4] P. Chatzipantelidies, "A finite volume method based on the Crouziex Raviart element for elliptic PDEs in two dimension," *Numerical Mathematics*, vol. 82, pp. 409–432, 1999.
[5] R. Ewing, R. Lazarov, and Y. Lin, "Finite volume element approximations of nonlocal reactive flows in porous media," *Numerical Methods for Partial Differential Equations*, vol. 16, no. 3, pp. 285–311, 2000.
[6] H. Guoliang and H. Yinnian, "The finite volume method based on stabilized finite element for the stationary Navier-Stokes problem," *Journal of Computational and Applied Mathematics*, vol. 205, no. 1, pp. 651–665, 2007.
[7] X. Ye, "A discontinuous finite volume method for the Stokes problems," *SIAM Journal on Numerical Analysis*, vol. 44, no. 1, pp. 183–198, 2006.
[8] S. H. Chou, "Analysis and convergence of a covolume method for the generalized Stokes problem," *Mathematics of Computation*, vol. 66, no. 217, pp. 85–104, 1997.
[9] M. Berggren, "A vertex-centered, dual discontinuous Galerkin method," *Journal of Computational and Applied Mathematics*, vol. 192, no. 1, pp. 175–181, 2006.
[10] S.-H. Chou and D. Y. Kwak, "Multigrid algorithms for a vertex-centered covolume method for elliptic problems," *Numerische Mathematik*, vol. 90, no. 3, pp. 441–458, 2002.
[11] R. Eymard, T. Gallouët, and R. Herbin, "Finite volume methods," in *Handbook of Numerical Analysis, Vol. VII*, pp. 713–1020, North-Holland, Amsterdam, The Netherlands, 2000.
[12] V. R. Voller, *Basic Control Volume Finite Element Methods for Fluids and Solids*, vol. 1, World Scientific, Hackensack, NJ, USA, 2009.
[13] J. Xu, "A novel two-grid method for semilinear elliptic equations," *SIAM Journal on Scientific Computing*, vol. 15, no. 1, pp. 231–237, 1994.
[14] J. Xu, "Two-grid discretization techniques for linear and nonlinear PDEs," *SIAM Journal on Numerical Analysis*, vol. 33, no. 5, pp. 1759–1777, 1996.
[15] V. Ervin, W. Layton, and J. Maubach, "A posteriori error estimators for a two-level finite element method for the Navier-Stokes equations," *Numerical Methods for Partial Differential Equations*, vol. 12, no. 3, pp. 333–346, 1996.
[16] V. Girault and J.-L. Lions, "Two-grid finite-element schemes for the steady Navier-Stokes problem in polyhedra," *Portugaliae Mathematica*, vol. 58, no. 1, pp. 25–57, 2001.
[17] Y. He, "Two-level method based on finite element and Crank-Nicolson extrapolation for the time-dependent Navier-Stokes equations," *SIAM Journal on Numerical Analysis*, vol. 41, no. 4, pp. 1263–1285, 2003.
[18] J. Li, "Investigations on two kinds of two-level stabilized finite element methods for the stationary Navier-Stokes equations," *Applied Mathematics and Computation*, vol. 182, no. 2, pp. 1470–1481, 2006.
[19] W. Layton, "A two-level discretization method for the Navier-Stokes equations," *Computers & Mathematics with Applications*, vol. 26, no. 2, pp. 33–38, 1993.

[20] W. Layton and W. Lenferink, "Two-level Picard, defect correction for the Navier-Stokes equations," *Applied Mathematics and Computation*, vol. 80, pp. 1–12, 1995.

[21] W. Layton and L. Tobiska, "A two-level method with backtracking for the Navier-Stokes equations," *SIAM Journal on Numerical Analysis*, vol. 35, no. 5, pp. 2035–2054, 1998.

[22] L. Zhu and Y. He, "Two-level Galerkin-Lagrange multipliers method for the stationary Navier-Stokes equations," *Journal of Computational and Applied Mathematics*, vol. 230, no. 2, pp. 504–512, 2009.

[23] C. Bi and V. Ginting, "Two-grid finite volume element method for linear and nonlinear elliptic problems," *Numerische Mathematik*, vol. 108, no. 2, pp. 177–198, 2007.

[24] C. Chen, M. Yang, and C. Bi, "Two-grid methods for finite volume element approximations of nonlinear parabolic equations," *Journal of Computational and Applied Mathematics*, vol. 228, no. 1, pp. 123–132, 2009.

[25] C. Chen and W. Liu, "Two-grid finite volume element methods for semilinear parabolic problems," *Applied Numerical Mathematics*, vol. 60, no. 1-2, pp. 10–18, 2010.

[26] P. Chatzipantelidis, R. D. Lazarov, and V. Thomée, "Error estimates for a finite volume element method for parabolic equations in convex polygonal domains," *Numerical Methods for Partial Differential Equations*, vol. 20, no. 5, pp. 650–674, 2004.

[27] R. Stenberg, "Analysis of mixed finite elements methods for the Stokes problem: a unified approach," *Mathematics of Computation*, vol. 42, no. 165, pp. 9–23, 1984.

[28] V. Girault and P.-A. Raviart, *Finite Element Methods for Navier-Stokes Equations*, vol. 5 of *Theory and Algorithms*, Springer, Berlin, Germany, 1986.

[29] R. Temam, *Navier-Stokes Equations*, vol. 2 of *Theory and Numerical Analysis*, North-Holland, Amsterdam, The Netherlands, 3rd edition, 1984.

[30] X. Ye, "On the relationship between finite volume and finite element methods applied to the Stokes equations," *Numerical Methods for Partial Differential Equations*, vol. 17, no. 5, pp. 440–453, 2001.

[31] N. Kechkar and D. Silvester, "Analysis of locally stabilized mixed finite element methods for the Stokes problem," *Mathematics of Computation*, vol. 58, no. 197, pp. 1–10, 1992.

[32] G. He, Y. He, and X. Feng, "Finite volume method based on stabilized finite elements for the nonstationary Navier-Stokes problem," *Numerical Methods for Partial Differential Equations*, vol. 23, no. 5, pp. 1167–1191, 2007.

6

New Approach for Solving a Class of Doubly Singular Two-Point Boundary Value Problems Using Adomian Decomposition Method

Randhir Singh, Jitendra Kumar, and Gnaneshwar Nelakanti

Department of Mathematics, Indian Institute of Technology Kharagpur, Kharagpur 721302, India

Correspondence should be addressed to Randhir Singh, randhir.math@gmail.com

Academic Editor: Norbert Heuer

We propose two new modified recursive schemes for solving a class of doubly singular two-point boundary value problems. These schemes are based on Adomian decomposition method (ADM) and new proposed integral operators. We use all the boundary conditions to derive an integral equation before establishing the recursive schemes for the solution components. Thus we develop recursive schemes without any undetermined coefficients while computing successive solution components, whereas several previous recursive schemes have done so. This modification also avoids solving a sequence of nonlinear algebraic or transcendental equations for the undetermined coefficients with multiple roots, which is required to complete calculation of the solution by several earlier modified recursion schemes using the ADM. The approximate solution is computed in the form of series with easily calculable components. The effectiveness of the proposed approach is tested by considering four examples and results are compared with previous known results.

1. Introduction

Consider the following class of doubly singular two-point boundary value problems:

$$(p(t)y'(t))' = q(t)f(t, y(t)), \quad 0 < t \leq 1, \tag{1.1}$$

with boundary conditions

$$y(0) = a_1, \qquad y(1) = c_1,$$
$$\text{or } y'(0) = 0, \qquad ay(1) + by'(1) = c, \tag{1.2}$$

where a_1, c_1, $a > 0$, $b \geq 0$, and c are any finite constants. The condition $p(0) = 0$ characterizes that problem (1.1) is singular and in addition to this if $q(t)$ is allowed to be discontinuous at $t = 0$, then problem (1.1) is called doubly singular Bobisud [1]. Consider problem (1.1) with the following conditions on $p(t)$, $q(t)$, and $f(t, y(t))$.

Type 1. Dirichlet boundary conditions: $y(0) = a_1$, $y(1) = c_1$

(E_1) $p \in C[0,1]$, $p \in C^1(0,1]$ with $p > 0$ in $(0,1]$ and $1/p \in L^1(0,1)$;

(E_2) $q > 0$ in $(0,1]$ and $\int_0^1 (1/p)(\int_x^1 q\,ds)dx < \infty$;

(E_3) (i) $f \in C([0,1] \times \mathbb{R})$; (ii) $\partial f / \partial y$ exists and continuous, $\partial f / \partial y \geq 0$ for all $0 \leq t \leq 1$ and all real y.

Type 2. Mixed type boundary conditions: $y'(0) = 0$, $ay(1) + by'(1) = c$

(E_1') $p \in C[0,1]$, $p \in C^1(0,1]$ with $p > 0$ in $(0,1]$;

(E_2') $q > 0$ in $(0,1]$, $q \in L^1(0,1)$ and $\int_0^1 (1/p)(\int_0^x q\,ds)dx < \infty$;

(E_3') the same as (E_3).

In recent years, the study of such singular boundary value problems (1.1) has attracted the attention of several researchers [2–9]. In particular, if $p(t) = 1$, $q(t) = t^{-1/2}$ and $f(t,y) = y^{3/2}$ problem (1.1) with Type 1 boundary conditions is known as Thomas-Fermi equation. Thomas [10] and Fermi [11] independently derived a boundary value problem for determining the electrical potential in an atom. The analysis leads to the nonlinear SBVP $y'' = t^{-1/2}y^{3/2}$ with boundary conditions given by $y(0) = 1$, $y(b_1) = 0$.

Chan and Hon [12] considered the generalized Thomas-Fermi equation: $(t^{b_2}y')' = cq(t)y^e$, $y(0) = 1$, $y(a) = 0$ with $q(t) = t^{b_2+d_2}$, where $0 \leq b_2 < 1$, $c > 0$, $d_2 > -2$, $e > 1$, which is doubly singular problem. Problem (1.1) with $q(t) = p(t) = t^\alpha$ where $\alpha = 2$ arises in the study of the distribution of heat sources in the human head [13] with $f(t,y) = -\delta e^{-\theta y}$, $\theta > 0, \delta > 0$.

There is a huge literature available on numerical methods for problem (1.1) with $q(t) = 1$, but very few numerical methods are available to tackle doubly singular boundary value problems. Reddien [14] studied the linear form of problem (1.1) and derived numerical methods for $q(t) \in L^2[0,1]$ which is stronger assumption than (E_2').

Chawla and Shivakumar [15] established the existence as well as uniqueness of solution for problem (1.1) where $q(t) = p(t) = t^\alpha$. The existence and uniqueness for problem (1.1) have also been discussed by Dunninger and Kurtz [16] and Bobisud [1]. Later Pandey and Verma [17] extended the results on the existence uniqueness for problem (1.1) with Types 1 and 2 boundary conditions.

1.1. Adomian Decomposition Method (ADM)

In this subsection, we briefly describe standard ADM for nonlinear second-order equation.

Recently, many researchers [4, 18–26] have shown interest to the study of ADM for different scientific models. Adomian [18] asserted that the ADM provides an efficient and computationally suitable method for generating approximate series solution for a large class of differential equations. Let us consider nonlinear second-order ordinary differential equation

$$Ly + Ry + Ny = g(t), \tag{1.3}$$

where $L \equiv d^2/dt^2$ is the second-order linear derivative operator, R is the linear remainder operator, N represents the nonlinear term, and $g(t)$ is a source term. The above equation can be rewritten as

$$Ly = g(t) - Ry - Ny. \tag{1.4}$$

The inverse operator of L is defined as $L^{-1}(\cdot) = \int_0^t \int_0^t (\cdot) dt dt$.

Operating the inverse linear operator $L^{-1}(\cdot)$ on both the sides of (1.4) yields

$$y = y(0) + ty'(0) + L^{-1}g(t) - L^{-1}Ry - L^{-1}Ny. \tag{1.5}$$

Next, we decompose the solution y and the nonlinear function Ny by an infinite series as

$$y = \sum_{n=0}^{\infty} y_n, \qquad Ny = \sum_{n=0}^{\infty} A_n, \tag{1.6}$$

where A_n are Adomian polynomials that can be constructed for various classes of nonlinear functions with the formula given by Adomian and Rach [19]

$$A_n = \frac{1}{n!} \frac{d^n}{d\lambda^n} \left[N \left(\sum_{k=0}^{\infty} y_k \lambda^k \right) \right]_{\lambda=0}, \qquad n = 0, 1, 2, \dots. \tag{1.7}$$

Substituting the series (1.6) into (1.5), we obtain

$$\sum_{n=0}^{\infty} y_n = y(0) + ty'(0) + L^{-1}g(t) - L^{-1}R \sum_{n=0}^{\infty} y_n - L^{-1} \sum_{n=0}^{\infty} A_n. \tag{1.8}$$

From (1.8), the various components y_n of the solution y can be determined by using the recursive relation

$$y_0 = y(0) + ty'(0) + L^{-1}g(t),$$
$$y_{k+1} = -L^{-1}Ry_k - L^{-1}A_k, \quad k \geq 0. \tag{1.9}$$

For numerical purpose, the n-term truncated series $\psi_n = \sum_{m=0}^{n-1} y_m$ may be used to give the approximate solution.

The ADM has been used to solve nonlinear boundary value problems (BVPs) for ordinary differential equations by several researchers [4, 20–26]. Solving nonlinear BVP by standard ADM is always a computationally involved task as it requires the computation of undetermined coefficients $y'(0)$ in (1.9) in a sequence of nonlinear equations which increases computational complexity.

It is important to note that the standard ADM (1.9) can not be applied directly to solve two-point boundary value problem (1.1) as the component y_0 in scheme (1.9) is not independent from undetermined coefficient. Many researchers [23–25] have proposed modified ADM to overcome the difficulty by setting undermined coefficient $y'(0) = c$, and then it will be determined by using second boundary condition satisfying $\psi_n(1) = y(1)$. In this case, it requires additional computational work to solve nonlinear equation $\psi_n(1) = y(1)$ for c and c may not be uniquely determined.

Benabidallah and Cherruault [24] have considered the following boundary value problem:

$$y'' + py' = f(t, y), \quad a < t < b,$$
$$y(a) = \alpha, \qquad y(b) = \beta. \tag{1.10}$$

The inverse operator $L^{-1} = \int_a^x \int_b^x dx dx$ was proposed, then unidentified constant $y'(b)$ is set, as $y'(b) = \sum_{n=0}^{\infty} c_n$, and each solution component y_n is obtained by the following scheme:

$$y_0 = y(a) + (x - a)[c_0 + y(b)p(b)],$$
$$y_{n+1} = c_{n+1}(t - a) - \int_a^t py_n dt + L^{-1}A_n + L^{-1}p'y_n, \quad n \geq 0. \tag{1.11}$$

In order to determine the unidentified constants c_n, it is also required n-term approximate solution $\psi_n = \sum_{m=0}^{n-1} y_m$ satisfying all the boundary conditions. It is clearly that $\psi_n(a) = y(a)$. For boundary condition $x = b$, the following scheme was proposed

$$y_0(b) = y(b), \qquad y_n(b) = 0, \quad n \geq 1. \tag{1.12}$$

However, since each constant c_n is determined by solving (1.12), this scheme also require additional computational work.

In order to avoid solving such nonlinear algebraic equations for a two-point boundary value problems, Jang [26] introduced extended ADM for nonsingular problems with Dirichlet boundary condition. Khuri and Sayfy [27] applied a novel approach based on the mixed decomposition spline for solving singular problems arising in physiology. The method proposed by Ebaid [4] is based on the modification of Lesnic's work [28]. Later, Duan and Rach [29] introduced special modified inverse linear integral operators for higher order boundary value problem.

To the best of our knowledge, no one has applied ADM to solve a class of doubly singular two-point boundary value problem (1.1) with Types 1 and 2 boundary conditions.

In this work, a new approach based on ADM and new integral operator are proposed for solving doubly singular two-point boundary value problems (1.1) with Types 1 and 2. To set up the modified scheme, we first consider DSBVP (1.1) and use all the boundary conditions and Adomian decomposition method to establish the recursion scheme. Thus a modified recursion scheme is developed which does not require the computation of undermined coefficients, whereas most of previous recursive schemes do require the computation of undermined coefficients (see [23–25]). In fact, the proposed recursion scheme is useful for solving problem (1.1) whether they are linear or nonlinear singular (nonsingular). The main advantage of the method is that it provides a direct scheme for solving the singular boundary value problem, that is, without linearization and discretization. Numerical results are presented to demonstrate the effectiveness of proposed recursive scheme. The symbolic and numerical computations have been performed using "MATHEMATICA" software.

The organization of the paper is as follows. Section 2 presents the inverse integral operator with modified Adomian decomposition method for doubly singular boundary value problems of (1.1) with Types 1 and 2 boundary conditions. In Section 3, the convergence of the method is discussed. In Section 4, we illustrate our method with numerical results along with graphical representation.

2. Modified Adomian Decomposition Method

In this section, we establish two modified recursive schemes for solving doubly singular boundary value problems (1.1) with Types 1 and 2 boundary conditions.

We again write (1.1) with Type 1 boundary conditions as:

$$(p(t)y'(t))' = q(t)f(t, y(t)), \quad 0 < t \le 1,$$
$$y(0) = a_1, \qquad y(1) = c_1, \tag{2.1}$$

which can be rewritten as

$$Ly(t) = q(t)Ny(t), \tag{2.2}$$

where $Ly(t) = (p(t)y'(t))'$ is the linear differential operator to be inverted and $Ny(t) = f(t, y(t))$ is an analytic nonlinear operator. It is also assumed that the solution of the problem (2.1) exits and unique. A twofold integral operator $L^{-1}(\cdot)$, regarded as the inverse operator of $L(\cdot)$, is proposed as

$$L^{-1}(\cdot) = \int_0^t \frac{1}{p(s)} \int_s^1 (\cdot)dxds. \tag{2.3}$$

To establish the recursive scheme, we operate $L^{-1}(\cdot)$ on the left hand side of (2.1) and use initial condition $y(0) = a_1$; we have

$$
\begin{aligned}
L^{-1}\big[(p(t)y'(t))'\big] &= \int_0^t \frac{1}{p(s)} \int_s^1 (p(x)y'(x))'dxds \\
&= c_2 \int_0^t \frac{1}{p(s)}ds - y(t) + a_1,
\end{aligned}
\tag{2.4}
$$

$$
L^{-1}\big[(p(t)y'(t))'\big] = c_2 \int_0^t \frac{1}{p(s)}ds - y(t) + a_1,
\tag{2.5}
$$

where $c_2 = p(1)y'(1)$.

We again operate the inverse operator $L^{-1}(\cdot)$ on both sides of (2.1) and use (2.5), yielding

$$
y(t) = a_1 + c_2 \int_0^t \frac{1}{p(s)}ds - \int_0^t \frac{1}{p(s)} \int_s^1 q(t)Ny(t)dxds.
\tag{2.6}
$$

For simplicity, we set $h(t) = \int_0^t (1/p(s))ds$ and $[L^{-1}(\cdot)]_{t=1} = \int_0^1 (1/p(s)) \int_s^1 (\cdot)dxds$.

Then (2.6) can be written as

$$
y(t) = a_1 + c_2 h(t) - \Big[L^{-1}q(t)Ny(t)\Big].
\tag{2.7}
$$

Using the boundary condition $y(1) = c_1$ in (2.7), we get

$$
c_2 = \frac{c_1 - a_1}{h(1)} + \frac{1}{h(1)}\Big[L^{-1}q(t)Ny(t)\Big]_{t=1}.
\tag{2.8}
$$

By using c_2 into (2.7), we obtain

$$
y(t) = a_1 + \frac{c_1 - a_1}{h(1)}h(t) + \frac{h(t)}{h(1)}\Big[L^{-1}q(t)Ny(t)\Big]_{t=1} - \Big[L^{-1}q(t)Ny(t)\Big].
\tag{2.9}
$$

It is important to note that the right hand side of (2.9) does not contain any unknown constants.

The equation (2.9) can be rewritten in the following operator equation form as

$$
y = f + M(y),
\tag{2.10}
$$

where

$$
f = a_1 + \frac{c_1 - a_1}{h(1)}h(t), \qquad M(y) = \frac{h(t)}{h(1)}\Big[L^{-1}q(t)Ny(t)\Big]_{t=1} - \Big[L^{-1}q(t)Ny(t)\Big].
\tag{2.11}
$$

Next, the solution $y(t)$ and the nonlinear function $Ny(t)$ are decomposed by infinite series of the form

$$y(t) = \sum_{n=0}^{\infty} y_n(t), \qquad Ny(t) = \sum_{n=0}^{\infty} A_n(y_0(t), y_1(t), \dots, y_n(t)), \qquad (2.12)$$

respectively, where $A_n(y_0(t), y_1(t), \dots, y_n(t))$ are Adomian polynomials [19].
    Substituting the series (2.12) into (2.9) gives

$$\sum_{n=0}^{\infty} y_n(t) = a_1 + \frac{c_1 - a_1}{h(1)} h(t) + \frac{h(t)}{h(1)} \left[L^{-1} q(t) \sum_{n=0}^{\infty} A_n \right]_{t=1} - \left[L^{-1} q(t) \sum_{n=0}^{\infty} A_n \right]. \qquad (2.13)$$

Using the above equation (2.13), the solution components $y_0, y_1, \dots, y_n$ are determined by the following recursive scheme:

$$y_0(t) = a_1 + \frac{c_1 - a_1}{h(1)} h(t),$$

$$y_{n+1}(t) = \frac{h(t)}{h(1)} \left[L^{-1} q(t) A_n \right]_{t=1} - \left[L^{-1} q(t) A_n \right], \quad n \ge 0,$$

$$\qquad (2.14)$$

or

$$y_{n+1}(t) = \frac{h(t)}{h(1)} \int_0^1 \frac{1}{p(s)} \int_s^1 q(t) A_n dx ds - \int_0^t \frac{1}{p(s)} \int_s^1 q(t) A_n dx ds, \quad n \ge 0.$$

We further modify the above algorithm (2.14) to get a more efficient and economic algorithm. To obtain modified algorithm, the zeroth component y_0 is divided into the sum of two parts, namely, $f_0 + f_1$, where $f_0 = a_1$, $f_1 = ((c_1 - a_1)/h(1))h(t)$. The first part, f_0, is kept in y_0 and the rest part, f_1, is added to y_1. Thus, the zeroth component will be $y_0 = a_1$ in modified algorithm as

$$y_0(t) = a_1,$$

$$y_1(t) = \frac{c_1 - a_1}{h(1)} h(t) + \frac{h(t)}{h(1)} \left[L^{-1} q(t) A_0 \right]_{t=1} - \left[L^{-1} q(t) A_0 \right],$$

$$\qquad (2.15)$$

$$\vdots$$

$$y_{n+1}(t) = \frac{h(t)}{h(1)} \left[L^{-1} q(t) A_n \right]_{t=1} - \left[L^{-1} q(t) A_n \right], \quad n \ge 0.$$

Note that new modified recursion scheme (2.15) does not require the computation of undermined coefficients, since this algorithm does not contain any undermined coefficient. This slight change plays a major role for minimizing the size of calculations.

Now, we again consider the same problem (1.1) with Type 2 boundary conditions:

$$(p(t)y'(t))' = q(t)f(t, y(t)), \quad 0 < t \le 1,$$
$$y'(0) = 0, \qquad ay(1) + by'(1) = c, \tag{2.16}$$

which can be rewritten as

$$Ly(t) = q(t)Ny(t), \tag{2.17}$$

where $Ly(t) = (p(t)y'(t))'$ is the linear differential operator to be inverted and $Ny(t) = f(t, y(t))$ is an analytic nonlinear operator. It is also assumed that the solution of problem (2.16) exits and is unique. In this case, we propose the following inverse integral operator:

$$L^{-1}(\cdot) = \int_t^1 \frac{1}{p(s)} \int_0^s (\cdot) dx ds. \tag{2.18}$$

To setup an algorithm, we operate the inverse operator $L^{-1}(\cdot)$ on the left hand side of (2.16) and using the boundary condition $y'(0) = 0$, we have

$$L^{-1}\left[(p(t)y'(t))'\right] = \int_t^1 \frac{1}{p(s)} \int_0^s (p(x)y'(x))' dx ds$$
$$= \int_t^1 y'(s) ds, \tag{2.19}$$

$$L^{-1}\left[(p(t)y'(t))'\right] = c_0 - y(t), \tag{2.20}$$

where $c_0 = y(1)$.

By operating the inverse operator $L^{-1}(\cdot)$ on both sides of (2.16) and using (2.20) we have

$$y(t) = c_0 - \int_t^1 \frac{1}{p(s)} \int_0^s q(t)Ny(t) dx ds. \tag{2.21}$$

Thus (2.21) can be written as

$$y(t) = c_0 - \left[L^{-1}q(t)Ny(t)\right]. \tag{2.22}$$

Using boundary condition $ay(1) + by'(1) = c$ into (2.22), we obtain

$$y(t) = \frac{c}{a} - \frac{b}{ap(1)} \int_0^1 q(t)Ny(t) dt - \left[L^{-1}q(t)Ny(t)\right]. \tag{2.23}$$

The equation (2.23) may be rewritten in operator equation form

$$y = f + M(y), \tag{2.24}$$

where

$$f = \frac{c}{a}, \qquad M(y) = -\frac{b}{ap(1)} \int_0^1 q(t) N y(t) dx - \left[L^{-1} q(t) N y(t) \right]. \tag{2.25}$$

Substituting the series (2.12) into (2.23) gives

$$\sum_{n=0}^{\infty} y_n(t) = \frac{c}{a} - \frac{b}{ap(1)} \int_0^1 q(t) \sum_{n=0}^{\infty} A_n dt - \left[L^{-1} q(t) \sum_{n=0}^{\infty} A_n \right]. \tag{2.26}$$

Upon matching both sides of (2.26), the solution components $y_0, y_1, \ldots, y_n$ can be determined by the following modified recursive scheme:

$$y_0(t) = \frac{c}{a},$$

$$y_{n+1}(t) = -\frac{b}{ap(1)} \int_0^1 q(t) A_n dt - \left[L^{-1} q(t) A_n \right], \quad n \geq 0. \tag{2.27}$$

Thus, we have established two modified recursive schemes which give the complete determination of solution components y_n, and hence the approximate series solution $\psi_n(t)$ can be obtained for the singular boundary value problems (1.1) with Types 1 and 2 boundary conditions. For numerical purpose, the truncated n-term approximate series solution is given by

$$\psi_n(t) = \sum_{m=0}^{n-1} y_m(t). \tag{2.28}$$

Note that the truncated n-term approximate series solution is obtained by adding n solutions components $y_0, y_1, \ldots, y_{n-1}$, that is, $\psi_n = \sum_{m=0}^{n-1} y_m$.

3. Convergence of Method

In this section, we discuss the convergence analysis of modified ADM for doubly singular boundary value problem (1.1).

Many authors [30–33] established the convergence of ADM for differential and integral equations. The first proof of convergence of ADM for the general functional equation was given by Cherruault [31]. Cherruault and Adomian [32] proposed a new convergence proof of Adomian decomposition method for the general nonlinear functional equation based on the properties of convergent series. Recently, Hosseini and Nasabzadeh [33] introduced a simple technique to determine the rate of convergence of ADM for initial value problem.

Now, we discuss the convergence of doubly singular two-point boundary value problems (1.1) with Types 1 and 2 boundary conditions. To do so, note that (2.9) and (2.23) may be written in the operator equation form

$$y = f + M(y), \tag{3.1}$$

where M is an operator from a Banach space $C[0,1]$ to $C[0,1]$, f is a given function in $C[0,1]$, and we are looking for $y \in C[0,1]$ satisfying (3.1).

For Type 1 boundary conditions, we have

$$f = a_1 + \frac{(c_1 - a_1)}{h(1)} h(t), \tag{3.2}$$

$$M(y) = \frac{h(t)}{h(1)} \left[L^{-1} q(t) N y(t) \right]_{t=1} - \left[L^{-1} q(t) N y(t) \right]. \tag{3.3}$$

For Type 2 boundary conditions, f and M are given as

$$f = \frac{c}{a}, \tag{3.4}$$

$$M(y) = -\frac{b}{ap(1)} \int_0^1 q(t) N y(t) dt - \left[L^{-1} q(t) N y(t) \right]. \tag{3.5}$$

Let the solution y and the nonlinear function Ny be considered as sum of series

$$y = \sum_{n=0}^{\infty} y_n, \qquad Ny = \sum_{n=0}^{\infty} A_n(y_0, y_1, \ldots, y_n), \tag{3.6}$$

where $A_n(y_0, y_1, \ldots, y_n)$ are Adomian polynomials [19].

Substituting the series (3.6) into the operator equation (3.1) yields

$$\sum_{n=0}^{\infty} y_n = f + \sum_{n=0}^{\infty} B_n(y_0, \ldots, y_n),$$
$$y_0 + y_1 + \cdots + y_n + \cdots = f + B_0(y_0) + \cdots + B_{n-1}(y_0, \ldots, y_{n-1}) + \cdots, \tag{3.7}$$

where

$$\sum_{n=0}^{\infty} B_n(y_0, \ldots, y_n) = \frac{h(t)}{h(1)} \left[L^{-1} q(t) \sum_{n=0}^{\infty} A_n \right]_{t=1} - \left[L^{-1} q(t) \sum_{n=0}^{\infty} A_n \right],$$

$$\sum_{n=0}^{\infty} B_n(y_0, \ldots, y_n) = -\frac{b}{ap(1)} \int_0^1 q(t) \sum_{n=0}^{\infty} A_n dt - \left[L^{-1} q(t) \sum_{n=0}^{\infty} A_n \right], \tag{3.8}$$

for Types 1 and 2 boundary conditions, respectively.

Upon matching both sides of (3.7), we obtain the following scheme:

$$y_0 = f,$$
$$y_1 = B_0(y_0),$$
$$\vdots$$
$$y_{n+1} = B_n(y_0, \ldots, y_n),$$
$$\vdots$$

(3.9)

Let $\psi_n = \sum_{m=0}^{n-1} y_m$ be approximate solution, then the modified ADM for (3.1) is equivalent to the following problem:

$$y_0 = f,$$
$$\psi_{n+1} = y_0 + M(\psi_n), \quad n \geq 0.$$

(3.10)

Finding the solution of (3.1) is equivalent to finding the sequence ψ_n such that $\psi_n = y_0 + y_1 + \cdots + y_{n-1}$ satisfies (3.10).

Theorem 3.1. *Let $M(y)$ be the nonlinear operator defined by (3.3) or (3.5) which satisfies the Lipschitz condition $\|M(\varphi) - M(\xi)\| \leq \delta\|\varphi - \xi\|$, for all $\varphi, \xi \in C[0,1]$ with Lipschitz constant δ, $0 \leq \delta < 1$. If $\|y_0\| < \infty$, then there holds $\|y_{k+1}\| \leq \delta\|y_k\|$, for all $k \in \mathbb{N} \cup \{0\}$ and the sequence $\{\psi_n\}$ defined by (3.10) converges to $y = \sum_{n=0}^{\infty} y_n = \lim_{n \to \infty} \psi_n = \psi$.*

Proof. Since

$$\psi_1 = y_0 + y_1, \qquad \psi_2 = y_0 + y_1 + y_2, \ldots, \qquad \psi_n = y_0 + y_1 + y_2 + \cdots + y_n, \ldots,$$

(3.11)

we have $y_{k+1} = \psi_{k+1} - \psi_k$, $k = 1, 2, \ldots$.

We now show that the sequence $\{\psi_n\}$ is convergent. To prove this, it is sufficient to show that $\{\psi_n\}$ is cauchy sequence in Banach space $C[0,1]$. Now using the nonlinearity of $M(y)$, we have

$$\|y_{n+1}\| = \|\psi_{n+1} - \psi_n\| = \|M(\psi_n) - M(\psi_{n-1})\| \leq \delta\|\psi_n - \psi_{n-1}\| = \delta\|y_n\|.$$

(3.12)

Thus we obtain

$$\|y_{n+1}\| \leq \delta\|y_n\| \leq \delta^2\|y_{n-1}\| \cdots \leq \delta^{n+1}\|y_0\|.$$

(3.13)

Now for every $n, m \in \mathbb{N}$, and $n \geq m$, we have

$$
\begin{aligned}
\|\psi_n - \psi_m\| &= \left\| (\psi_n - \psi_{n-1}) + (\psi_{n-1} - \psi_{n-2}) + \cdots + (\psi_{m+1} - \psi_m) \right\| \\
&\leq \|\psi_n - \psi_{n-1}\| + \|\psi_{n-1} - \psi_{n-2}\| + \cdots + \|\psi_{m+1} - \psi_m\| \\
&\leq \delta^n \|y_0\| + \delta^{n-1} \|y_0\| + \cdots + \delta^{m+1} \|y_0\| \\
&\leq \delta^{m+1} \left(1 + \delta + \delta^2 + \cdots + \delta^{n-m-1} \right) \|y_0\|.
\end{aligned}
\tag{3.14}
$$

As we know $\sum_{j=0}^{\infty} \delta^j$ is geometric series with common ration $\delta, 0 \leq \delta < 1$. Therefore, we have

$$
\|\psi_n - \psi_m\| \leq \frac{\delta^{m+1}}{1 - \delta} \|y_0\|,
\tag{3.15}
$$

which converges to zero, that is, $\|\psi_n - \psi_m\| \to 0$, as $m \to \infty$. This implies that there exits ψ such that $\lim_{n \to \infty} \psi_n = \psi$. But, we have $y = \sum_{n=0}^{\infty} y_n = \lim_{n \to \infty} \psi_n$, that is, $y = \psi$ which is solution of (3.1). Hence, we have convergence in norm. $\square$

4. Numerical Examples and Discussion

In this section, we consider four examples to demonstrate the effectiveness of proposed recursive schemes (2.15) and (2.27). The numerical results are compared with known results and maximum absolute error is also calculated. We also plot approximate and exact solutions to show how approximate solutions converge to exact solution.

Example 4.1. Consider linear two-point boundary value problem [6]

$$
(t^\alpha y')' = \beta t^{\alpha+\beta-2} \left(\beta t^\beta + \alpha + \beta - 1 \right) y, \quad 0 < t \leq 1,
\tag{4.1}
$$

$$
y(0) = 1, \qquad y(1) = e,
$$

with exact solution $y(t) = e^{t^\beta}$. For every $\alpha > 0$ and $\beta > 0$, (4.1) is called singular.

In particular, if $\alpha = 0.5$, $\beta = 1$, that is, $p(t) = t^{0.5}$ and $q(t) = t^{-0.5}$ then problem (4.1) is called doubly singular boundary value problem. Applying modified recursive scheme (2.15) to (4.1), the solution components are obtained as

$$y_0 = 1,$$

$$y_1 = 0.384948t^{0.5} + t + 0.333333t^2,$$

$$y_2 = -0.551338t^{0.5} + 0.128316t^{1.5} + 0.166667t^2 + 0.076989t^{2.5} + 0.155556t^3 + 0.023809t^4,$$

$$y_3 = 0.224316t^{0.5} - 0.183779t^{1.5} - 0.097435t^{2.5} + 0.011111t^3 + 0.015886t^{3.5} + 0.017460t^4$$

$$+ 0.004277t^{4.5} + 0.007442t^5 + 0.000721t^6,$$ (4.2)

$$y_4 = -0.077011t^{0.5} + 0.074771t^{1.5} + 0.026485t^{2.5} - 0.022142t^{3.5} + 0.000396t^4$$

$$- 0.004971t^{4.5} + 0.000881t^5 + 0.000655t^{5.5} + 0.0006418t^6 + 0.0001096t^{6.5}$$

$$+ 0.0001715t^7 + 0.000012t^8,$$

$$\vdots$$

For numerical purpose, the approximate series solution is

$$\psi_6 = 1 - 0.002003t^{0.5} + t + 0.002065t^{1.5} + 0.5t^2 + 0.000603t^{2.5} + 0.166667t^3$$

$$- 0.0006957t^{3.5} + 0.041666t^4 - 0.0000457t^{4.5} + 0.0083333t^5 + 0.0000802t^{5.5}$$

$$+ 0.001388t^6 + 1.0684407 \times 10^{-6}t^{6.5} + 0.000198t^7 - 4.663181 \times 10^{-6}t^{7.5}$$

$$+ 0.000024t^8 - 3.885381 \times 10^{-8}t^{8.5} + 2.745287 \times 10^{-6}t^9 + 1.676726 \times 10^{-7}t^{9.5}$$ (4.3)

$$+ 2.664448 \times 10^{-7}t^{10} + 1.5360215 \times 10^{-8}t^{10.5} + 2.063867 \times 10^{-8}t^{11}$$

$$+ 9.172396 \times 10^{-10}t^{12}.$$

Now, we define error function as $E_n(t) = \psi_n(t) - y(t)$ and the maximum absolute errors as

$$\|E_n\|_\infty = \max_{0 < t \leq 1} |E_n(t)|.$$ (4.4)

For different values of α and β, the maximum absolute error $\|E_n\|_\infty$, for $n = 4, 6$, and 8, is shown in Tables 1 and 2, and we observe that when n increases the error decreases. On the other hand, Chawla and Katti [6] applied M_3 method based on three evaluations of f to solve the same problem in which a huge amount of computation work is needed to obtain the numerical solution. The results are compared with our approximate series solution in Table 3. Furthermore, the approximate series solutions ψ_2, ψ_4 and exact solution y are plotted in Figures 1 and 2. It can be seen from the figures that series solution is very close to exact solution.

Table 1: Maximum error of Example 4.1, when $\beta = 1$.

α	$\|E_4\|_\infty$	$\|E_6\|_\infty$	$\|E_8\|_\infty$
0.25	2.3211×10^{-3}	4.2210×10^{-5}	2.4843×10^{-6}
0.5	2.3151×10^{-3}	6.1226×10^{-5}	3.2320×10^{-6}
0.75	3.5315×10^{-3}	7.0126×10^{-5}	2.2117×10^{-6}

Table 2: Maximum error of Example 4.1, when $\beta = 5$.

α	$\|E_4\|_\infty$	$\|E_6\|_\infty$	$\|E_8\|_\infty$
0.25	2.2419×10^{-3}	3.3304×10^{-5}	3.5127×10^{-6}
0.5	1.3307×10^{-3}	5.0264×10^{-5}	4.6572×10^{-6}
0.75	4.2104×10^{-3}	4.3231×10^{-5}	2.1260×10^{-6}

Example 4.2. Consider nonlinear singular two-point boundary value problem

$$\left(t^\alpha y'\right)' = t^{\alpha+\beta-2}e^y\left(t^\beta e^y - \alpha - \beta + 1\right), \quad 0 < t \le 1,$$

$$y(0) = \ln\left(\frac{\beta}{4}\right), \qquad y(1) = \ln\left(\frac{\beta}{5}\right), \tag{4.5}$$

with exact solution $y(t) = \ln(\beta/(4 + t^\beta))$ and for any real $\alpha > 0$ and $\beta > 0$, this equation is called singular.

If $\alpha = 0.25$, $\beta = 1$, that is, $p(t) = t^{0.25}$, $q(t) = t^{-0.75}$, then problem (4.5) is called doubly singular boundary value problem.

Now we apply recursive scheme (2.15) to obtain the solution components as

$$y_0 = -1.38629,$$

$$y_1 = 0.001856t^{0.75} - 0.25t + 0.025t^2,$$

$$y_2 = -0.001605t^{0.75} - 0.000066t^{1.75} + 0.00625t^2 + 0.000042t^{2.75} - 0.004861t^3 + 0.000240t^4,$$

$$y_3 = -0.000283t^{0.75} + 0.000057t^{1.75} - 2.461715 \times 10^{-8}t^{2.5} - 0.000030t^{2.75} - 0.000347t^3$$

$$+ 4.475846 \times 10^{-8}t^{3.5} - 0.000011t^{3.75} + 0.000714t^4 + 8.882529 \times 10^{-7}t^{4.75}$$

$$- 0.000103t^5 + 3.434065 \times 10^{-6}t^6,$$

$$\vdots$$

$$\tag{4.6}$$

Table 3: Maximum error of Example 4.1.

α, β	Proposed method		Chawla and Katti [6]	
	$\|E_8\|_\infty$	N		error
$\alpha = 0.50$, $\beta = 4.0$	2.4003×10^{-6}	128		1.80×10^{-4}
$\alpha = 0.75$, $\beta = 3.75$	5.1431×10^{-6}	128		1.80×10^{-4}

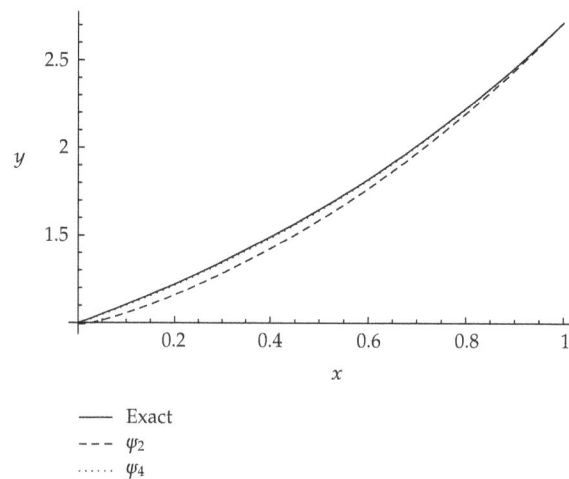

Figure 1: Comparison of exact y and approximate ψ_2, ψ_4 solutions of Example 4.1, $\alpha = 0.50$, $\beta = 1$.

To compare the results numerically, the 6-term truncated series solution is given by

$$
\begin{aligned}
\psi_6 = {}& -1.38629 - 3.372935 \times 10^{-7} x^{0.75} - 0.25t - 1.108619 \times 10^{-7} t^{1.75} + 0.03125t^2 \\
& + 7.059037 \times 10^{-9} t^{2.5} - 3.299920 \times 10^{-8} t^{2.75} - 0.005208t^3 + 1.307442 \times 10^{-11} t^{3.25} \\
& - 1.557418 \times 10^{-8} t^{3.5} + 1.289457 \times 10^{-7} t^{3.75} + 0.000976t^4 - 5.530529 \times 10^{-11} t^{4.25} \\
& + 1.078097 \times 10^{-8} t^{4.5} + 3.090869 \times 10^{-7} t^{4.75} - 0.000195t^5 - 1.480660 \times 10^{-11} t^{5.25} \\
& + 1.980188 \times 10^{-9} t^{5.5} - 3.668865 \times 10^{-7} t^{5.75} + 0.0000405t^6 + 2.567699 \times 10^{-12} t^{6.25} \\
& - 1.009792 \times 10^{-9} t^{6.5} + 1.315036 \times 10^{-7} t^{6.75} - 7.607217 \times 10^{-6} t^7 \\
& + 5.654798 \times 10^{-11} t^{7.5} - 1.378289 \times 10^{-8} t^{7.75} + 8.830360 \times 10^{-7} t^8 + 4.154631 \\
& \times 10^{-10} t^{8.75} - 4.867214 \times 10^{-8} t^9 + 9.674748 \times 10^{-10} t^{10}.
\end{aligned}
\tag{4.7}
$$

New Approach for Solving a Class of Doubly Singular Two-Point Boundary Value Problems
Using Adomian Decomposition Method

111

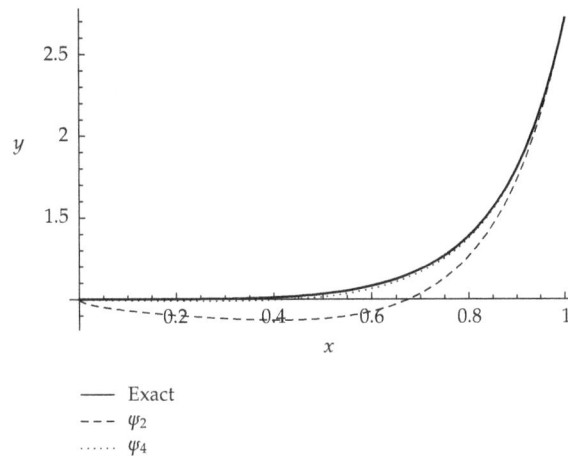

Figure 2: Comparison of exact y and approximate ψ_2, ψ_4 solutions of Example 4.1, $\alpha = 0.50$, $\beta = 5$.

Table 4: Maximum error for Example 4.2, when $\beta = 1$.

α	$\|E_4\|_\infty$	$\|E_6\|_\infty$	$\|E_8\|_\infty$
0.25	1.3889×10^{-4}	2.0435×10^{-6}	2.5693×10^{-8}
0.5	5.2987×10^{-4}	1.1481×10^{-6}	1.1616×10^{-8}
0.75	2.4545×10^{-4}	3.7571×10^{-7}	2.1654×10^{-9}

Similarly, for $\alpha = 0.5$, $\beta = 1$, that is, $p(t) = t^{0.5}$ and $q(t) = t^{-0.5}$, the solution components are obtained using the scheme (2.15) as

$$y_0 = -1.38629,$$

$$y_1 = 0.006023t^{0.5} - 0.25t + 0.020833t^2,$$

$$y_2 = -0.005737t^{0.5} - 0.000501t^{1.5} + 0.010416t^2 + 0.000150t^{2.5} - 0.004513t^3 + 0.000186t^4,$$

$$y_3 = -0.000302t^{0.5} + 0.000478t^{1.5} - 7.557899 \times 10^{-7}t^2 - 0.000093t^{2.5} - 0.000693t^3$$

$$- 0.000045t^{3.5} + 0.0007378t^4 + 2.788479 \times 10^{-6}t^{4.5} - 0.0000851t^5 + 2.348635 \times 10^{-6}t^6,$$

$$\vdots$$

$$(4.8)$$

The maximum absolute error $\|E_n\|_\infty$, for $n = 4, 6$, and 8, is listed in Tables 4 and 5. As the value of n increases, the maximum absolute error decreases. In addition, to verify how close is our approximate series solutions ψ_2, ψ_4 to exact solution y, we have plotted approximate solution and exact solution in Figures 3 and 4. It is shown from figures that the approximate series solution ψ_4 is very near to the exact solution y.

Table 5: Maximum error for Example 4.2, when $\beta = 5$.

α	$\|E_4\|_\infty$	$\|E_6\|_\infty$	$\|E_8\|_\infty$
0.25	6.5109×10^{-4}	2.6137×10^{-7}	6.0161×10^{-8}
0.5	8.8947×10^{-4}	1.1610×10^{-6}	7.1746×10^{-8}
0.75	7.6596×10^{-4}	5.2644×10^{-7}	6.3210×10^{-8}

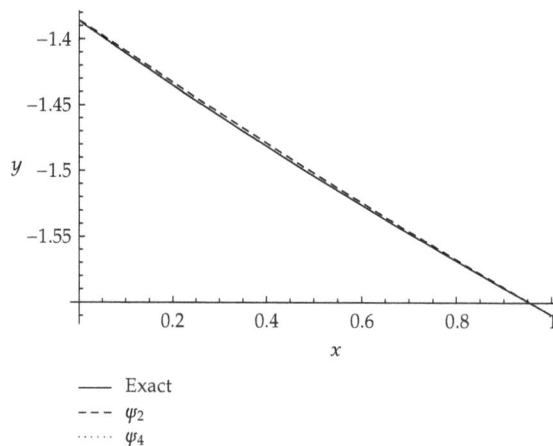

Figure 3: Comparison of exact y and approximate ψ_2, ψ_4 solutions of Example 4.2, when $\alpha = 0.50$, $\beta = 1$.

Example 4.3. The heat conduction model in human head problem Duggan and Goodman [13] is

$$\left(t^2 y'\right)' = -t^2 e^{-y}, \quad 0 < t \leq 1,$$

$$y'(0) = 0, \qquad 2y(1) + y'(1) = 0. \tag{4.9}$$

We apply the recursive scheme (2.27) to (4.9), where $p(t) = q(t) = t^2$, $f = e^{-y}$, $a = 2$, $b = 1$, and $c = 0$ then scheme (2.27) for (4.9) becomes as

$$y_0 = 0,$$

$$y_{n+1} = \frac{1}{2} \int_0^1 t^2 A_n dt + \int_t^1 s^{-2} \left(\int_0^s x^2 A_n dx \right) ds, \quad n \geq 0. \tag{4.10}$$

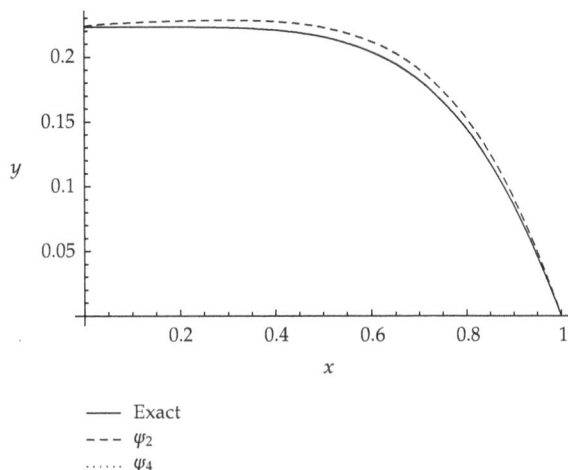

Figure 4: Comparison of exact y and approximate ψ_2, ψ_4 solutions of Example 4.2, when $\alpha = 0.50$, $\beta = 5$.

We can easily determine the solution components from above scheme as

$$y_0 = 0,$$

$$y_1 = 0.333333 - 0.166667t^2,$$

$$y_2 = -0.086111 + 0.055555t^2 - 0.008333t^4,$$

$$y_3 = 0.032672 - 0.023611t^2 + 0.005555t^4 - 0.000529t^6,$$

$$y_4 = -0.014584 + 0.011258t^2 - 0.003287t^4 + 0.000529t^6 - 0.0000373t^8,$$

$$\vdots$$

(4.11)

The truncated series solution ψ_8 is obtained by adding $y_0, y_1, \ldots, y_6$, given by

$$\psi_8 = 0.270259 - 0.127429t^2 - 0.0047677t^{4.} - 0.0002650t^6 - 4.7439600 \times 10^{-6}t^8$$

$$- 2.4024518 \times 10^{-6}t^{10} + 2.2847637 \times 10^{-7}t^{12} - 3.485344 \times 10^{-8}t^{14}$$

$$+ 2.384089 \times 10^{-9}t^{16} - 1.1874334 \times 10^{-10}t^{18}.$$

(4.12)

The comparison of numerical results of solutions with known solution is given in Table 6. It can be noted that the approximate solution ψ_8 is very close to known solution [13]. However, Duggan and Goodman [13] utilized the maximum principle based on lower and an upper solutions technique in which a lot of calculation is required as we need to solve sequence of linear differential equation analytically.

Table 6: Comparison of numerical results with known results of Example 4.3.

t	ψ_6	ψ_8	ψ_{10}	Solution in [13]
0	0.27073627	0.27025895	0.27010891	0.27035006
0.2	0.26561546	0.26515412	0.26500927	0.26525434
0.4	0.25016332	0.24974709	0.24961677	0.24986712
0.6	0.22410365	0.22375398	0.22364491	0.22388597
0.8	0.18695255	0.18668071	0.18659620	0.18679895
1	0.13798246	0.13778979	0.13772998	0.13787263

Example 4.4. The nonlinear singular boundary problem considered by Chawla et al. [34] is

$$y'' + \frac{1}{t}y' = -ve^y, \quad 0 < t < 1,$$

$$y'(0) = 0, \qquad y(1) = 0, \tag{4.13}$$

with exact solution $y(t) = 2\ln((B + 1)/(Bt^2 + 1))$, where $B = (4 - v - 2(4 - 2v)^{1/2})/v$, $v = 1$.

In the above problem, we have $p(t) = q(t) = t$ and $f = -e^y$, $a = 1$, $b = 0$, $c = 0$, by applying recursive scheme (2.27), we have following recursive scheme:

$$y_0 = 0,$$

$$y_{n+1} = \frac{0}{1}\int_0^1 tA_n dt + \int_t^1 s^{-1}\left(\int_0^s xA_n dx\right)ds, \quad n \geq 0. \tag{4.14}$$

We have successive solution components y_n as

$$y_0 = 0,$$

$$y_1 = 0.25 - 0.25t^2,$$

$$y_2 = 0.046875 - 0.0625t^2 + 0.015625t^4, \tag{4.15}$$

$$y_3 = 0.0130208 - 0.0195313t^2 + 0.0078125t^4 - 0.00130208t^6,$$

$$\vdots$$

and then the truncated series is obtained as

$$\psi_6 = 0.316294 - 0.342438t^2 + 0.0289497t^4 - 0.003107t^6 + 0.0003280x^8$$

$$- 0.0000274t^{10} + 1.2715657 \times 10^{-6}t^{12}. \tag{4.16}$$

In order to check whether the approximate solution converges to exact solution y, we plot approximate solution ψ_n, for $n = 2, 4$ and exact solution y in Figure 5. It is clear from the figure that the approximate solution ψ_4 converges rapidly to exact solution. The maximum

Table 7: Numerical results and maximum error of Example 4.4.

n	$\|E_n\|_\infty$	n	Error in [34]
5	6.2129×10^{-4}	16	2.52×10^{-3}
8	1.0919×10^{-5}	32	1.83×10^{-4}
10	3.0638×10^{-6}	64	1.28×10^{-5}

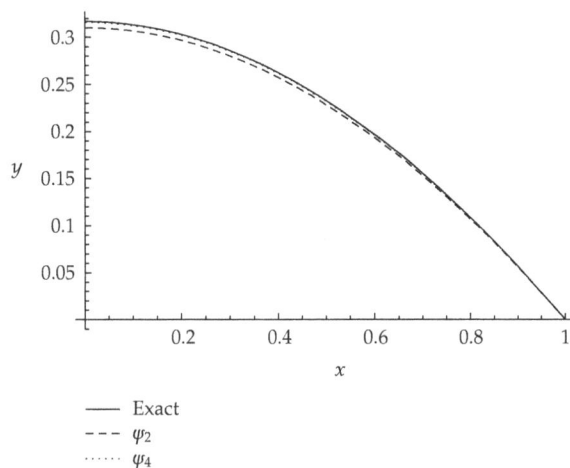

Figure 5: Comparison of exact y and approximate ψ_2, ψ_4 solutions of Example 4.4.

absolute error $\|E_n\|_\infty$, for $n = 5, 8$, and 10, is presented in Table 7. It can be noted that Chawla et al. [34] employed fourth-order finite difference method to obtain the numerical solution in which a lot of computation work is required. However, the maximum absolute error obtained was 1.28×10^{-5} using $N = 64$.

5. Conclusion

In this paper, we we have illustrated how the modified recursive schemes (2.15) and (2.27) can be used to solve a class of doubly singular two-point boundary value problems (1.1) with Types 1 and 2 boundary conditions. The accuracy of the numerical results indicates that the method is well suited for the solution of this type problem. The main advantage of this approach is that it provides a direct scheme to obtain approximate solutions, and we have also shown graphically that these approximate solutions are almost identical to the analytic solution. Another advantage of a modified recursion scheme is that this does not require the computation of undermined coefficients, whereas most of previous recursive schemes do require the computation of undermined coefficients (see [23–25]). The method provides a reliable technique which requires less work compared to the traditional techniques such as finite difference method, Cubic spline method, and standard ADM. The numerical results of the examples are presented and only a few terms are required to obtain accurate solutions. By comparing the results with other existing methods, it has been proved that proposed modified ADM is a more powerful method for solving the singular problems (1.1).

References

[1] L. E. Bobisud, "Existence of solutions for nonlinear singular boundary value problems," *Applicable Analysis*, vol. 35, no. 1–4, pp. 43–57, 1990.

[2] T. Aziz and M. Kumar, "A fourth-order finite-difference method based on non-uniform mesh for a class of singular two-point boundary value problems," *Journal of Computational and Applied Mathematics*, vol. 136, no. 1-2, pp. 337–342, 2001.

[3] M. M. Chawla and C. P. Katti, "A uniform mesh finite difference method for a class of singular two-point boundary value problems," *SIAM Journal on Numerical Analysis*, vol. 22, no. 3, pp. 561–565, 1985.

[4] A. Ebaid, "A new analytical and numerical treatment for singular two-point boundary value problems via the Adomian decomposition method," *Journal of Computational and Applied Mathematics*, vol. 235, no. 8, pp. 1914–1924, 2011.

[5] M. Kumar, "A new finite difference method for a class of singular two-point boundary value problems," *Applied Mathematics and Computation*, vol. 143, no. 2-3, pp. 551–557, 2003.

[6] M. M. Chawla and C. P. Katti, "Finite difference methods and their convergence for a class of singular two-point boundary value problems," *Numerische Mathematik*, vol. 39, no. 3, pp. 341–350, 1982.

[7] M. M. Chawla and C. P. Katti, "A finite-difference method for a class of singular two-point boundary value problems," *IMA Journal of Numerical Analysis*, vol. 4, no. 4, pp. 457–466, 1984.

[8] M. M. Chawla, S. McKee, and G. Shaw, "Order 2 method for a singular two-point boundary value problem," *BIT Numerical Mathematics*, vol. 26, no. 3, pp. 318–326, 1986.

[9] S. R. K. Iyengar and P. Jain, "Spline finite difference methods for singular two point boundary value problems," *Numerische Mathematik*, vol. 50, no. 3, pp. 363–376, 1987.

[10] L. Thomas, "The calculation of atomic fields," in *Mathematical Proceedings of the Cambridge Philosophical Society*, vol. 23, pp. 542–548, Cambridge University Press, 1927.

[11] E. Fermi, "Un metodo statistico per la determinazione di alcune priorieta dell'atome," *Rendicondi Accademia Nazionale de Lincei*, vol. 32, no. 6, pp. 602–607, 1927.

[12] C. Y. Chan and Y. C. Hon, "A constructive solution for a generalized Thomas-Fermi theory of ionized atoms," *Quarterly of Applied Mathematics*, vol. 45, no. 3, pp. 591–599, 1987.

[13] R. Duggan and A. Goodman, "Pointwise bounds for a nonlinear heat conduction model of the human head," *Bulletin of Mathematical Biology*, vol. 48, no. 2, pp. 229–236, 1986.

[14] G. W. Reddien, "Projection methods and singular two point boundary value problems," *Numerische Mathematik*, vol. 21, pp. 193–205, 1974.

[15] M. M. Chawla and P. N. Shivakumar, "On the existence of solutions of a class of singular nonlinear two-point boundary value problems," *Journal of Computational and Applied Mathematics*, vol. 19, no. 3, pp. 379–388, 1987.

[16] D. R. Dunninger and J. C. Kurtz, "Existence of solutions for some nonlinear singular boundary value problems," *Journal of Mathematical Analysis and Applications*, vol. 115, no. 2, pp. 396–405, 1986.

[17] R. K. Pandey and A. K. Verma, "A note on existence-uniqueness results for a class of doubly singular boundary value problems," *Nonlinear Analysis: Theory, Methods & Applications*, vol. 71, no. 7-8, pp. 3477–3487, 2009.

[18] G. Adomian, *Solving Frontier Problems of Physics: The Decomposition Method*, vol. 60 of *Fundamental Theories of Physics*, Kluwer Academic, Dordrecht, The Netherlands, 1994.

[19] G. Adomian and R. Rach, "Inversion of nonlinear stochastic operators," *Journal of Mathematical Analysis and Applications*, vol. 91, no. 1, pp. 39–46, 1983.

[20] G. Adomian and R. Rach, "A new algorithm for matching boundary conditions in decomposition solutions," *Applied Mathematics and Computation*, vol. 58, no. 1, pp. 61–68, 1993.

[21] G. Adomian and R. Rach, "Modified decomposition solution of linear and nonlinear boundary-value problems," *Nonlinear Analysis: Theory, Methods & Applications*, vol. 23, no. 5, pp. 615–619, 1994.

[22] A.-M. Wazwaz, "Approximate solutions to boundary value problems of higher order by the modified decomposition method," *Computers & Mathematics with Applications*, vol. 40, no. 6-7, pp. 679–691, 2000.

[23] A. M. Wazwaz, "A reliable algorithm for obtaining positive solutions for nonlinear boundary value problems," *Computers & Mathematics with Applications*, vol. 41, no. 10-11, pp. 1237–1244, 2001.

[24] M. Benabidallah and Y. Cherruault, "Application of the Adomian method for solving a class of boundary problems," *Kybernetes*, vol. 33, no. 1, pp. 118–132, 2004.

[25] M. Inc and D. J. Evans, "The decomposition method for solving of a class of singular two-point boundary value problems," *International Journal of Computer Mathematics*, vol. 80, no. 7, pp. 869–882, 2003.

[26] B. Jang, "Two-point boundary value problems by the extended Adomian decomposition method," *Journal of Computational and Applied Mathematics*, vol. 219, no. 1, pp. 253–262, 2008.

[27] S. A. Khuri and A. Sayfy, "A novel approach for the solution of a class of singular boundary value problems arising in physiology," *Mathematical and Computer Modelling*, vol. 52, no. 3-4, pp. 626–636, 2010.

[28] D. Lesnic, "A computational algebraic investigation of the decomposition method for time-dependent problems," *Applied Mathematics and Computation*, vol. 119, no. 2-3, pp. 197–206, 2001.

[29] J.-S. Duan and R. Rach, "A new modification of the Adomian decomposition method for solving boundary value problems for higher order nonlinear differential equations," *Applied Mathematics and Computation*, vol. 218, no. 8, pp. 4090–4118, 2011.

[30] K. Abbaoui and Y. Cherruault, "Convergence of Adomian's method applied to differential equations," *Computers & Mathematics with Applications*, vol. 28, no. 5, pp. 103–109, 1994.

[31] Y. Cherruault, "Convergence of Adomian's method," *Kybernetes*, vol. 18, no. 2, pp. 31–38, 1989.

[32] Y. Cherruault and G. Adomian, "Decomposition methods: a new proof of convergence," *Mathematical and Computer Modelling*, vol. 18, no. 12, pp. 103–106, 1993.

[33] M. M. Hosseini and H. Nasabzadeh, "On the convergence of Adomian decomposition method," *Applied Mathematics and Computation*, vol. 182, no. 1, pp. 536–543, 2006.

[34] M. M. Chawla, R. Subramanian, and H. L. Sathi, "A fourth order method for a singular two-point boundary value problem," *BIT Numerical Mathematics*, vol. 28, no. 1, pp. 88–97, 1988.

7

A Note on Fourth Order Method for Doubly Singular Boundary Value Problems

R. K. Pandey and G. K. Gupta

Department of Mathematics, Indian Institute of Technology, Kharagpur 721302, India

Correspondence should be addressed to R. K. Pandey, rkp@maths.iitkgp.ernet.in

Academic Editor: Hassan Safouhi

We present a fourth order finite difference method for doubly singular boundary value problem $(p(x)y'(x))' = q(x)f(x,y)$, $0 < x \leq 1$ with boundary conditions $y(0) = A$ (or $y'(0) = 0$ or $\lim_{x \to 0} p(x)y'(x) = 0$) and $\alpha y(1) + \beta y'(1) = \gamma$, where $\alpha(> 0)$, $\beta(\geq 0)$, γ and A are finite constants. Here $p(0) = 0$ and $q(x)$ is allowed to be discontinuous at the singular point $x = 0$. The method is based on uniform mesh. The accuracy of the method is established under quite general conditions and also corroborated through one numerical example.

1. Introduction

Consider the following class of singular two point boundary value problems:

$$(p(x)y')' = q(x)f(x,y), \quad 0 < x \leq 1, \tag{1.1}$$

with boundary conditions

$$y(0) = A, \quad \alpha y(1) + \beta y'(1) = \gamma \tag{1.2}$$

$$\text{or,} \quad y'(0) = 0 \quad \left(\text{or } \lim_{x \to 0} p(x)y'(x) = 0\right), \quad \alpha y(1) + \beta y'(1) = \gamma, \tag{1.3}$$

where $\alpha > 0$, $\beta \geq 0$ and A, γ are finite constants. Here $p(0) = 0$ and $q(x)$ is unbounded near $x = 0$. The condition $p(0) = 0$ says that the problem is singular and as $q(x)$ is unbounded near $x = 0$, so the problem is doubly singular [1]. Let $p(x)$, $q(x)$, and $f(x,y)$ satisfy the following conditions:

(A-1)

(i) $p(0) = 0$, $p(x) > 0$ in $(0,1]$,
(ii) $p(x) \in C[0,1] \cap C^1(0,1]$,
(iii) $p(x) = x^{b_0} g(x)$, $(0 \leq b_0 < 1)$, $g(x) > 0$, and
(iv) $G \in C^4[0,1]$ and $G^v(x)$ exists on $(0,1)$,

where $G(x) = 1/g(x)$ on $[0,1]$.

(A-2)

(i) $q(x) > 0$ in $(0,1]$, $q(x)$ is unbounded near $x = 0$,
(ii) $q(x) \in L^1(0,1)$,
(iii) $q(x) = x^{a_0} H(x)$, $H(0) > 0$ with $(a_0 > -1)$,
(iv) $H(x) \in C^4[0,1]$ and $H^v(x)$ exists on $(0,1)$, and
(v) $1 + a_0 - b_0 \geq 0$.

(A-3) $f(x,y)$ is continuous on $\{[0,1] \times \mathbb{R}\}$, $\partial f/\partial y$ exists, and is continuous and $\partial f/\partial y \geq 0$ for all $0 \leq x \leq 1$ and for all real y.

Thomas [2] and Fermi [3] independently derived a boundary value problem for determining the electrical potential in an atom. The analysis leads to the nonlinear singular second order problem

$$y'' = x^{-1/2} y^{3/2} \tag{1.4}$$

with a set of boundary conditions. The following are of our interest:

(i) the neutral atom with Bohr radius b given by $y(0) = 1$, $by'(b) - y(b) = 0$;
(ii) the ionized atom given by $y(0) = 1$, $y(b) = 0$.

Furthermore, Chan and Hon [4] have considered the generalized Thomas-Fermi equation

$$\left(x^b y'\right)' = cq(x)y^e, \qquad y(0) = 1, \qquad y(a) = 0 \tag{1.5}$$

for parameter values $0 \leq b < 1$, $c > 0$, $d > -2$, $e > 1$, and $q(x) = x^{b+d}$.

Such singular problems have been the concern of several researchers [5–8]. The existence-uniqueness of the solution of the boundary value problem (1.1) with boundary condition (1.2) or (1.3) is established in [1, 9–13]. Bobisud [1] has mentioned that in case $\lim_{x \to 0^+} (q(x)/p'(x)) \neq 0$, the condition $y'(0) = 0$ is quite severe, that is, it is sufficient but not necessary for forcing the solution to be differentiable at $x = 0$. In fact if $\lim_{x \to 0^+} p(x)y'(x) = 0$, then

$$\lim_{x \to 0^+} y'(x) = \lim_{x \to 0^+} \frac{q(x)}{p'(x)} \cdot \lim_{x \to 0^+} f(x, y(x)). \tag{1.6}$$

Thus $\lim_{x \to 0^+} y'(x) = 0$ if either $\lim_{x \to 0^+} (q(x)/p'(x)) = 0$ or $f(0, y(0)) = 0$. But $\lim_{x \to 0^+} (q(x)/p'(x)) \neq 0$ if $q(x)$ has discontinuity at $x = 0$; thus it is natural to consider the weaker boundary condition $\lim_{x \to 0^+} p(x)y'(x) = 0$.

There is a considerable literature on numerical methods for $q(x) = 1$ but to the best of our knowledge very few numerical methods are available to tackle doubly singular boundary value problems. Reddien [14] has considered the linear form of (1.1) and derived numerical methods for $q(x) \in L^2[0,1]$ which is stronger assumption than (A-2)(ii).

Some second order methods (Chawla and Katti [15], Pandey and Singh [16, 17]) as well as fourth order methods (Chawla et al. [18–21], Pandey and Singh [22–24]) have been developed for $q(x) = 1$, $p(x)$. Most of the researchers have developed methods for the function $p(x) = x^{b_0}$, $0 \le b_0 < 1$ and the boundary conditions $y(0) = A$ and $y(1) = B$.

Chawla [19] has given fourth order method for the problem (1.1)-(1.2) with $q(x) = 1$ and $p(x) = x^{b_0}$, $0 \le b_0 < 1$. The method is extended by Pandey and Singh [22] to a class of function $p(x) = x^{b_0}g(x)$ satisfying (A-1)(i–iii) with $G(x) = 1/g(x)$ analytic in the neighborhood of the singular point.

In this work we extend the fourth order accuracy method developed in [22] to doubly singular boundary value problems (1.1)-(1.2) and (1.1), (1.3) both. That is, we allow the data function to be discontinuous at the singular point $x = 0$. Further, we do not require analyticity of the function $G(x) = 1/g(x)$. The convergence of the method is established under quite general conditions on the functions $p(x)$, $q(x)$, and $f(x, y)$. Fourth order convergence of the method is corroborated through one example and maximum absolute errors are displayed in Table 1.

The work is organized in the following manner. In Section 2, first the method is described and then its construction is explained. In Section 3, convergence of the method is established and in Section 4, the order of the method is corroborated through one example.

2. Finite Difference Method

This section is divided in to two parts: (i) description of the method and (ii) derivation of the method. The coefficients not specified explicitly in this section are specified in the appendices.

2.1. Description of the Method

In this section we first state the method; detailed derivation is given in Section 2.2.

For positive integer $N \ge 2$, we consider the mesh $w_h = \{x_k\}_{k=0}^N$ over $[0, 1] : 0 = x_0 < x_1 < x_2 < \cdots x_N = 1$, $h = 1/N$. For uniform mess $x_k = kh$, where b_0 is mentioned in condition (A-1). Let $r(x) := f(x, y(x))$ ($y(x)$ is the solution), $y_k = y(x_k)$, and so forth. Now we approximate the differential equation (1.1) with boundary condition (1.2) on the grid w_h by the following difference equations:

$$\frac{\tilde{y}_{k-1}}{J_{k-1}} - \left(\frac{1}{J_k} + \frac{1}{J_{k-1}}\right)\tilde{y}_k + \frac{\tilde{y}_{k+1}}{J_k} = a_{0,k}\tilde{r}_k + a_{1,k}\tilde{r}_{k+1} - a_{-1,k}\tilde{r}_{k-1}, \quad k = 1(1)(N-1),$$

$$(2.1)$$

$$\frac{\tilde{y}_{N-1}}{J_{N-1}} - \left(\frac{1}{J_{N-1}} + \frac{\alpha}{(\beta G_N)}\right)\tilde{y}_N + \frac{\gamma}{(\beta G_N)} = a_{0,N}\tilde{r}_N - a_{-1,N}\tilde{r}_{N-1} + a_{1,N}\tilde{r}_{N-\theta}, \quad (2.2)$$

for boundary value problem (1.1)-(1.2). In the case of boundary condition (1.3) we use the following difference equation:

$$\frac{-\tilde{y}_1}{J_1} + \frac{\tilde{y}_2}{J_1} = a_{-1,1}\tilde{r}_0 + a_{0,1}\tilde{r}_1 + a_{1,1}\tilde{r}_2. \tag{2.3}$$

Here $\tilde{y} = (\tilde{y}_k)$ denotes the approximate solution, $\tilde{y}_k \approx y_k$, $G_k = G(x_k)$, and

$$J_k = \int_{x_k}^{x_{k+1}} (p(\tau))^{-1} d\tau. \tag{2.4}$$

Note that the functions $p(x)$ and $G(x)$ are defined in assumption (A-1). Thus the method for boundary value problem (1.1)-(1.2) is given by difference equations (2.1)-(2.2). Further, for boundary value problem (1.1) and (1.3) the method is given by difference equations (2.1) (with $k = 2(1)(N-1)$) and (2.2)-(2.3).

2.2. Construction of the Method

In this section we describe the derivation of the method. Local truncation errors are mentioned without proof.

With $z(x) = p(x)y'(x)$ the differential equation (1.1) becomes $z'(x) = q(x)f(x, y(x))$. Now integrating $z'(x) = q(x)f(x, y(x))$ twice, first from x_k to x and then from x_k to x_{k+1} and changing the order of integration we get

$$y_{k+1} - y_k = z_k J_k + \int_{x_k}^{x_{k+1}} \left(\int_t^{x_{k+1}} (p(\tau))^{-1} d\tau \right) q(t) r(t) dt, \tag{2.5}$$

where $z_k = z(x_k)$, $r(x) := f(t, y(x))$ and $J_k = \int_{x_k}^{x_{k+1}} (p(\tau))^{-1} d\tau$. In an analogous manner, we get

$$y_k - y_{k-1} = z_k J_{k-1} - \int_{x_{k-1}}^{x_k} \left(\int_{x_{k-1}}^t (p(\tau))^{-1} d\tau \right) q(t) r(t) dt. \tag{2.6}$$

Eliminating z_k from (2.5) and (2.6) we obtain the Chawla's identity

$$\frac{y_{k+1} - y_k}{J_k} - \frac{y_k - y_{k-1}}{J_{k-1}} = \frac{I_k^+}{J_k} + \frac{I_k^-}{J_{k-1}}, \quad k = 1(1)(N-1), \tag{2.7}$$

where

$$I_k^+ = \int_{x_k}^{x_{k+1}} \left(\int_t^{x_{k+1}} (p(\tau))^{-1} d\tau \right) q(t) r(t) dt,$$

$$I_k^- = \int_{x_{k-1}}^{x_k} \left(\int_{x_{k-1}}^t (p(\tau))^{-1} d\tau \right) q(t) r(t) dt. \tag{2.8}$$

Via Taylor expansion of $G(x)$, $r(x)$ and $H(x)$ about x_k in $I_k^{\pm}$ and taking the following approximations for r_k' and r_k''

$$r_k' = \frac{r_{k+1} - r_{k-1}}{2h} - \frac{1}{6}h^2 r'''(\eta_k), \quad x_{k-1} < \eta_k < x_{k+1},$$

$$r_k'' = \frac{r_{k+1} - 2r_k + r_{k-1}}{h^2} - \frac{1}{12}h^2 r^{iv}(\xi_k), \quad x_{k-1} < \xi_k < x_{k+1}, \tag{2.9}$$

we get the approximation for the smooth solution $y(x)$ as:

$$\frac{\tilde{y}_{k-1}}{J_{k-1}} - \left(\frac{1}{J_k} + \frac{1}{J_{k-1}}\right)\tilde{y}_k + \frac{\tilde{y}_{k+1}}{J_k} = a_{0,k}r_k + a_{1,k}r_{k+1} - a_{-1,k}r_{k-1} + t_k, \quad k = 1(1)(N-1), \tag{2.10}$$

where t_k is local truncation error given by

$$\begin{aligned}
t_k = {}& \left(H_k G_k''' b_{03,k} + H_k G^{iv}(\xi_k)b_{04,k}\right)f_k + \left(H_k f_k' + H_k' f_k\right)\left(G_k'' b_{12,k} + G_k''' b_{13,k}\right) \\
& + \left(\frac{1}{2}H_k f_k'' + H_k' f_k' + \frac{1}{2}H_k'' f_k\right)\left(G_k' b_{21,k} + G_k'' b_{22,k}\right) \\
& + \left(\frac{1}{6}H_k f_k''' + \frac{1}{2}H_k' f_k'' + \frac{1}{2}H_k'' f_k' + \frac{1}{6}H_k''' f_k\right)\left(G_k b_{30,k} + G_k' b_{31,k}\right) \\
& + \left(\frac{1}{24}H_k f^{iv}(\xi_k) + \frac{1}{6}H_k' f_k''' + \frac{1}{4}H_k'' f_k'' + \frac{1}{6}H_k''' f_k' + \frac{1}{24}H^{iv}(\eta_k)f_k\right)G_k b_{40,k} \\
& - \frac{1}{6}h^2 f'''(\eta_k)\left(b_{10,k}G_k H_k + b_{11,k}G_k' H_k + b_{20,k}G_k H_k'\right) \\
& - \frac{1}{24}b_{20,k}h^2 G_k H_k f^{iv}(\xi_k), \quad k = 1(1)(N-1), \ x_{k-1} < \xi_k, \ \eta_k < x_{k+1}.
\end{aligned} \tag{2.11}$$

The coefficients $b_{03,k}$, $b_{04,k}$, $b_{10,k}$, $b_{11,k}$, $b_{12,k}$, $b_{13,k}$, $b_{20,k}$, $b_{21,k}$, $b_{22,k}$, $b_{30,k}$, $b_{31,k}$, $b_{40,k}$, and so forth, are specified in the appendices and the functions $G(x)$, $H(x)$ are defined in the assumptions (A-1), (A-2).

2.2.1. Discretization of the Boundary Condition at $x = 1$

We write (2.6) for $k = N$

$$\frac{y_N}{J_{N-1}} - \frac{y_{N-1}}{J_{N-1}} = p_N y_N' - \frac{I_N^-}{J_{N-1}}. \tag{2.12}$$

Now, we use the boundary condition at $x = 1$, Taylor expansion of $r(x)$ and following approximations for r'_N and r''_N

$$r'_N = \frac{1}{h}\left[\frac{(1+\theta)}{\theta}r_N + \frac{\theta}{(1-\theta)}r_{N-1} + \frac{1}{\theta(\theta-1)}r_{N-\theta}\right] - \frac{h^2\theta}{6}r'''(\eta_N), \quad x_{N-1} < \eta_N < x_N,$$

$$r''_N = \frac{2}{h^2}\left[\frac{1}{\theta}r_N + \frac{1}{(1-\theta)}r_{N-1} + \frac{1}{\theta(\theta-1)}r_{N-\theta}\right] - \frac{h(1+\theta)}{3}r'''(\eta'_N), \quad x_{N-1} < \eta'_N < x_N;\ 0 < \theta < 1$$

$$(2.13)$$

in (2.12) and get the discretization at $x = 1$ as follows:

$$\frac{\tilde{y}_{N-1}}{J_{N-1}} - \left(\frac{1}{J_{N-1}} + \frac{\alpha}{(\beta G_N)}\right)\tilde{y}_N + \frac{\gamma}{(\beta G_N)} = a_{0,N}r_N - a_{-1,N}r_{N-1} + a_{1,N}r_{N-\theta} + t_N, \quad (2.14)$$

where the local truncation error t_N is given by

$$t_N = \frac{1}{J_{N-1}}\left[H_N f_N G'''_N a^-_{03,N} + H_N f_N G(\xi^-_N)^{iv} a^-_{04,N} + (H_N f'_N + H'_N f_N)\left\{G''_N a^-_{12,N} + G'''_N a^-_{13,N}\right\}\right.$$

$$+ \left(\frac{1}{2}H_N f''_N + H'_N f'_N + \frac{1}{2}H''_N f_N\right)\left(G'_N a^-_{21,N} + G''_N a^-_{22,N}\right)$$

$$+ \left(\frac{1}{6}H_N f'''_N + \frac{1}{2}H'_N f''_N + \frac{1}{2}H''_N f'_N + \frac{1}{6}H'''_N f_N\right)\left(G_N a^-_{30,N} + G'_N a^-_{31,N}\right)$$

$$+ \left(\frac{1}{24}H_N f(\xi_N)^{iv} + \frac{1}{6}H'_N f'''_N + \frac{1}{4}H''_N f''_N + \frac{1}{6}H'''_N f'_N + \frac{1}{24}H(\eta_N)^{iv} f_N\right)G_N a^-_{40,k}$$

$$- \frac{\theta}{6}h^2 f'''(\eta^-_N)\left(a^-_{10,N}G_N H_N + a^-_{11,N}G'_N H_N + a^-_{20,N}G_N H'_N\right)$$

$$\left. - \frac{h(1+\theta)}{3}a^-_{20,N}h^2 G_N H_N f'''(\eta^-_N)\right], \quad x_{N-1} < \xi^-_N,\ \eta^-_N < x_N.$$

$$(2.15)$$

The coefficients $a^-_{ij,N}$ are specified in the appendices.

The discretization (2.14) involves unknown $y_{N-\theta}$, which we approximate in the following way:

$$y_{N-\theta} = \overline{y}_{N-\theta} - \frac{h^3\theta^2(1-\theta)}{6}y'''_N, \quad (2.16)$$

where

$$\overline{y}_{N-\theta} = \theta(1-\theta)\left[\frac{\theta}{(1-\theta)}y_{N-1} + \left(\frac{(1+\theta)}{\theta} + \frac{h\alpha}{\beta}\right) - \frac{h\gamma}{\beta}\right]. \quad (2.17)$$

Now, let $\bar{r}_{N-\theta} = f(x_{N-\theta}, \bar{y}_{N-\theta})$ then replacing $r_{N-\theta}$ by $\bar{r}_{N-\theta}$ in (2.14) we obtain the following discretization for a smooth solution $y(x)$ at $k = N$

$$\frac{\tilde{y}_{N-1}}{J_{N-1}} - \left(\frac{1}{J_{N-1}} + \frac{\alpha}{(\beta G_N)}\right)\tilde{y}_N + \frac{\gamma}{(\beta G_N)} = a_{0,N}r_N - a_{-1,N}r_{N-1} + a_{1,N}\bar{r}_{N-\theta} + \overline{t_N}, \qquad (2.18)$$

where

$$\overline{t_N} = t_N^{(1)} - \frac{h^3}{6}\theta^2(1-\theta)a_{1,N}y_N'''\frac{\partial f}{\partial y}(x_{N-\theta}, y_{N-\theta}^*),$$

$$(2.19)$$

$$y_{N-\theta}^* \in (\min\{y_{N-\theta}, \bar{y}_{N-\theta}\}, \max\{y_{N-\theta}, \bar{y}_{N-\theta}\}).$$

2.2.2. Discretization of the Boundary Condition at $x = 0$

Integrating the differential equation (1.1) twice, first from x_1 to x; then from x_1 to x_2 and interchanging the order of integration we get

$$y_2 - y_1 = J_1 \int_0^{x_1} q(t)r(t)dt + \int_{x_1}^{x_2}\left(\int_t^{x_2}(p(\tau))^{-1}d\tau\right)q(t)r(t)dt. \qquad (2.20)$$

Via Taylor expansion of $r(x)$, $H(x)$, and $G(x)$ about $x = x_1$ we get the following discretization for $k = 1$:

$$\frac{-\tilde{y}_1}{J_1} + \frac{\tilde{y}_2}{J_1} = a_{-1,1}r_0 + a_{0,1}r_1 + a_{1,1}r_2 + t_1, \qquad (2.21)$$

where the local truncation error t_1 is given by

$$t_1 = \left(\frac{1}{6}H_1f_1''' + \frac{1}{2}H_1'f_1'' + \frac{1}{2}H_1''f_1' + \frac{1}{6}H_1'''f_1\right)\left(-\frac{6x_1^{4+a_0}}{\psi(4)}\right)$$

$$+ \left(\frac{1}{24}H_1f_1^{iv} + \frac{1}{6}H_1'f_1''' + \frac{1}{4}H_1''f_1'' + \frac{1}{6}H_1'''f_1' + \frac{1}{24}H_1^{iv}f_1\right)\left(\frac{24x_1^{4+a_0}}{\psi(5)}\right)$$

$$+ \frac{1}{J_1}\left\{H_1f_1\left(\frac{1}{6}a_{03,1}^+G_1''' + \frac{1}{24}a_{04,1}^+G_1^{iv}\right)\right.$$

$$+ (H_1f_1' + H_1'f_1)\left(\frac{1}{2}a_{12,1}^+G_1'' + \frac{1}{6}a_{13,1}^+G_1'''\right)$$

$$+ \left(\frac{1}{2} H_1 f_1'' + H_1' f_1' + \frac{1}{2} H_1'' f_1 \right) \left(a_{21,1}^+ G_1' + \frac{1}{2} a_{22,1}^+ G_1'' \right)$$

$$+ \left(\frac{1}{6} H_1 f_1''' + \frac{1}{2} H_1' f_1'' + \frac{1}{2} H_1'' f_1' + \frac{1}{6} H_1''' f_1 \right) \left(a_{30,1}^+ G_1 + a_{31,1}^+ G_1' \right)$$

$$+ \left(\frac{1}{24} H_1 f_1^{iv} + \frac{1}{6} H_1' f_1''' + \frac{1}{4} H_1'' f_1'' + \frac{1}{6} H_1''' f_1' + \frac{1}{24} H_1^{iv} f_1 \right) a_{40,1}^+ G_1 \Bigg\}$$

$$+ - \frac{h^2}{6} f'''(\eta_1) \left[-\frac{x_1^{2+a_0}}{\psi(2)} H_1 + \frac{2 x_1^{3+a_0}}{\psi(3)} H_1' \right.$$

$$+ \frac{1}{J_1} \left\{ \left(a_{10,1}^+ G_1 + a_{11,1}^+ G_1' \right) H_1 + a_{20,1}^+ G_1 H_1' \right\} \Bigg]$$

$$- \frac{h^2}{12} f^{iv}(\xi_1) \left[\frac{x_1^{3+a_0}}{\psi(3)} H_1 + \frac{1}{2 J_1} a_{20,1}^+ G_1 H_1 \right], \quad 0 < \xi_1 < x_2.$$

$$(2.22)$$

The coefficients $a_{i,j}$ are specified in the appendices and the functions $G(x)$, $H(x)$ are defined in the assumptions (A-1) and (A-2).

The discretization (2.22) involves y_0; so for y_0 we use the following approximation:

$$\overline{y}_0 = y_1 + \frac{x_1^{2+a_0-b_0}}{(1+a_0)(2+a_0-b_0)} H_1 G_1 f_1$$

$$+ x_1^{3+a_0-b_0} \left\{ \left(-\frac{1}{(1+a_0)(2-b_0)} + \frac{1}{(2+a_0-b_0)(3+a_0-b_0)} + \frac{1}{(2-b_0)(3+a_0-b_0)} \right) G_1' H_1 f_1 \right.$$

$$+ \left(-\frac{1}{(1+a_0)(2+a_0)} + \frac{1}{(3+a_0-b_0)} \right) G_1 H_1' f_1 \Bigg\}$$

$$+ \frac{x_1^{2+a_0-b_0}}{2} \left(-\frac{1}{(1+a_0)(2+a_0)} + \frac{1}{(3+a_0-b_0)} \right) (f_2 - f_0),$$

$$(2.23)$$

where

$$y_0 = \overline{y}_0 + \tau_0,$$

$$(2.24)$$

$$\tau_0 = c^* h^{4+a_0-b_0}, \quad c^* \text{ is some constant.}$$

Let $\overline{r}_0 = f(x_0, \overline{y}_0)$; then in (2.21) replacing r_0 by $\overline{r}_0$ we obtain the following discretization for $k = 1$:

$$\frac{1}{J_1} y_1 - \frac{1}{J_1} y_2 + b_{-1,1} \overline{r}_0 + b_{0,1} r_1 + b_{1,1} r_2 + \overline{t}_1^{(1)} = 0, \qquad (2.25)$$

where $\bar{t}_1 = t_1 + b_{-1,1}\tau_0(\partial f/\partial y)(x_0, y_0^*)$, $y_0^* \in (\min\{y_0, \bar{y}_0\}, \max\{y_0, \bar{y}_0\})$, and coefficients $b_{i,j}$, are specified in appendices.

3. Convergence of the Method

In this section we show that under suitable conditions the method (2.1)-(2.2) for the boundary value problems (1.1)-(1.2) described in Section 2 is of fourth order accuracy.

Let $\tilde{Y} = (\tilde{y}_1, \ldots, \tilde{y}_N)^T$, $R(\tilde{Y}) = (\tilde{r}_1, \ldots, \tilde{r}_N)^T$, $Q = (q_1, \ldots, q_N)^T$, and $T = (t_1, \ldots, t_N)^T$, then the finite difference method given by (2.1)-(2.2) for the boundary value problems (1.1)-(1.2) can be expressed in matrix form as

$$B\tilde{Y} + PR(\tilde{Y}) + G\tilde{Y} = Q, \tag{3.1}$$

and $Y = (y_1, \ldots, y_N)^T$ corresponding to smooth solution $y(x)$ satisfies the perturbed system

$$BY + PR(Y) + GY + T = Q, \tag{3.2}$$

where $B = (b_{ij})$ and $P = (p_{ij})$ are $(N \times N)$ tridiagonal matrices.

Let $E = \tilde{Y} - Y = (e_1, \ldots, e_N)^T$, then from Mean Value Theorem $R(\tilde{Y}) - R(Y) = ME$, $M = \text{diag}\{U_1, \ldots, U_N\}(U_k = \partial f_k/\partial y_k \geq 0)$, then from (3.1) and (3.2) we get the error equation as

$$(B + PM + W)E = T. \tag{3.3}$$

We recall that by the notation $Z \geq 0$ we mean that all components z_i of the vector Z satisfy $z_i \geq 0$. Similarly by $B \geq 0$ we mean all the elements b_{ij} of the matrix B satisfy $b_{ij} \geq 0$.

From Corollary of Theorems 7.2 and 7.4 in [25] an (irreducible) tridiagonal matrix B is monotone if

$$b_{k,k+1} \leq 0, \quad b_{k,k-1} \leq 0, \quad \sum_{j=1}^{N} b_{k,j} \begin{cases} \geq 0, & k = 1, 2, \ldots, N, \\ > 0, & \text{for at least one } i. \end{cases} \tag{3.4}$$

Furthermore, from Theorem 7.3 in [25], B^{-1} exists and the elements of B^{-1} are nonnegative. Now, if $B + PM + W$ is monotone and $B + PM + W \geq B$, then from Theorem 7.5 in [25]

$$(B + PM + W)^{-1} \leq B^{-1}. \tag{3.5}$$

Let vector norm $\|E\|$ and matrix norm $\|B\|$ be defined by

$$\|E\| = \max_{1 \leq k \leq N} |e_k|, \qquad \|B\| = \max_{1 \leq k \leq N} \sum_{j=1}^{N} |b_{kj}|, \tag{3.6}$$

then from (3.3) we get

$$\|E\| = \left\| (B + PM + W)^{-1} |T| \right\|, \tag{3.7}$$

where $|T| \equiv (|t_1|, |t_2|, \ldots, |t_N|)$.

Furthermore, if $B + PM + W \geq B$ and B is also monotone matrix, then from (3.5) we get

$$\|E\| \leq \left\| B^{-1} |T| \right\|. \tag{3.8}$$

Now from [22], $B^{-1} = (b_{i,j}^{-1})$ of matrix B is given by

$$b_{i,j}^{-1} = \begin{cases} \dfrac{P(x_i)\left[P(x_N) - P(x_j) + \beta G_N / \alpha\right]}{\left[P(x_N) + \beta G_N / \alpha\right]}, & i \leq j, \\[4mm] \dfrac{P(x_j)\left[P(x_N) - P(x_i) + \beta G_N / \alpha\right]}{\left[P(x_N) + \beta G_N / \alpha\right]}, & i \geq j, \end{cases} \tag{3.9}$$

where

$$P(x) = \int_0^x \left(p(\tau)\right)^{-1} d\tau. \tag{3.10}$$

Let xr^{iv}, $r^{(i)}$ for $i = 0\,(1)\,3$, and y''' be bounded on $(0, 1]$. Now, as b_0 is fixed in $[0, 1)$ and a_0 is chosen such that $1 + a_0 - b_0 > 0$, for sufficiently small h we get

$$|t_k| \leq C h^5 x_k^{-1+a_0}, \quad \left|\bar{t}_N\right| \leq \tilde{C} h^4, \tag{3.11}$$

for suitable constants C and $\tilde{C}$.

Now from (3.8)–(3.11) we get

$$|e_i| \leq C^* h^5 \left[\left\{ 1 - \frac{P(x_i)}{\left[P(x_N) + \beta G_N / \alpha\right]} \right\} \sum_{j=1}^{i} P(x_j) x_j^{-1+a_0} \right.$$

$$\left. + \frac{P(x_i)}{\left[P(x_N) + \beta G_N / \alpha\right]} \left\{ \sum_{j=i+1}^{N-1} \left[P(x_N) - P(x_j) + \frac{\beta G_N}{\alpha} \right] x_j^{-1+a_0} + \frac{\beta G_N}{h\alpha} \right\} \right], \tag{3.12}$$

where $C^* = \max\{C, \overline{C}\}$. It is easy to see that

$$h\sum_{j=1}^{i} P(x_j)x_j^{-1+a_0} < \int_0^{x_i} P(x)x^{-1+a_0}\,dx,$$

$$h\sum_{j=i+1}^{N-1} \left[P(x_N) - P(x_j) + \frac{\beta G_N}{\alpha}\right]x_j^{-1+a_0} < \int_{x_i}^{x_N}\left[P(x_N) - P(x) + \frac{\beta G_N}{\alpha}\right]x^{-1+a_0}\,dx.$$

(3.13)

From (3.12)-(3.13) we obtain

$$|e_i| \leq \frac{C^*h^4}{(1-b_0)}\begin{cases}\dfrac{(1-b_0)x_i^{1-b_0}}{-a_0(1-b_0+a_0)}\left[x_i^{a_0} - \dfrac{\alpha + (1-b_0+a_0)(1-a_0)\beta}{\alpha + (1-b_0)\beta}\right]\\[2mm]\qquad \times\sup_{[0,1]}|G(x)| + (1-b_0)d_1, \qquad\qquad\qquad a_0 \neq 0\\[3mm]x_i^{1-b_0}\left[\ln\left(\dfrac{1}{x_i}\right) + \dfrac{(2-b_0)\beta}{\alpha + (1-b_0)\beta}\right]\sup_{[0,1]}|G(x)| + d_2, \qquad a_0 = 0,\end{cases}$$

(3.14)

where

$$d_1 = \frac{\alpha(1-b_0+a_0)^{(1-b_0+a_0)/-a_0}}{\left(\alpha + (1-b_0)\beta\right)(1-b_0)^{(1-b_0)/-a_0}}$$

$$\times\left[\frac{(2-b_0+a_0)^{(1-b_0+a_0)/-a_0}}{(2-b_0)^{(1-b_0)/-a_0}}\sup_{[0,1]}|G'(x)| + \frac{(3-b_0+a_0)^{(1-b_0+a_0)/-a_0}}{2(3-b_0)^{(1-b_0)/-a_0}}\sup_{[0,1]}|G''(x)|\right.$$

$$\left.+ \frac{(4-b_0+a_0)^{(1-b_0+a_0)/-a_0}}{3!(4-b_0)^{(1-b_0)/-a_0}}\sup_{[0,1]}|G'''(x)| + \frac{(5-b_0+a_0)^{(1-b_0+a_0)/-a_0}}{4!(5-b_0)^{(1-b_0)/-a_0}}\sup_{[0,1]}\left|G^{v}(x)\right|\right]$$

$$d_2 = \frac{\alpha}{(\alpha + (1-b_0)\beta)e^2}\left[\frac{1}{(2-b_0)}e^{1/(2-b_0)}\sup_{[0,1]}|G'(x)| + \frac{1}{2(3-b_0)}e^{2/(3-b_0)}\sup_{[0,1]}|G''(x)|\right.$$

$$\left.+ \frac{1}{3!(4-b_0)}e^{3/(4-b_0)}\sup_{[0,1]}|G'''(x)| + \frac{1}{4!(5-b_0)}e^{4/(5-b_0)}\sup_{[0,1]}\left|G^{v}(x)\right|\right].$$

(3.15)

In view of the following inequalities

$$x^{1-b_0}\left(x^{a_0} - \frac{\alpha + (1-b_0+a_0)(1-a_0)\beta}{\alpha + (1-b_0)\beta}\right) \leq \left(\frac{\alpha + (1-b_0)\beta}{\alpha + (1-b_0+a_0)(1-a_0)\beta}\right)^{(1-b_0+a_0)/-a_0},$$

$$x^{1-b_0}\left(\ln\left(\frac{1}{x}\right) + \frac{\beta}{\alpha + (1-b_0)\beta}\right) \leq \frac{1}{(1-b_0)}e^{-[\alpha-(1-b_0)\beta]/[\alpha+(1-b_0)\beta]}$$

(3.16)

for $x \in (0,1]$, $(B + PM + W)^{-1} \leq B^{-1}$ and from (3.14) it is easy to establish that

$$\|E\|_{\infty} = O\left(h^4\right). \tag{3.17}$$

Thus we have established the following result.

Theorem 3.1. *Assume that $p(x)$, $q(x)$, and $f(x, y)$ satisfy assumptions given in (A-1), (A-2), and (A-3), respectively. Let $r(x) := f(x, y(x))$, then the method is of fourth order accuracy for sufficiently small mesh size h provided, $xr^{(iv)}(x)$, $r^{(i)}$, $i = 0$ (1) 3, and y''' are bounded on $(0,1]$.*

Remark 3.2. We have developed the numerical method for the boundary condition $y'(0) = 0$ but could not establish the fourth order convergence although the order of the accuracy is verified through one example in the next section.

4. Numerical Illustrations

To illustrate the convergence of the method and to corroborate their order of the accuracy, we apply the method to following example. The maximum absolute errors are displayed in Table 1.

Example 4.1. Consider

$$\left(x^{b_0}\frac{1}{1 + x^{3.5}}y'\right)' = (1.9 - b_0)x^{-0.1}\frac{\left((1.9 - b_0)x^{1.9-b_0}e^y - 0.9 + 3.5x^{3.5}/(1 + x^{3.5})\right)}{(4 + x^{1.9-b_0})(1 + x^{3.5})},$$

$$y(0) = \ln\left(\frac{1}{4}\right) \quad \text{or} \quad (y'(0) = 0), \qquad y(1) + 5y'(1) = \ln\left(\frac{1}{5}\right) - (1.9 - b_0), \tag{4.1}$$

with exact solution $y(x) = \ln(1/(4 + x^{1.9-b_0}))$.

The method is applied on the example for $b_0 = 0.1$, 0.5. Maximum absolute errors and order of accuracy for Example 4.1 are displayed in Table 1.

We use the following approximation of J_k in our numerical program:

$$J_k = \frac{x_{k+1}^{1-b_0} - x_k^{1-b_0}}{1 - b_0}G_k + \left(\frac{x_{k+1}^{2-b_0} - x_k^{2-b_0}}{2 - b_0} - \frac{G_k\left(x_{k+1}^{1-b_0} - x_k^{1-b_0}\right)}{1 - b_0}\right)G_k'$$

$$+ \frac{1}{2}G_k''\left(\frac{x_{k+1}^{3-b_0} - x_k^{3-b_0}}{3 - b_0} + x_k^2\frac{x_{k+1}^{1-b_0} - x_k^{1-b_0}}{1 - b_0} - 2x_k\frac{x_{k+1}^{2-b_0} - x_k^{2-b_0}}{2 - b_0}\right) \tag{4.2}$$

$$+ \frac{1}{6}G_k'''\left(\frac{x_{k+1}^{4-b_0} - x_k^{4-b_0}}{4 - b_0} - 3x_k\frac{x_{k+1}^{3-b_0} - x_k^{3-b_0}}{3 - b_0} + 3x_k^2\frac{x_{k+1}^{2-b_0} - x_k^{2-b_0}}{2 - b_0} - x_k^3\frac{x_{k+1}^{1-b_0} - x_k^{1-b_0}}{1 - b_0}\right).$$

Table 1: Maximum absolute errors and order of the methods for Example 4.1.

N	$b_0 = 0.1$	Order	$b_0 = 0.5$	Order	$b_0 = 0.1$	Order	$b_0 = 0.5$	Order
	$y(0) = \ln(1/4)$ and $\theta = 0.01$				$y(0) = \ln(1/4)$ and $\theta = 0.98$			
32	8.23 (−7)[a]		1.10 (−6)		6.11 (−7)		8.11 (−7)	
64	5.05 (−8)	4.03	6.80 (−8)	4.02	3.50 (−8)	4.13	4.60 (−8)	4.14
128	3.13 (−9)		4.22 (−9)		2.09 (−9)		2.72 (−9)	
256	1.94 (−10)	4.01	2.64 (−10)	4.00	1.27 (−10)	4.04	1.67 (−10)	4.03
	$y'(0) = 0$ and $\theta = 0.01$				$y'(0) = 0$ and $\theta = 0.98$			
32	3.37 (−7)		4.03 (−7)		1.72 (−7)		1.95 (−7)	
64	2.04 (−8)	4.05	2.46 (−8)	4.03	9.13 (−9)	4.24	9.80 (−9)	4.31
128	1.25 (−9)		1.52 (−9)		5.20 (−10)		5.39 (−10)	
266	7.74 (−11)	4.02	9.56 (−11)	3.99	3.09 (−11)	4.08	3.23 (−11)	4.06

[a]8.23 (−7) = 8.23×10^{-7}.

4.1. Numerical Results for Uniform Mesh Case

Remark 4.2. In the discretization (2.18), θ may take any value in $(0, 1)$ but from the Table 1, maximum absolute errors are less for the value of θ close to one. So it is better to take value of θ close to one.

Appendices

The coefficients involved in the method and their approximations are mentioned below.

A. Coefficients for Method

$$a_{\pm 1,k}^{(1)} = \frac{1}{2h}\left[\pm\left(b_{10,k}G_kH_k + b_{11,k}G_k'H_k + b_{20,k}G_kH_k'\right) + \frac{1}{h}b_{20,k}G_kH_K\right],$$

$$a_{0,k}^{(1)} = \left(b_{00,k} - \frac{1}{h^2}b_{20,k}\right)G_kH_k + b_{01,k}H_kG_k' + \frac{1}{2}b_{02,k}G_k''H_k$$

$$+ b_{10,k}G_kH_k' + b_{11,k}G_k'H_k' + \frac{1}{2}b_{20,k}G_kH_k'',$$

$$a_{-1,N}^{(1)} = \frac{1}{J_{N-1}}\left[\frac{\theta}{h(1-\theta)}\left\{a_{10,N}^-G_NH_N + a_{11,N}^-G_N'H_N + a_{20,N}^-G_NH_N'\right\}\right.$$

$$\left. + \frac{1}{h^2(1-\theta)}a_{20,N}^-G_NH_N\right],$$

$$a_{0,N}^{(1)} = \frac{1}{J_{N-1}}\left[a_{00,N}^-G_NH_N + a_{10,N}^-G_NH_N' + a_{01,N}^-G_N'H_N + \frac{1}{2}a_{02,N}^-G_N''H_N + a_{11,N}^-G_N'H_N'\right.$$

$$\left. + \frac{1}{2}a_{20,N}^-G_NH_N'' + \frac{(1+\theta)}{h\theta}\left\{a_{10,N}^-G_NH_N + a_{11,N}^-G_N'H_N + a_{20,N}^-G_NH_N'\right\}\right]$$

$$+ \frac{1}{\theta h^2} a_{20,N}^- G_N H_N \Bigg],$$

$$a_{1,N}^{(1)} = \frac{1}{J_{N-1}} \Bigg[\frac{1}{h\theta(1-\theta)} \Big\{ a_{10,N}^- G_N H_N + a_{11,N}^- G_N' H_N + a_{20,N}^- G_N H_N' \Big\}$$

$$+ \frac{1}{h^2\theta(1-\theta)} a_{20,N}^- G_N H_N \Bigg],$$

$$(A.1)$$

where

$$a_{00,k}^{\pm} = \frac{1}{(1-b_0)} \left[-\frac{\left(x_{k\pm1}^{2+a_0-b_0} - x_k^{2+a_0-b_0} \right)}{\phi(2)} + \frac{x_{k\pm1}}{\psi(1)} \left(x_{k\pm1}^{1+a_0} - x_k^{1+a_0} \right) \right],$$

$$a_{01,k}^{\pm} = \frac{1}{(1-b_0)} \left[\frac{(x_k - x_{k\pm1}) x_{k\pm1}^{2+a_0-b_0}}{\phi(2)} + \frac{1}{\phi(3)} \left(x_{k\pm1}^{3+a_0-b_0} - x_k^{3+a_0-b_0} \right) \right.$$

$$+ \frac{(x_{k\pm1} - x_k) x_{k\pm1}^{1-b_0}}{\psi(1)} \left(x_{k\pm1}^{1+a_0} - x_k^{1+a_0} \right) - \frac{x_{k\pm1}^{2-b_0}}{\psi(1)\varphi(2)} \left(x_{k\pm1}^{1+a_0} - x_k^{1+a_0} \right)$$

$$+ \left. \frac{1}{\varphi(2)(3+a_0-b_0)} \left(x_{k\pm1}^{3+a_0-b_0} - x_k^{3+a_0-b_0} \right) \right],$$

$$a_{02,k}^{\pm} = \frac{1}{(1-b_0)} \left[-\frac{(x_{k\pm1} - x_k)^2 x_{k\pm1}^{2+a_0-b_0}}{\phi(2)} + \frac{1}{\phi(3)} \left(2(x_{k\pm1} - x_k) x_{k\pm1}^{3+a_0-b_0} \right) \right.$$

$$- \frac{2}{\phi(4)} \left(x_{k\pm1}^{4+a_0-b_0} - x_k^{4+a_0-b_0} \right) + \frac{(x_{k\pm1} - x_k)^2 x_{k\pm1}^{1-b_0}}{(1+a_0)} \left(x_{k\pm1}^{1+a_0} - x_k^{1+a_0} \right)$$

$$+ \frac{2(x_{k\pm1} - x_k)}{\varphi(2)(3+a_0-b_0)} x_{k\pm1}^{3+a_0-b_0} - \frac{2\phi(2)}{\varphi(2)\phi(4)} \left(x_{k\pm1}^{4+a_0-b_0} - x_k^{4+a_0-b_0} \right)$$

$$- \frac{2(x_{k\pm1} - x_k)}{\varphi(2)\psi(1)} x_{k\pm1}^{2-b_0} \left(x_{k\pm1}^{1+a_0} - x_k^{1+a_0} \right) + \frac{2}{\varphi(3)\psi(1)} \left(x_{k\pm1}^{1+a_0} - x_k^{1+a_0} \right)$$

$$- \left. \frac{2}{\varphi(3)(4+a_0-b_0)} \left(x_{k\pm1}^{4+a_0-b_0} - x_k^{4+a_0-b_0} \right) \right],$$

$$a_{03,k}^{\pm} = \frac{1}{1-b_0} \left[\pm \frac{h^3}{(1+a_0)} x_{k\pm1}^{1-b_0} \left(x_{k\pm1}^{1+a_0} - x_k^{1+a_0} \right) \mp \frac{h^3}{\phi(2)} x_{k\pm1}^{2+a_0-b_0} + \frac{3h^2}{\phi(3)} x_{k\pm1}^{3+a_0-b_0} \right.$$

$$\mp \frac{6h}{\phi(4)} x_{k\pm1}^{4+a_0-b_0} + \frac{6}{\phi(5)} \left(x_{k\pm1}^{5+a_0-b_0} - x_k^{5+a_0-b_0} \right)$$

$$- \frac{3h^2}{(1+a_0)(2-b_0)} x_{k\pm1}^{2-b_0} \left(x_{k\pm1}^{1+a_0} - x_k^{1+a_0} \right)$$

$$+ \frac{3}{(2-b_0)} \left\{ \frac{h^2}{3+a_0-b_0} x_{k\pm1}^{3+a_0-b_0} \mp \frac{h}{(3+a_0-b_0)(4+a_0-b_0)} x_{k\pm1}^{4+a_0-b_0} \right.$$

$$\left. + \frac{2\phi(2)}{\phi(5)} \left(x_{k\pm1}^{5+a_0-b_0} - x_k^{5+a_0-b_0} \right) \right\}$$

$$\pm \frac{6}{(2-b_0)(3-b_0)(1+a_0)} x_{k\pm1}^{3-b_0} \left(x_{k\pm1}^{1+a_0} - x_k^{1+a_0} \right) - \frac{6}{(2-b_0)(3-b_0)}$$

$$\left\{ \pm \frac{h}{(4+a_0-b_0)} x_{k\pm1}^{4+a_0-b_0} - \frac{1}{(4+a_0-b_0)(5+a_0-b_0)} \left(x_{k\pm1}^{5+a_0-b_0} - x_k^{5+a_0-b_0} \right) \right\}$$

$$- \frac{6}{\varphi(4)(1+a_0)} x_{k\pm1}^{4-b_0} \left(x_{k\pm1}^{1+a_0} - x_k^{1+a_0} \right)$$

$$\left. \frac{6}{\varphi(4)(5+a_0-b_0)} \left(x_{k\pm1}^{5+a_0-b_0} - x_k^{5+a_0-b_0} \right) \right],$$

$$a_{04,k}^{\pm} = \frac{1}{1-b_0} \left[\frac{h^4}{\psi(1)} x_{k\pm1}^{1-b_0} \left(x_{k\pm1}^{1+a_0} - x_k^{1+a_0} \right) - \frac{h^4}{\phi(2)} x_{k\pm1}^{2+a_0-b_0} \pm \frac{4h^3}{\phi(3)} x_{k\pm1}^{3+a_0-b_0} \right.$$

$$- \frac{12h^2}{\phi(4)} x_{k\pm1}^{4+a_0-b_0} \pm \frac{24h}{\phi(5)} x_{k\pm1}^{5+a_0-b_0} + \frac{24}{\phi(6)} \left(x_{k\pm1}^{6+a_0-b_0} - x_k^{6+a_0-b_0} \right)$$

$$\mp \frac{4h^3}{\varphi(2)\psi(1)} x_{k\pm1}^{2-b_0} \left(x_{k\pm1}^{1+a_0} - x_k^{1+a_0} \right)$$

$$+ \frac{4}{(2-b_0)} \left\{ \mp \frac{h^3}{(3+a_0-b_0)} x_{k\pm1}^{3+a_0-b_0} - \frac{3h^2\phi(2)}{\phi(4)} x_{k\pm1}^{4+a_0-b_0} \right.$$

$$\left. \pm \frac{6h\phi(2)}{\phi(5)} x_{k\pm1}^{5+a_0-b_0} - \frac{6\phi(2)}{\phi(6)} \left(x_{k\pm1}^{6+a_0-b_0} - x_k^{6+a_0-b_0} \right) \right\}$$

$$+ \frac{12h^2}{\varphi(2)\psi(1)} x_{k\pm1}^{3-b_0} \left(x_{k\pm1}^{1+a_0} - x_k^{1+a_0} \right)$$

$$- \frac{12}{\varphi(3)} \left\{ \frac{h^2}{(4+a_0-b_0)} x_{k\pm1}^{4+a_0-b_0} \mp \frac{2h\phi(3)}{\phi(5)} x_{k\pm1}^{5+a_0-b_0} \right.$$

$$\left. + \frac{2\phi(3)}{\phi(5)} \left(x_{k\pm1}^{6+a_0-b_0} - x_k^{6+a_0-b_0} \right) \right\}$$

$$\mp \frac{24h}{\varphi(4)\psi(1)} x_{k\pm1}^{4-b_0} \left(x_{k\pm1}^{1+a_0} - x_k^{1+a_0} \right)$$

$$+ \frac{24}{\varphi(4)} \left\{ \pm \frac{h}{(5+a_0-b_0)} x_{k\pm1}^{5+a_0-b_0} \right.$$

$$\left. - \frac{1}{(5+a_0-b_0)(6+a_0-b_0)} \left(x_{k\pm1}^{6+a_0-b_0} - x_k^{6+a_0-b_0} \right) \right\}$$

$$+ \frac{24}{\varphi(5)\psi(1)} x_{k\pm1}^{5-b_0} \left(x_{k\pm1}^{1+a_0} - x_k^{1+a_0} \right)$$

$$- \frac{24}{\varphi(6)(6+a_0-b_0)} \left(x_{k\pm1}^{6+a_0-b_0} - x_k^{6+a_0-b_0} \right) \Bigg],$$

$$a_{10,k}^{\pm} = \frac{1}{(1-b_0)} \left[-\frac{(x_{k\pm1}-x_k)}{\phi(1)} x_{k\pm1}^{2+a_0-b_0} + \frac{(x_{k\pm1}-x_k)}{\psi(1)} x_{k\pm1}^{2+a_0-b_0} - \frac{1}{\psi(2)} x_{k\pm1}^{1-b_0} \left(x_{k\pm1}^{2+a_0} - x_k^{2+a_0} \right) \right.$$

$$\left. + \frac{1}{\phi(3)} \left(x_{k\pm1}^{3+a_0-b_0} - x_k^{3+a_0-b_0} \right) \right],$$

$$a_{11,k}^{\pm} = \frac{1}{(1-b_0)} \left[-\frac{(x_{k\pm1}-x_k)^2 x_{k\pm1}^{2+a_0-b_0}}{\phi(2)} + \frac{(x_{k\pm1}-x_k)^2}{\psi(1)} x_{k\pm1}^{2+a_0-b_0} \right.$$

$$- \frac{(x_{k\pm1}-x_k)}{\psi(2)} x_{k\pm1}^{1-b_0} \left(x_{k\pm1}^{1-b_0} - x_k^{1-b_0} \right)$$

$$+ \frac{2(x_{k\pm1}-x_k)}{\phi(3)} x_{k\pm1}^{3+a_0-b_0} - \frac{2}{\phi(4)} \left(x_{k\pm1}^{4+a_0-b_0} - x_k^{4+a_0-b_0} \right)$$

$$+ \frac{1}{(2-b_0)} \left[\frac{(x_{k\pm1}-x_k)}{(3+a_0-b_0)} x_{k\pm1}^{3+a_0-b_0} - \frac{(x_{k\pm1}-x_k)}{\psi(1)} x_{k\pm1}^{3+a_0-b_0} + \frac{1}{\psi(2)} x_{k\pm1}^{2-b_0} \right.$$

$$\left. \left. \times \left(x_{k\pm1}^{2+a_0} - x_k^{2+a_0} \right) - \frac{\phi(2)}{\phi(4)} \left(x_{k\pm1}^{4+a_0-b_0} - x_k^{4+a_0-b_0} \right) \right] \right],$$

$$a_{12,k}^{\pm} = \frac{1}{1-b_0} \left[h^2 x_{k\pm1}^{1-b_0} \left\{ \pm \frac{h}{\psi(1)} x_{k\pm1}^{1+a_0} - \frac{1}{\psi(2)} \left(x_{k\pm1}^{2+a_0} - x_k^{2+a_0} \right) \right\} \right.$$

$$\mp \frac{h^3}{\phi(2)} x_{k\pm1}^{2+a_0-b_0} + \frac{3h^2}{\phi(3)} x_{k\pm1}^{3+a_0-b_0} \mp \frac{6h}{\phi(4)} x_{k\pm1}^{4+a_0-b_0}$$

$$+ \frac{6}{\phi(5)} \left(x_{k\pm1}^{5+a_0-b_0} - x_k^{5+a_0-b_0} \right)$$

$$\mp \frac{2h}{\varphi(2)} x_{k\pm1}^{2-b_0} \left\{ \frac{h}{\psi(1)} x_{k\pm1}^{1+a_0} - \frac{1}{\psi(2)} \left(x_{k\pm1}^{2+a_0} - x_k^{2+a_0} \right) \right\}$$

$$+ \frac{2}{\varphi(2)} \left\{ \frac{h^2}{(3+a_0-b_0)} x_{k\pm1}^{3+a-0-b_0} \mp \frac{2h\phi(2)}{\phi(4)} x_{k\pm1}^{4+a-0-b_0} \right.$$

$$\left. + \frac{2\phi(2)}{\phi(5)} \left(x_{k\pm1}^{5+a_0-b_0} - x_k^{5+a_0-b_0} \right) \right\}$$

$$+ \frac{2}{\varphi(3)} \left\{ x_{k\pm1}^{3-b_0} \left\{ \pm \frac{h}{\psi(1)} x_{k\pm1}^{1+a_0} - \frac{1}{\psi(2)} \left(x_{k\pm1}^{2+a_0} - x_k^{2+a_0} \right) \right\} \right.$$

$$\mp \frac{h}{(4+a_0-b_0)} x_{k\pm1}^{4+a_0-b_0} + \frac{1}{(4+a_0-b_0)(5+a_0-b_0)} \left(x_{k\pm1}^{5+a_0-b_0} - x_k^{5+a_0-b_0} \right) \bigg],$$

$$a_{13,k}^{\pm} = \frac{1}{1-b_0}\left[\pm h^3 x_{k\pm 1}^{1-b_0}\left\{\pm\frac{h}{\psi(1)}x_{k\pm 1}^{1+a_0} - \frac{1}{\psi(2)}\left(x_{k\pm 1}^{2+a_0} - x_k^{2+a_0}\right)\right\}\right.$$

$$-\frac{h^4}{\phi(2)}x_{k\pm 1}^{2+a_0-b_0} \pm \frac{4h^3}{\phi(3)}x_{k\pm 1}^{3+a_0-b_0} - \frac{12h^2}{\phi(4)}x_{k\pm 1}^{4+a_0-b_0} \pm \frac{24h}{\phi(5)}x_{k\pm 1}^{5+a_0-b_0}$$

$$-\frac{24}{\phi(6)}\left(x_{k\pm 1}^{6+a_0-b_0} - x_k^{6+a_0-b_0}\right)$$

$$-\frac{3h^2}{\varphi(2)}x_{k\pm 1}^{2-b_0}\left\{\pm\frac{h}{\psi(1)}x_{k\pm 1}^{1+a_0} - \frac{1}{\psi(2)}\left(x_{k\pm 1}^{2+a_0} - x_k^{2+a_0}\right)\right\}$$

$$+\frac{3\phi(2)}{\varphi(2)}\left\{\pm\frac{h^3}{\phi(3)}x_{k\pm 1}^{3+a_0-b_0} - \frac{3h^2}{\phi(4)}x_{k\pm 1}^{4+a_0-b_0} \pm \frac{6h}{\phi(5)}x_{k\pm 1}^{5+a_0-b_0}\right.$$

$$\left.-\frac{6}{\phi(6)}\left(x_{k\pm 1}^{6+a_0-b_0} - x_k^{6+a_0-b_0}\right)\right\}$$

$$\pm\frac{6h}{\varphi(3)}x_{k\pm 1}^{3-b_0}\left\{\pm\frac{h}{\psi(1)}x_{k\pm a}^{1+a_0} - \frac{1}{\psi(2)}\left(x_{k\pm 1}^{2+a_0} - x_k^{2+a_0}\right)\right\}$$

$$-\frac{6\phi(3)}{\varphi(3)}\left\{\frac{h^2}{\phi(4)}x_{k\pm 1}^{4+a_0-b_0} \mp \frac{2h}{\phi(5)}x_{k\pm 1}^{5+a_0-b_0} + \frac{2}{\phi(6)}\left(x_{k\pm 1}^{6+a_0-b_0} - x_k^{6+a_0-b_0}\right)\right\}$$

$$-\frac{6}{\varphi(4)}\left\{x_{k\pm 1}^{4-b_0}\left\{\pm\frac{h}{\psi(1)}x_{k\pm 1}^{1+a_0} - \frac{1}{\psi(2)}\left(x_{k\pm 1}^{2+a_0} - x_k^{2+a_0}\right)\right\} \mp \frac{h}{(5+a_0-b_0)}x_{k\pm 1}^{5+a_0-b_0}\right.$$

$$\left.\left.+\frac{1}{(5+a_0-b_0)(6+a_0-b_0)}\left(x_{k\pm 1}^{6+a_0-b_0} - x_k^{6+a_0-b_0}\right)\right\}\right],$$

$$a_{20,k}^{\pm} = \frac{1}{(1-b_0)}\left[-\frac{(x_{k\pm 1} - x_k)^2}{\phi(2)}x_{k\pm 1}^{2+a_0-b_0} + \frac{2(x_{k\pm 1} - x_k)}{\phi(3)}x_{k\pm 1}^{3+a_0-b_0} + \frac{(x_{k\pm 1} - x_k)^2}{\psi(1)}x_{k\pm 1}^{2+a_0-b_0}\right.$$

$$-\frac{2}{\phi(4)}\left(x_{k\pm 1}^{4+a_0-b_0} - x_k^{4+a_0-b_0}\right) - \frac{2(x_{k\pm 1} - x_k)}{\psi(2)}x_{k\pm 1}^{3+a_0-b_0}$$

$$\left.+\frac{2}{\psi(3)}x_{k\pm 1}^{1-b_0}\left(x_{k\pm 1}^{3+a_0} - x_k^{3+a_0}\right)\right],$$

$$a_{21,k}^{\pm} = \frac{1}{1-b_0}\left[\pm h x_{k\pm 1}^{1-b_0}\left\{\frac{h^2}{\psi(1)}x_{k\pm 1}^{1+a_0} \mp \frac{2}{\psi(2)}x_{k\pm 1}^{2+a_0} + \frac{2}{\psi(3)}\left(x_{k\pm 1}^{3+a_0} - x_k^{3+a_0}\right)\right\}\right.$$

$$\mp\frac{h^3}{\phi(2)}x_{k\pm 1}^{2+a_0-b_0} + \frac{3h^2}{\phi(3)}x_{k\pm 1}^{3+a_0-b_0} \mp \frac{6h}{\phi(4)}x_{k\pm 1}^{4+a_0-b_0}$$

$$+\frac{6}{\phi(5)}\left(x_{k\pm 1}^{5+a_0-b_0} - x_k^{5+a_0-b_0}\right)$$

$$-\frac{1}{\varphi(2)}\left\{x_{k\pm 1}^{2-b_0}\left\{\frac{h^2}{\psi(1)}x_{k\pm 1}^{1+a_0} \mp \frac{2h}{\psi(2)}x_{k\pm 1}^{2+a_0} + \frac{2}{\psi(3)}\left(x_{k\pm 1}^{3+a_0} - x_k^{3+a_0}\right)\right\}\right.$$

$$+ \phi(2)\left\{-\frac{h^2}{\phi(3)}x_{k\pm1}^{3+a_0-b_0}\right.$$

$$\left.\pm\frac{2h}{\phi(4)}x_{k\pm1}^{4+a_0-b_0} - \frac{2}{\phi(5)}\left(x_{k\pm1}^{5+a_0-b_0} - x_k^{5+a_0-b_0}\right)\right\}\Bigg\}\Bigg],$$

$$a_{22,k}^{\pm} = \frac{1}{1-b_0}\left[x_{k\pm1}^{1-b_0}\left(h^2 + \frac{\mp hx + 2x^2}{(2-b_0)}\right)\left\{\frac{h^2}{\psi(1)}x_{k\pm1}^{1+a_0} \mp \frac{2h}{\psi(2)}x_{k\pm1}^{2+a_0}\right.\right.$$

$$\left.+ \frac{1}{\psi(3)}\left(x_{k\pm1}^{3+a_0} - x_k^{3+a_0}\right)\right\}$$

$$-\frac{h^4}{\phi(2)}x_{k\pm1}^{2+a_0-b_0} \pm \frac{4h^3}{\phi(3)}x_{k\pm1}^{3+a_0-b_0} + \frac{12h^2}{\phi(4)}x_{k\pm1}^{4+a_0-b_0} \mp \frac{24h}{\phi(5)}x_{k\pm1}^{5+a_0-b_0}$$

$$+\frac{24}{\phi(6)}\left(x_{k\pm1}^{6+a_0-b_0} - x_k^{6+a_0-b_0}\right)$$

$$+\frac{2\phi(2)}{(2-b_0)}\left\{\pm\frac{h^3}{\phi(3)}x_{k\pm1}^{3+a_0-b_0} - \frac{3h^2}{\phi(4)}x_{k\pm1}^{4+a_0-b_0}\right.$$

$$\left.\mp\frac{6h}{\phi(5)}x_{k\pm1}^{5+a_0-b_0} - \frac{6}{\phi(6)}\left(x_{k\pm1}^{6+a_0-b_0} - x_k^{6+a_0-b_0}\right)\right\}$$

$$\left.-\frac{2\varphi(1)}{\varphi(3)}\phi(3)\left\{\frac{h^2}{\phi(4)}x_{k\pm1}^{4+a_0-b_0} \mp \frac{h}{\phi(5)}x_{k\pm1}^{5+a_0-b_0} + \frac{1}{\phi(6)}\left(x_{k\pm1}^{6+a_0-b_0} - x_k^{6+a_0-b_0}\right)\right\}\right],$$

$$a_{30,k}^{\pm} = \frac{1}{1-b_0}\left[x_{k\pm1}^{1-b_0}\left\{\pm\frac{h^3}{\psi(1)}x_{k\pm1}^{1+a_0} - \frac{3h^2}{\psi(2)}x_{k\pm1}^{2+a_0} \pm \frac{6h}{\psi(3)}x_{k\pm1}^{3+a_0}\right.\right.$$

$$\left.-\frac{1}{\psi(4)}\left(x_{k\pm1}^{4+a_0} - x_k^{4+a_0}\right)\right\}$$

$$\mp\frac{h^3}{\phi(2)}x_{k\pm1}^{2+a_0-b_0} + \frac{3h^2}{\phi(3)}x_{k\pm1}^{3+a_0-b_0} \mp \frac{6h}{\phi(4)}x_{k\pm1}^{4+a_0-b_0}$$

$$\left.+\frac{6}{\phi(5)}\left(x_{k\pm1}^{5+a_0-b_0} - x_k^{5+a_0-b_0}\right)\right],$$

$$a_{31,k}^{\pm} = \frac{1}{1-b_0}\left[\left(\pm hx_{k\pm1}^{1-b_0} - \frac{1}{(2-b_0)}x_{k\pm1}^{2-b_0}\right)\left\{\pm\frac{h^3}{\psi(1)}x_{k\pm1}^{1+a_0} - \frac{3h^2}{\psi(2)}x_{k\pm1}^{2+a_0} \pm \frac{6h}{\psi(3)}x_{k\pm1}^{3+a_0}\right.\right.$$

$$\left.-\frac{6}{\psi(4)}\left(x_{k\pm1}^{4+a_0} - x_k^{4+a_0}\right)\right\}$$

$$-\frac{h^4}{\phi(2)}x_{k\pm1}^{2+a_0-b_0} \pm \frac{4h^3}{\phi(3)}x_{k\pm1}^{3+a_0-b_0} - \frac{12h^2}{\phi(4)}x_{k\pm1}^{4+a_0-b_0} \pm \frac{24h}{\phi(5)}x_{k\pm1}^{5+a_0-b_0}$$

$$-\frac{24}{\phi(6)}\left(x_{k\pm1}^{6+a_0-b_0} - x_k^{6+a_0-b_0}\right)$$

$$+\frac{\phi(2)}{(2-b_0)}\left\{\pm\frac{h^3}{\phi(3)}x_{k\pm1}^{3+a_0-b_0} - \frac{3h^2}{\phi(4)}x_{k\pm1}^{4+a_0-b_0}\right.$$

$$\left.\pm\frac{6h}{\phi(5)}x_{k\pm1}^{5+a_0-b_0} - \frac{6}{\phi(6)}\left(x_{k\pm1}^{6+a_0-b_0} - x_k^{6+a_0-b_0}\right)\right\}\Bigg],$$

$$a_{40,k}^{\pm} = \frac{1}{1-b_0}\left[x_{k\pm1}^{1-b_0}\left\{\frac{h^4}{\psi(1)}x_{k\pm1}^{1+a_0} \mp \frac{4h^3}{\psi(2)}x_{k\pm1}^{2+a_0} + \frac{12h^2}{\psi(3)}x_{k\pm1}^{3+a_0} \mp \frac{24h}{\psi(4)}x_{k\pm1}^{4+a_0}\right.\right.$$

$$\left.\left.+\frac{24}{\psi(5)}\left(x_{k\pm1}^{5+a_0} - x_k^{5+a_0}\right)\right\}\right.$$

$$-\frac{h^4}{\phi(2)}x_{k\pm1}^{2+a_0-b_0} \pm \frac{4h^3}{\phi(3)}x_{k\pm1}^{3+a_0-b_0} - \frac{12h^2}{\phi(4)}x_{k\pm1}^{4+a_0-b_0}$$

$$\left.\pm\frac{24h}{\phi(5)} - \frac{24}{\phi(6)}\left(x_{k\pm1}^{6+a_0-b_0} - x_k^{6+a_0-b_0}\right)\right],$$

$$b_{ij,k} = \left(\frac{a_{ij,k}^{+}}{J_k} + \frac{a_{ij,k}^{-}}{J_{k-1}}\right),$$

$$\phi(i) = \prod_{j=2}^{i}(j + a_0 - b_0), \qquad \psi(i) = \prod_{j=1}^{i}(j + a_0), \qquad \varphi(i) = \prod_{j=2}^{i}(j - b_0).$$

$$(A.2)$$

B. Approximations of the Coefficients

For uniform mesh $x_k = kh$, $k = 0(1)N$ and $h = 1/N$ then for fixed x_k as $h \to 0$, we get the following approximations for the coefficients:

$$b_{00,k} \sim \frac{hx_k^{a_0}}{G_k}, \quad b_{01,k} \sim (a_0 - 2b_0)\frac{h^3 x_k^{a_0-1}}{4G_k}, \quad b_{02,k} \sim \frac{h^3 x_k^{a_0}}{2G_k}, \quad b_{10,k} \sim (2a_0 - 3b_0)\frac{h^3 x_k^{a_0-1}}{12G_k},$$

$$b_{11,k} \sim \frac{h^3 x_k^{a_0}}{4G_k}, \quad b_{20,k} \sim \frac{h^3 x_k^{a_0}}{6G_k}, \quad b_{03,k} \sim (2a_0 - 3b_0)\frac{h^3 x_k^{a_0-1}}{3G_k}, \quad b_{04,k} \sim \frac{2h^5 x_k^{a_0}}{3G_k},$$

$$b_{12,k} \sim (3a_0 - 4b_0)\frac{h^5 x_k^{a_0-1}}{9G_k}, \quad b_{13,k} \sim \frac{h^5 x_k^{a_0}}{3G_k}, \quad b_{21,k} \sim (5a_0 - 6b_0)\frac{h^5 x_k^{a_0-1}}{18G_k}, \quad b_{22,k} \sim \frac{2h^5 x_k^{a_0}}{9G_k},$$

$$b_{30,k} \sim (6a_0 - 7b_0)\frac{h^5 x_k^{a_0-1}}{30G_k}, \quad b_{31,k} \sim \frac{h^5 x_k^{a_0}}{6G_k}, \quad b_{40,k} \sim \frac{2h^5 x_k^{a_0}}{15G_k},$$

$$\left|a_{00,N}^{-}\right| < \frac{h^2}{2}, \quad \left|a_{01,N}^{-}\right| < \frac{2h^3}{3}, \quad \left|a_{02,N}^{-}\right| < \frac{h^4}{6}, \quad \left|a_{10,N}^{-}\right| < \frac{h^3}{3}, \quad \left|a_{11,N}^{-}\right| < \frac{h^4}{4}, \quad \left|a_{20,N}^{-}\right| < \frac{h^4}{6},$$

$$\left|a_{03,N}^{-}\right| < \frac{2h^5}{5}, \quad \left|a_{04,N}^{-}\right| < \frac{h^6}{3}, \quad \left|a_{12,N}^{-}\right| < \frac{h^5}{5}, \quad \left|a_{13,N}^{-}\right| < \frac{h^6}{6}, \quad \left|a_{21,N}^{-}\right| < \frac{2h^5}{15}, \quad \left|a_{22,N}^{-}\right| < \frac{h^6}{9},$$

$$\left|a_{30,N}^{-}\right| < \frac{h^5}{10}, \quad \left|a_{31,N}^{-}\right| < \frac{h^6}{12}, \quad \left|a_{40,N}^{-}\right| < \frac{h^6}{15}.$$

$$(B.1)$$

Acknowledgments

The authors are thankful to the referee for the valuable suggestions. This work is supported by Department of Science and Technology, New Delhi, India.

References

[1] L. E. Bobisud, "Existence of solutions for nonlinear singular boundary value problems," *Applicable Analysis*, vol. 35, no. 1–4, pp. 43–57, 1990.

[2] L. H. Thomas, "The calculation of atomic fields," *Mathematical Proceedings of the Cambridge Philosophical Society*, vol. 23, no. 5, pp. 542–548, 1927.

[3] E. Fermi, "Un methodo statistico per la determinazione di alcune proprieta dell'atomo," *Atti dell'Accademia Nazionale dei Lincei. Classe di Scienze Fisiche, Matematiche e Naturali*, vol. 6, pp. 602–607, 1927.

[4] C. Y. Chan and Y. C. Hon, "A constructive solution for a generalized Thomas-Fermi theory of ionized atoms," *Quarterly of Applied Mathematics*, vol. 45, no. 3, pp. 591–599, 1987.

[5] L. E. Bobisud, D. O'Regan, and W. D. Royalty, "Singular boundary value problems," *Applicable Analysis*, vol. 23, no. 3, pp. 233–243, 1986.

[6] L. E. Bobisud, D. O'Regan, and W. D. Royalty, "Solvability of some nonlinear boundary value problems," *Nonlinear Analysis: Theory, Methods & Applications*, vol. 12, no. 9, pp. 855–869, 1988.

[7] E. Hille, "Some aspects of the Thomas-Fermi equation," *Journal d'Analyse Mathématique*, vol. 23, no. 1, pp. 147–170, 1970.

[8] A. Granas, R. B. Guenther, and J. Lee, "Nonlinear boundary value problems for ordinary differential equations," *Dissertationes Mathematicae*, vol. 244, p. 128, 1985.

[9] D. R. Dunninger and J. C. Kurtz, "Existence of solutions for some nonlinear singular boundary value problems," *Journal of Mathematical Analysis and Applications*, vol. 115, no. 2, pp. 396–405, 1986.

[10] R. K. Pandey and A. K. Verma, "Existence-uniqueness results for a class of singular boundary value problems arising in physiology," *Nonlinear Analysis: Real World Applications*, vol. 9, no. 1, pp. 40–52, 2008.

[11] R. K. Pandey and A. K. Verma, "A note on existence-uniqueness results for a class of doubly singular boundary value problems," *Nonlinear Analysis: Theory, Methods & Applications*, vol. 71, no. 7-8, pp. 3477–3487, 2009.

[12] R. K. Pandey and A. K. Verma, "On solvability of derivative dependent doubly singular boundary value problems," *Journal of Applied Mathematics and Computing*, vol. 33, no. 1-2, pp. 489–511, 2010.

[13] Y. Zhang, "A note on the solvability of singular boundary value problems," *Nonlinear Analysis: Theory, Methods & Applications*, vol. 26, no. 10, pp. 1605–1609, 1996.

[14] G. W. Reddien, "Projection methods and singular two point boundary value problems," *Numerische Mathematik*, vol. 21, pp. 193–205, 1973.

[15] M. M. Chawla and C. P. Katti, "Finite difference methods and their convergence for a class of singular two-point boundary value problems," *Numerische Mathematik*, vol. 39, no. 3, pp. 341–350, 1982.

[16] R. K. Pandey and A. K. Singh, "On the convergence of a finite difference method for a class of singular boundary value problems arising in physiology," *Journal of Computational and Applied Mathematics*, vol. 166, no. 2, pp. 553–564, 2004.

[17] R. K. Pandey and A. K. Singh, "On the convergence of finite difference methods for weakly regular singular boundary value problems," *Journal of Computational and Applied Mathematics*, vol. 205, no. 1, pp. 469–478, 2007.

[18] M. M. Chawla and C. P. Katti, "A uniform mesh finite difference method for a class of singular two-point boundary value problems," *SIAM Journal on Numerical Analysis*, vol. 22, no. 3, pp. 561–565, 1985.

[19] M. M. Chawla, "A fourth-order finite difference method based on uniform mesh for singular two-point boundary-value problems," *Journal of Computational and Applied Mathematics*, vol. 17, no. 3, pp. 359–364, 1987.

[20] M. M. Chawla, R. Subramanian, and H. L. Sathi, "A fourth-order spline method for singular two-point boundary value problems," *Journal of Computational and Applied Mathematics*, vol. 21, no. 2, pp. 189–202, 1988.

[21] M. M. Chawla, R. Subramanian, and H. L. Sathi, "A fourth order method for a singular two-point boundary value problem," *BIT Numerical Mathematics*, vol. 28, no. 1, pp. 88–97, 1988.

[22] R. K. Pandey and A. K. Singh, "On the convergence of fourth-order finite difference method for weakly regular singular boundary value problems," *International Journal of Computer Mathematics*, vol. 81, no. 2, pp. 227–238, 2004.

[23] R. K. Pandey and A. K. Singh, "A new high-accuracy difference method for a class of weakly nonlinear singular boundary-value problems," *International Journal of Computer Mathematics*, vol. 83, no. 11, pp. 809–817, 2006.

[24] R. K. Pandey and A. K. Singh, "On the convergence of a fourth-order method for a class of singular boundary value problems," *Journal of Computational and Applied Mathematics*, vol. 224, no. 2, pp. 734–742, 2009.

[25] P. Henrici, *Discrete Variable Methods in Ordinary Differential Equations*, John Wiley & Sons, New York, NY, USA, 1962.

8

A Class of Numerical Methods for the Solution of Fourth-Order Ordinary Differential Equations in Polar Coordinates

Jyoti Talwar and R. K. Mohanty

Department of Mathematics, Faculty of Mathematical Sciences, University of Delhi, 110007 Delhi, India

Correspondence should be addressed to R. K. Mohanty, rmohanty@maths.du.ac.in

Academic Editor: Alfredo Bermudez De Castro

In this piece of work using only three grid points, we propose two sets of numerical methods in a coupled manner for the solution of fourth-order ordinary differential equation $u^{iv}(x) = f(x, u(x), u'(x), u''(x), u'''(x))$, $a < x < b$, subject to boundary conditions $u(a) = A_0$, $u'(a) = A_1$, $u(b) = B_0$, and $u'(b) = B_1$, where A_0, A_1, B_0, and B_1 are real constants. We do not require to discretize the boundary conditions. The derivative of the solution is obtained as a byproduct of the discretization procedure. We use block iterative method and tridiagonal solver to obtain the solution in both cases. Convergence analysis is discussed and numerical results are provided to show the accuracy and usefulness of the proposed methods.

1. Introduction

Consider the fourth-order boundary value problem

$$u^{iv}(x) = f(x, u(x), u'(x), u''(x), u'''(x)), \quad a < x < b, \tag{1.1}$$

subject to the prescribed natural boundary conditions

$$u(0) = A_0, \qquad u'(0) = A_1, \qquad u(1) = B_0, \qquad u'(1) = B_1, \tag{1.2}$$

or equivalently, for $u'(x) = v(x)$,

$$u^{iv}(x) = f(x, u(x), v(x), u''(x), v''(x)), \quad a < x < b, \tag{1.3}$$

subject to the natural boundary conditions

$$u(0) = A_0, \qquad v(0) = A_1, \qquad u(1) = B_0, \qquad v(1) = B_1, \tag{1.4}$$

where A_0, A_1, B_0, and B_1 are real constants and $-\infty < a \le x \le b < \infty$.

Fourth-order differential equations occur in a number of areas of applied mathematics, such as in beam theory, viscoelastic and inelastic flows, and electric circuits. Some of them describe certain phenomena related to the theory of elastic stability. A classical fourth-order equation arising in the beam-column theory is the following (see Timoshenko [1]):

$$EI\frac{d^4u}{dx^4} + P\frac{d^2u}{dx^2} = q, \tag{1.5}$$

where u is the lateral deflection, q is the intensity of a distributed lateral load, P is the axial compressive force applied to the beam, and EI represents the flexural rigidity in the plane of bending. Various generalizations of the equation describing the deformation of an elastic beam with different types of two-point boundary conditions have been extensively studied via a broad range of methods.

The existence and uniqueness of solutions of boundary value problems are discussed in the papers and book of Agarwal and Krishnamoorthy, Agarwal and Akrivis (see [2–5]). Several authors have investigated solving fourth-order boundary value problem by some numerical techniques, which include the cubic spline method, Ritz method, finite difference method, multiderivative methods, and finite element methods (see [6–16]). In the 1980s, Usmani et al. (see [17–19]) worked on finite difference methods for solving $[p(x)y'']'' + q(x)y = r(x)$ and finite difference methods for computing eigenvalues of fourth-order linear boundary value problem. In 1984, Twizell and Tirmizi (see [20]) developed multi-derivative methods for linear fourth-order boundary value problems. In 1984, Agarwal and Chow (see [21]) developed iterative methods for a fourth-order boundary value problem. In 1991, O'Regan (see [13]) worked on the solvability of some fourth-(and higher) order singular boundary value problems. In 1994, Cabada (see [22]) developed the method of lower and upper solutions for fourth- and higher-order boundary value problems. In 2005 Franco et al. (see [23]) dealt with some fourth-order problems with nonlinear boundary conditions. In 2006, Noor and Mohyud-Din (see [12]) used the variational iteration method to solve fourth-order boundary value problems and further developed the homotopy perturbation method for solving fourth order boundary value problems. Some of these methods use transformation in order to reduce the equation into more simple equation or system of equations and some other methods give the solution in a series form which converges to the exact solution. Later, Han and Li (see [24]) worked on some fourth-order boundary value problems.

In this paper we present the finite difference methods for the solution of fourth-order boundary value problem. We discretize the interval [a,b] into $N + 1$ subintervals each of width $h = (b - a)/(N + 1)$, where N is a positive integer. We seek the solution of (la) or (2a) at the grid points, $x_k = kh$, $k = 1, 2, \ldots, N$. Let u_k and u'_k denote the approximate solutions, and let $U_k = u(x_k)$ and $U'_k = u'(x_k)$ be the exact solution values of $u(x)$ and $u'(x)$ at the grid point $x = x_k$, respectively. Also, $x_0 = a$ and $x_{N+1} = b$.

Using the second-order central differences, we obtain a five-point difference formula for (1.1), which requires the use of fictitious points outside [a,b]. The accuracy of the numerical solution depends upon the boundary approximation used. The finite difference method discussed here is based only on three grid points for second-and fourth-order methods. Therefore, no fictitious points are required for incorporating the boundary conditions. Here, we use a combination of the value of the solution $u(x)$ and its derivative $u'(x)$ to derive the difference scheme using three grid points.

Since we need to solve a coupled system of equations at each mesh point, the block successive overrelaxation (BSOR) iterative method is used.

2. The Finite Difference Method

The method is described as follows: for $k = 1(1)N$, let

$$\overline{u}_k'' = \frac{(u'_{k+1} - u'_{k-1})}{2h},$$

$$\overline{u}_k''' = \frac{(u'_{k+1} - 2u'_k + u'_{k-1})}{h^2}, \tag{2.1}$$

$$\overline{f}_k = f(x_k, u_k, u'_k, \overline{u}_k'', \overline{u}_k''').$$

Then, the difference method of order two for the given differential (1.1) is given by,

$$-2(u_{k+1} - 2u_k + u_{k-1}) + h(u'_{k+1} - u'_{k-1}) = \frac{h^4}{6}\overline{f}_k, \tag{2.2a}$$

and the corresponding difference method for the derivative $u'(x)$ is given by

$$-3(u_{k+1} - u_{k-1}) + h(u'_{k+1} + 4u'_k + u'_{k-1}) = 0. \tag{2.2b}$$

Also, let

$$\overline{u}_{k\pm1}'' = \frac{(\pm 3u'_{k\pm1} \mp 4u'_k \pm u'_{k\mp1})}{2h}, \tag{2.3}$$

$$\overline{\overline{u}}_k''' = \frac{15(u_{k+1} - u_{k-1})}{2h^3} - \frac{3(u'_{k+1} + 8u'_k + u'_{k-1})}{2h^2}, \tag{2.4}$$

$$\overline{\overline{u}}_{k\pm1}'' = \frac{-(11u_{k\pm1} - 16u_k + 5u_{k\mp1})}{2h^2} \pm \frac{(4u'_{k\pm1} - u'_{k\mp1})}{h}, \tag{2.5}$$

$$\overline{\overline{u}}_k'' = \frac{2(u_{k+1} - 2u_k + u_{k-1})}{h^2} - \frac{(u'_{k+1} - u'_{k-1})}{2h}, \tag{2.6}$$

$$\overline{u}_{k\pm1}''' = \mp\frac{(27u_{k\pm1} - 48u_k + 21u_{k\mp1})}{2h^3} + \frac{(15u'_{k\pm1} - 9u'_{k\mp1})}{2h^2}, \tag{2.7}$$

$$\overline{\overline{u}}_{k\pm1}''' = \mp\frac{(99u_{k\pm1} - 48u_k - 51u_{k\mp1})}{2h^3} + \frac{(39u'_{k\pm1} + 96u'_k + 15u'_{k\mp1})}{2h^2}, \tag{2.8}$$

and set

$$\overline{f}_{k\pm1} = f\left(x_{k\pm1}, u_{k\pm1}, u'_{k\pm1}, \overline{u}''_{k\pm1}, \overline{u}'''_{k\pm1}\right), \tag{2.9a}$$

$$\overline{\overline{f}}_{k\pm1} = f\left(x_{k\pm1}, u_{k\pm1}, u'_{k\pm1}, \overline{\overline{u}}''_{k\pm1}, \overline{\overline{u}}'''_{k\pm1}\right), \tag{2.9b}$$

$$\overline{\overline{f}}_k = f\left(x_k, u_k, u'_k, \overline{\overline{u}}''_k, \overline{\overline{u}}'''_k\right). \tag{2.9c}$$

Then, the difference method of order four for the differential equation and the corresponding difference method for the derivative u'_k are given by

$$-2(u_{k+1} - 2u_k + u_{k-1}) + h(u'_{k+1} - u'_{k-1}) = \frac{h^4}{90}\left(\overline{\overline{f}}_{k+1} + \overline{\overline{f}}_{k-1} + 13\overline{\overline{f}}_k\right), \tag{2.10}$$

$$-3(u_{k+1} - u_{k-1}) + h(u'_{k+1} + 4u'_k + u'_{k-1}) = \frac{h^4}{60}\left(\overline{f}_{k+1} - \overline{f}_{k-1}\right). \tag{2.11}$$

Note that u_0, u'_0, u_{N+1} and u'_{N+1}, are prescribed. It is convenient to express the above finite difference schemes in block tridiagonal matrix form. If the differential equation is linear, the resulting block tridiagonal linear system can be solved using the block Gauss-Seidel (BGS) iterative method. If the differential equation is nonlinear, the system can be solved using the Newton nonlinear block successive overrelaxation (NBSOR) method (see [25, 26]).

3. Derivation of the Difference Scheme

For the derivation of the method we follow the approaches given by Chawla [27] and Mohanty [10]. In this section we discuss the derivation of the difference methods and the block iterative methods.

At the grid point x_k, the given differential equation can be written as

$$u_k^{iv} = f\left(x_k, u_k, u'_k, u''_k, u'''_k\right) = f_k, \quad k = 1(1)N. \tag{3.1a}$$

Similarly,

$$f_{k\pm1} = f\left(x_{k\pm1}, u_{k\pm1}, u'_{k\pm1}, u''_{k\pm1}, u'''_{k\pm1}\right), \quad k = 1(1)N. \tag{3.1b}$$

Using the Taylor expansion about the grid point x_k, we first obtain

$$-2(u_{k+1} - 2u_k + u_{k-1}) + h(u'_{k+1} - u'_{k-1}) = \frac{h^4}{90}(f_{k+1} + f_{k-1} + 13f_k) + T_1, \tag{3.2}$$

where $T_1 = O(h^8)$.

Now, we need the $O(h^2)$ approximation for $u'''_{k\pm1}$. Let

$$\overline{u}'''_{k+1} = \frac{1}{h^3}[a_{10}u_k + a_{11}u_{k+1} + a_{12}u_{k-1}] + \frac{1}{h^2}[b_{10}u'_k + b_{11}u'_{k+1} + b_{12}u'_{k-1}]. \tag{3.3}$$

Expanding each term on the right-hand side of (3.3) in the Taylor series about the point x_k and equating the coefficients of h^p ($p = -3, -2, -1,\ 0$, and 1) to zero, we get

$$(a_{10}, a_{11}, a_{12}, b_{10}, b_{11}, b_{12}) = \left(24, \frac{-27}{2}, \frac{-21}{2}, 0, \frac{15}{2}, \frac{-9}{2}\right). \tag{3.4}$$

Thus, we obtain

$$\overline{u}'''_{k+1} = -\frac{1}{2h^3}[27u_{k+1} - 48u_k + 21u_{k-1}] + \frac{1}{2h^2}[15u'_{k+1} - 9u'_{k-1}]$$

$$= u'''_{k+1} - \frac{2h^2}{5}u^v_k + O\!\left(h^3\right). \tag{3.5a}$$

Similarly, we obtain

$$\overline{u}'''_{k-1} = \frac{1}{2h^3}[27u_{k-1} - 48u_k + 21u_{k+1}] + \frac{1}{2h^2}[15u'_{k-1} - 9u'_{k+1}]$$

$$= u'''_{k-1} - \frac{2h^2}{5}u^v_k - O\!\left(h^3\right). \tag{3.5b}$$

Further, from (2.3) we obtain:

$$\overline{u}''_{k\pm1} = u''_{k\pm1} - \frac{h^2}{3}u^{iv}_k \pm O\!\left(h^3\right). \tag{3.6}$$

Hence, we can verify that $\overline{f}_{k\pm1} = f_{k\pm1} + O(h^2)$.

Next, we obtain the $O(h^4)$ approximation for u''_k. Let

$$\overline{\overline{u}}''_k = \frac{1}{h^2}[a_{20}u_k + a_{21}u_{k+1} + a_{22}u_{k-1}] + \frac{1}{h}[b_{20}u'_k + b_{21}u'_{k+1} + b_{22}u'_{k-1}]. \tag{3.7}$$

Using the Taylor series expansion and equating the coefficients of h^p ($p = -2,\ -1,\ 0,\ 1,\ 2$, and 3) to zero, we get

$$(a_{20}, a_{21}, a_{22}, b_{20}, b_{21}, b_{22}) = \left(-4, 2, 2, 0, \frac{-1}{2}, \frac{1}{2}\right). \tag{3.8}$$

Therefore,

$$\bar{\bar{u}}_k'' = \frac{2}{h^2}[u_{k+1} - 2u_k + u_{k-1}] - \frac{1}{2h}[u_{k+1}' - u_{k-1}'] = u_k'' + O\left(h^4\right). \tag{3.9}$$

Similarly,

$$\bar{\bar{u}}_k''' = \frac{15}{2h^3}[u_{k+1} - u_{k-1}] - \frac{3}{2h^2}[u_{k+1}' + 8u_k' + u_{k-1}'] = u_k''' + O\left(h^4\right), \tag{3.10}$$

$$\bar{\bar{u}}_{k\pm1}'' = \frac{-1}{2h^2}[11u_{k\pm1} - 16u_k + 5u_{k\mp1}] \pm \frac{1}{h}[4u_{k\pm1}' - u_{k\mp1}'] = u_{k\pm1}'' \mp \frac{h^3}{15}u_k^v + O\left(h^4\right). \tag{3.11}$$

Next, we obtain the $O(h^3)$ approximation for u_{k+1}'''. Let

$$\bar{\bar{u}}_{k+1}''' = \frac{1}{h^3}[a_{30}u_k + a_{31}u_{k+1} + a_{32}u_{k-1}] + \frac{1}{h^2}[b_{30}u_k' + b_{31}u_{k+1}' + b_{32}u_{k-1}']. \tag{3.12}$$

Equating the coefficients of h^p ($p = -3,-2,-1,0,\ 1$, and 2) to zero, we get

$$(a_{30}, a_{31}, a_{32}, b_{30}, b_{31}, b_{32}) = \left(24, \frac{-99}{2}, \frac{51}{2}, 48, \frac{39}{2}, \frac{15}{2}\right). \tag{3.13}$$

Thus, we obtain

$$\bar{\bar{u}}_{k+1}''' = \frac{-1}{2h^3}[99u_{k+1} - 48u_k - 51u_{k-1}] + \frac{1}{2h^2}[39u_{k+1}' + 96u_k' + 15u_{k-1}']$$
$$= u_{k+1}''' - \frac{h^3}{10}u_k^{vi} + O\left(h^4\right). \tag{3.14}$$

Similarly,

$$\bar{\bar{u}}_{k-1}''' = \frac{1}{2h^3}[99u_{k-1} - 48u_k - 51u_{k+1}] + \frac{1}{2h^2}[39u_{k-1}' + 96u_k' + 15u_{k+1}']$$
$$= u_{k-1}''' + \frac{h^3}{10}u_k^{vi} + O\left(h^4\right). \tag{3.15}$$

Let

$$\alpha_k = \frac{\partial f}{\partial u_k''}, \qquad \beta_k = \frac{\partial f}{\partial u_k'''}. \tag{3.16}$$

From (3.5a), (3.5b), (3.6), and (2.9a) it follows that $\overline{f}_{k\pm1}$ provides the $O(h^2)$ approximation for $f_{k\pm1}$ and

$$\overline{f}_{k\pm1} = f_{k\pm1} - \frac{h^2}{3}u_k^{iv}\alpha_k - \frac{2h^2}{5}u_k^{v}\beta_k \pm O\left(h^3\right). \tag{3.17}$$

Also, from (2.9b), (3.11), (3.14), and (3.15) we obtain the $O(h^3)$ approximation for $f_{k\pm1}$:

$$\overline{\overline{f}}_{k\pm1} = f_{k\pm1} \mp \frac{h^3}{15}u_k^{v}\alpha_k \mp \frac{h^3}{10}u_k^{vi}\beta_k + O\left(h^4\right). \tag{3.18}$$

From (2.9c), (3.9), and (3.10) it follows that $\overline{\overline{f}}_k$ provides an $O(h^4)$ approximation for f_k

$$\overline{\overline{f}}_k = f_k + O\left(h^4\right). \tag{3.19}$$

From (3.18) and (3.19) we get

$$\overline{\overline{f}}_{k+1} + \overline{\overline{f}}_{k-1} + 13\overline{\overline{f}}_k = f_{k+1} + f_{k-1} + 13f_k + O\left(h^4\right). \tag{3.20}$$

With the help of (3.2) and (3.20), from (2.10), we obtain that the local truncation error associated with the difference scheme (2.10) is of $O(h^8)$. Similarly, we can verify that the local truncation error associated with the difference scheme (2.11) is of $O(h^7)$.

Thus, we have the following result.

Theorem 3.1. *Consider the fourth-order nonlinear ordinary differential equation (1.1) along with the boundary conditions (1.2). Let $u \in C^8[a,b]$, and the function f is sufficiently differentiable with respect to its arguments. Then, the difference methods (2.10) and (2.11) with the approximations of u'' and u''' listed in (2.3)–(2.8) are of $O(h^4)$.*

If the differential (1.1) is linear, then the difference method (2.10) and (2.11) in the matrix form can be written as

$$\begin{bmatrix} \mathbf{A}_{11} & \mathbf{A}_{12} \\ \mathbf{A}_{21} & \mathbf{A}_{22} \end{bmatrix}\begin{bmatrix} \mathbf{u} \\ \mathbf{u}' \end{bmatrix} = \begin{bmatrix} \mathbf{d}_1 \\ \mathbf{d}_2 \end{bmatrix}, \tag{3.21}$$

where $\mathbf{A}_{11}$, $\mathbf{A}_{12}$, $\mathbf{A}_{21}$, and $\mathbf{A}_{22}$ are the Nth-order tri-diagonal matrices and $\mathbf{d}_1$ and $\mathbf{d}_2$ are vectors consisting of right-hand side functions and some boundary conditions associated with the block system.

The block successive over relaxation (BSOR) method (See Mohanty and Evans [26, 28]) is given by

$$\mathbf{A}_{11}\mathbf{u}^{(n+1)} = \omega\left[-\mathbf{A}_{12}(\mathbf{u}')^{(n)} + \mathbf{d}_1\right] + (1-\omega)\mathbf{A}_{11}\mathbf{u}^{(n)}, \quad n = 0,1,2,\ldots,$$

$$\mathbf{A}_{22}(\mathbf{u}')^{(n+1)} = \omega\left[-\mathbf{A}_{21}\mathbf{u}^{(n+1)} + \mathbf{d}_2\right] + (1-\omega)\mathbf{A}_{22}(\mathbf{u}')^{(n)}, \quad n = 0,1,2,\ldots, \tag{3.22}$$

where ω is a parameter known as relaxation parameter. With $\omega = 1$, the BSOR method reduces to block Gauss-Seidel (BGS) method. If $\omega > 1$ or $\omega < 1$, we have overrelaxation or underrelaxation, respectively.

If $f(x, u(x), v(x), u'(x), v'(x))$ is nonlinear, the difference equations (2.2a), (2.2b) or (2.10), (2.11) form a coupled nonlinear system. To solve the coupled nonlinear system we apply the Newton NBSOR method.

We first write the difference equations (2.2a), (2.2b) or (2.10), (2.11) as

$$\boldsymbol{\Phi}(\mathbf{u}, \mathbf{v}) = 0,$$

$$\boldsymbol{\Psi}(\mathbf{u}, \mathbf{v}) = 0,$$

$$(3.23)$$

where

$$\mathbf{u} = \begin{bmatrix} u_1 \\ u_2 \\ \vdots \\ u_N \end{bmatrix}, \quad \mathbf{v} = \begin{bmatrix} v_1 \\ v_2 \\ \vdots \\ v_N \end{bmatrix}, \quad \boldsymbol{\Phi}(\mathbf{u}, \mathbf{v}) = \begin{bmatrix} \phi_1(\mathbf{u}, \mathbf{v}) \\ \phi_2(\mathbf{u}, \mathbf{v}) \\ \vdots \\ \phi_N(\mathbf{u}, \mathbf{v}) \end{bmatrix}, \quad \boldsymbol{\Psi}(\mathbf{u}, \mathbf{v}) = \begin{bmatrix} \psi_1(\mathbf{u}, \mathbf{v}) \\ \psi_2(\mathbf{u}, \mathbf{v}) \\ \vdots \\ \psi_N(\mathbf{u}, \mathbf{v}) \end{bmatrix}.$$

$$(3.24)$$

Let

$$\mathbf{J} = \begin{bmatrix} \mathbf{T}_{11} & \mathbf{T}_{12} \\ \mathbf{T}_{21} & \mathbf{T}_{22} \end{bmatrix} \qquad (3.25)$$

be the Jacobian of $\boldsymbol{\Phi}$ and $\boldsymbol{\Psi}$, which is the $2N$th-order block tridiagonal matrix, where

$$\mathbf{T}_{11} = \frac{\partial(\phi_1, \phi_2, \ldots, \phi_N)}{\partial(u_1, u_2, \ldots, u_N)} = \begin{bmatrix} \dfrac{\partial \phi_1}{\partial u_1} & \dfrac{\partial \phi_1}{\partial u_2} & & & \\ \dfrac{\partial \phi_2}{\partial u_1} & \dfrac{\partial \phi_2}{\partial u_2} & \dfrac{\partial \phi_2}{\partial u_3} & & \\ & 0 & & \dfrac{\partial \phi_N}{\partial u_{N-1}} & \dfrac{\partial \phi_N}{\partial u_N} \end{bmatrix},$$

$$\mathbf{T}_{12} = \frac{\partial(\phi_1, \phi_2, \ldots, \phi_N)}{\partial(v_1, v_2, \ldots, v_N)},$$

$$T_{21} = \frac{\partial(\psi_1, \psi_2, \ldots, \psi_N)}{\partial(u_1, u_2, \ldots, u_N)},$$

$$T_{22} = \frac{\partial(\psi_1, \psi_2, \ldots, \psi_N)}{\partial(v_1, v_2, \ldots, v_N)}$$

$$(3.26)$$

are the Nth-order tridiagonal matrices.

In the NBSOR method starting with any initial approximation $(\mathbf{u}^{(0)}, \mathbf{v}^{(0)})$ of $(\mathbf{u}^{(k)}, \mathbf{v}^{(k)})$, $k = 1(1)N$, we define

$$\mathbf{u}^{(n+1)} = \mathbf{u}^{(n)} + \Delta\mathbf{u}^{(n)}, \quad n = 0, 1, 2, \ldots,$$

$$\mathbf{v}^{(n+1)} = \mathbf{v}^{(n)} + \Delta\mathbf{v}^{(n)}, \quad n = 0, 1, 2, \ldots,$$

$$(3.27)$$

where $\Delta\mathbf{u}$ and $\Delta\mathbf{v}$ are intermediate values obtained by solving the matrix equation for NBSOR method given by

$$\begin{bmatrix} \mathbf{T}_{11} & \mathbf{T}_{12} \\ \mathbf{T}_{21} & \mathbf{T}_{22} \end{bmatrix} \begin{bmatrix} \Delta\mathbf{u} \\ \Delta\mathbf{v} \end{bmatrix} = \begin{bmatrix} -\boldsymbol{\Phi} \\ -\boldsymbol{\Psi} \end{bmatrix}.$$

$$(3.28)$$

The above system can be solved for $\Delta\mathbf{u}^{(n)}$ and $\Delta\mathbf{v}^{(n)}$ by using the block SOR method (inner iterative method) as follows:

$$\mathbf{T}_{11}\Delta\mathbf{u}^{(n+1)} = \omega\left[-\boldsymbol{\Phi}\left(\mathbf{u}^{(n)}, \mathbf{v}^{(n)}\right) - \mathbf{T}_{12}\Delta\mathbf{v}^{(n)}\right] + (1 - \omega)\mathbf{T}_{11}\Delta\mathbf{u}^{(n)}, \quad n = 0, 1, 2, \ldots,$$

$$\mathbf{T}_{22}\Delta\mathbf{v}^{(n+1)} = \omega\left[-\boldsymbol{\Psi}\left(\mathbf{u}^{(n)}, \mathbf{v}^{(n)}\right) - \mathbf{T}_{21}\Delta\mathbf{u}^{(n+1)}\right] + (1 - \omega)\mathbf{T}_{22}\Delta\mathbf{v}^{(n)}, \quad n = 0, 1, 2, \ldots,$$

$$(3.29)$$

where $\omega \in (0, 2)$ is a relaxation parameter and $n = 0, 1, 2, \ldots$. The above system of equations can be solved by using the tridiagonal solver. In order for this method to converge it is sufficient that the initial approximation $(\mathbf{u}^{(0)}, \mathbf{v}^{(0)})$ be close to the solution.

4. Convergence and Stability Analysis

Consider the model problem $u^{iv} = f(x)$, where f is a function of x only. Applying the fourth-order difference methods (2.10) and (2.11) to the above equation, we get

$$(u_{k-1} - 2u_k + u_{k+1}) + \frac{h}{2}(v_{k-1} - v_{k+1}) = \frac{-h^4}{180}(f_{k+1} + f_{k-1} + 13f_k),$$

$$\frac{3}{h}(u_{k-1} - u_{k+1}) + (v_{k-1} + 4v_k + v_{k+1}) = \frac{h^3}{60}(f_{k+1} - f_{k-1}),$$

$$(4.1)$$

where we denote $u' = v$.

Let us denote by $\mathbf{P} = [1, 0, 1]$, $\mathbf{L} = [1, 1, 1]$, and $\mathbf{M} = [1, 0, -1]$ the Nth-order tridiagonal matrices.

The system of (4.1) can be written in the block form as

$$
\begin{bmatrix}
(\mathbf{L} - 3\mathbf{I}) & \dfrac{h}{2}\mathbf{M} \\
\dfrac{3}{h}\mathbf{M} & (\mathbf{L} + 3\mathbf{I})
\end{bmatrix}
\begin{bmatrix}
\mathbf{u} \\
\mathbf{v}
\end{bmatrix}
=
\begin{bmatrix}
\mathbf{d}_1 \\
\mathbf{d}_2
\end{bmatrix},
\tag{4.2}
$$

where $\mathbf{u}$ and $\mathbf{v}$ are two $\mathbf{N}$-dimensional solution vectors and $\mathbf{d}_1$, $\mathbf{d}_2$ are vectors consisting of right-hand side functions and some boundary values associated with (4.1). The BSOR method for the scheme is

$$
\mathbf{u}^{(n+1)} = (1 - \omega)\mathbf{u}^{(n)} - \frac{\omega h}{2}(\mathbf{L} - 3\mathbf{I})^{-1}\mathbf{M}(\mathbf{v})^{(n)} + \omega(\mathbf{L} - 3\mathbf{I})^{-1}\mathbf{d}_1,
$$

$$
(\mathbf{v})^{(n+1)} = (1 - \omega)(\mathbf{v})^{(n)} - \frac{3\omega}{h}(\mathbf{L} + 3\mathbf{I})^{-1}\mathbf{M}\mathbf{u}^{(n+1)} + \omega(\mathbf{L} + 3\mathbf{I})^{-1}\mathbf{d}_2,
\tag{4.3}
$$

The associated block SOR and block Jacobi iteration matrices are given by

$$
\mathbf{L}_\omega =
\begin{bmatrix}
(1 - \omega)\mathbf{I} & \dfrac{-\omega h}{2}(\mathbf{L} - 3\mathbf{I})^{-1}\mathbf{M} \\
\dfrac{-3\omega}{h}(\mathbf{L} + 3\mathbf{I})^{-1}\mathbf{M} & (1 - \omega)\mathbf{I}
\end{bmatrix},
$$

$$
\mathbf{B} =
\begin{bmatrix}
0 & \dfrac{-h}{2}(\mathbf{L} - 3\mathbf{I})^{-1}\mathbf{M} \\
\dfrac{-3}{h}(\mathbf{L} + 3\mathbf{I})^{-1}\mathbf{M} & 0
\end{bmatrix},
\tag{4.4}
$$

and λ and η are the eigenvalues associated with the corresponding matrices $\mathbf{L}_\omega$ and $\mathbf{B}$, which are related by the equation

$$
(\lambda + \omega - 1)^2 = \lambda \omega^2 \eta^2.
\tag{4.5}
$$

Let $\begin{bmatrix} \mathbf{v}_1 \\ \mathbf{v}_2 \end{bmatrix}$ be the eigenvector associated with the eigenvalue η so that

$$
\begin{bmatrix}
0 & \dfrac{-h}{2}(\mathbf{L} - 3\mathbf{I})^{-1}\mathbf{M} \\
\dfrac{-3}{h}(\mathbf{L} + 3\mathbf{I})^{-1}\mathbf{M} & 0
\end{bmatrix}
\begin{bmatrix}
\mathbf{v}_1 \\
\mathbf{v}_2
\end{bmatrix}
=
\begin{bmatrix}
\mathbf{v}_1 \\
\mathbf{v}_2
\end{bmatrix}
\eta.
\tag{4.6}
$$

That is,

$$
\frac{-h}{2}(\mathbf{L} - 3\mathbf{I})^{-1}\mathbf{M}\mathbf{v}_2 = \eta \mathbf{v}_1,
$$

$$
\frac{-3}{h}(\mathbf{L} + 3\mathbf{I})^{-1}\mathbf{M}\mathbf{v}_1 = \eta \mathbf{v}_2.
\tag{4.7}
$$

On eliminating $\mathbf{v}_2$, we obtain

$$\frac{3}{2}(\mathbf{L} - 3\mathbf{I})^{-1}\mathbf{M}(\mathbf{L}+3\mathbf{I})^{-1}M\mathbf{v}_1 = \eta^2\mathbf{v}_1. \tag{4.8}$$

The rate of convergence of the BSOR method is given by $-\mathbf{log}(\rho(\mathbf{L}_\omega))$.

The rate of convergence of the BSOR method is dependent on the eigenvalues of B (through relation (4.5)), which are given by $(3\tau/2) = \eta^2$, where τ are the eigenvalues of $(\mathbf{L} - 3\mathbf{I})^{-1}\mathbf{M}(\mathbf{L} + 3\mathbf{I})^{-1}\mathbf{M}$.

Hence, we can determine the optimal parameter as

$$\omega_0 = \frac{2}{1 + \sqrt{1 - (3/2)\bar{\tau}}}, \tag{4.9}$$

where $\bar{\tau} = S((\mathbf{L} - 3\mathbf{I})^{-1}\mathbf{M}(\mathbf{L} + 3\mathbf{I})^{-1}\mathbf{M})$.

Thus, we can determine the convergence factor

$$\bar{\lambda} = \omega_0 - 1 = \frac{1 - \sqrt{1 - (3/2)\bar{\tau}}}{1 + \sqrt{1 - (3/2)\bar{\tau}}}. \tag{4.10}$$

For convergence, we must have $|\bar{\lambda}| < 1$ to give the range

$$0 < \bar{\tau} < \frac{2}{3} \tag{4.11}$$

Hence, we get the convergence.

Thus, we have the following result.

Theorem 4.1. *The iteration method of the form (4.3) for the solution of $u^{iv} = f(x)$ converges if $0 < \bar{\tau} < 2/3$, where $\bar{\tau} = S((\mathbf{L} - 3\mathbf{I})^{-1}\mathbf{M}(\mathbf{L} + 3\mathbf{I})^{-1}\mathbf{M})$, $\mathbf{L} = [1, 1, 1]$, and $\mathbf{M} = [1, 0, -1]$ are the $N \times N$ tridiagonal matrices.*

Now, we discuss the stability analysis.

An iterative method for (4.1) can be written as

$$\mathbf{u}^{(k+1)} = \frac{1}{2}\mathbf{P}\mathbf{u}^{(k)} + \frac{h}{4}\mathbf{M}\mathbf{v}^{(k)} + \mathbf{RHU},$$

$$\mathbf{v}^{(k+1)} = \frac{-3}{4h}\mathbf{M}\mathbf{u}^{(k)} - \frac{1}{4}\mathbf{P}\mathbf{v}^{(k)} + \mathbf{RHV}, \tag{4.12}$$

where $\mathbf{u}^{(k)}$, $\mathbf{v}^{(k)}$ are solution vectors at the kth iteration and $\mathbf{RHU}$, $\mathbf{RHV}$ are right-hand side vectors consisting of boundary and homogenous function values.

The above iterative method in matrix form can be written as

$$\begin{bmatrix} \mathbf{u}^{(k+1)} \\ \mathbf{v}^{(k+1)} \end{bmatrix} = \mathbf{G}\begin{bmatrix} \mathbf{u}^{(k)} \\ \mathbf{v}^{(k)} \end{bmatrix} + \mathbf{RH}, \tag{4.13}$$

where

$$
\mathbf{G} = \begin{bmatrix} \dfrac{1}{2}\mathbf{P} & \dfrac{h}{4}\mathbf{M} \\ \dfrac{-3}{4h}\mathbf{M} & \dfrac{-1}{4}\mathbf{P} \end{bmatrix}, \quad \mathbf{RH} = \begin{bmatrix} \mathbf{RHU} \\ \mathbf{RHV} \end{bmatrix}. \tag{4.14}
$$

The eigenvalues of $\mathbf{P}$ and $\mathbf{M}$ are $2\cos k\pi/(N+1)$ and $2i\cos k\pi/(N+1)$, respectively, where $k = 1, 2, \ldots, N$. The characteristic equation of the matrix $\mathbf{G}$ is given by

$$
\det \begin{bmatrix} \dfrac{1}{2}\mathbf{P} - \xi\mathbf{I}_N & \dfrac{h}{4}\mathbf{M} \\ \dfrac{-3}{4h}\mathbf{M} & \dfrac{-1}{4}\mathbf{P} - \xi\mathbf{I}_N \end{bmatrix} = 0. \tag{4.15}
$$

Thus the eigenvalues of $\mathbf{G}$ are given by

$$
\det\left[\frac{-1}{4}\mathbf{P} - \xi\mathbf{I}_N\right] \times \det\left[\left(\frac{1}{2}\mathbf{P} - \xi\mathbf{I}_N\right) + \frac{3}{16}\mathbf{M}\left(\frac{-1}{4}\mathbf{P} - \xi\mathbf{I}_N\right)^{-1}\mathbf{M}\right] = 0. \tag{4.16}
$$

The proposed iterative method (4.13) is stable as long as the maximum absolute eigenvalues of the iteration matrix are less than or equal to one. It has been verified computationally that all eigenvalues of the system (4.16) are less than one. Hence, the iterative method (4.1) is stable.

5. Application to Singular Equation

Consider a singular fourth-order linear ordinary differential equation of the form

$$
\Delta^4 u \equiv \left(\frac{d^2}{dr^2} + \frac{1}{r}\frac{d}{dr}\right)^2 u = f(r), \quad 0 < r < 1, \tag{5.1}
$$

or equivalently

$$
u^{iv} = b(r)u''' + c(r)u'' + d(r)u' + f(r), \quad 0 < r < 1, \tag{5.2}
$$

where

$$
b(r) = -2/r, \quad c(r) = 1/r^2, \quad d(r) = -1/r^3. \tag{5.3}
$$

The above equation represents fourth-order ordinary differential equation in cylindrical polar coordinates.

The boundary conditions are given by

$$
u(0) = A_0, \quad u'(0) = A_1, \quad u(1) = B_0, \quad u'(1) = B_1, \tag{5.4}
$$

where A_0, A_1, B_0, and B_1 are constants.

Applying the difference scheme (2.2a), (2.2b) to the singular equation (5.2), we obtain a second-order difference method

$$-2(u_{k+1} - 2u_k + u_{k-1}) + h(u'_{k+1} - u'_{k-1}) = \frac{h^4}{6}(b_k\overline{u}'''_k + c_k\overline{u}''_k + d_k u'_k + f_k),$$

$$-3(u_{k+1} - u_{k-1}) + h(u'_{k+1} + 4u'_k + u'_{k-1}) = 0, \qquad k = 1(1)N.$$

(5.5)

Applying the fourth-order difference scheme (2.10) to the singular equation (5.2), we obtain

$$
\begin{aligned}
&- 2(u_{k+1} - 2u_k + u_{k-1}) + h(u'_{k+1} - u'_{k-1})\\
&= \frac{h^4}{90}\Big[\Big(b_{k+1}\overline{\overline{u}}'''_{k+1} + c_{k+1}\overline{\overline{u}}''_{k+1} + d_{k+1}u'_{k+1} + f_{k+1}\Big)\\
&\qquad + \Big(b_{k-1}\overline{\overline{u}}'''_{k-1} + c_{k-1}\overline{\overline{u}}''_{k-1} + d_{k-1}u'_{k-1} + f_{k-1}\Big)\\
&\qquad + \Big(13b_k\overline{\overline{u}}'''_k + 13c_k\overline{\overline{u}}''_k + 13d_k u'_k + 13f_k\Big)\Big], \qquad k = 1(1)N,
\end{aligned}
$$

(5.6)

where

$$b_{k\pm1} = b(r_{k\pm1}), \qquad c_{k\pm1} = c(r_{k\pm1}), \qquad d_{k\pm1} = d(r_{k\pm1}), \qquad f_{k\pm1} = f(r_{k\pm1}).$$

(5.7)

Note that the scheme fails when the solution is to be determined at $k = 1$. We overcome the difficulty by modifying the method in such a way that the solutions retain the order and accuracy even in the vicinity of the singularity $r = 0$ (see [29]). We consider the following approximations:

$$b_{k\pm1} = b_k \pm hb'_k + \frac{h^2}{2!}b''_k + O\big(\pm h^3\big),$$

$$c_{k\pm1} = c_k \pm hc'_k + \frac{h^2}{2!}c''_k + O\big(\pm h^3\big),$$

(5.8)

$$d_{k\pm1} = d_k \pm hd'_k + \frac{h^2}{2!}d''_k + O\big(\pm h^3\big),$$

$$f_{k\pm1} = f_k \pm hf'_k + \frac{h^2}{2!}f''_k + O\big(\pm h^3\big).$$

Using the approximation (5.8) in (5.6) and neglecting higher-order terms, we can rewrite (5.6) in compact operator form as

$$
\begin{aligned}
&- 2(u_{k+1} - 2u_k + u_{k-1}) + h(u'_{k+1} - u'_{k-1})\\
&= \frac{h^2}{90}\big(-24b'_k + 18c_k - 4h^2 c''_k\big)\delta_x^2 u_k
\end{aligned}
$$

$$+ \frac{h}{180}\left(45b_k - 75h^2 b_k'' - 6h^2 c_k'\right)(2\mu_x \delta_x)u_k$$

$$+ \frac{h^4}{90}\left(\frac{-45b_k}{h^2} + 75b_k'' + 6c_k' + 15d_k + h^2 d_k''\right)u_k'$$

$$+ \frac{h^2}{180}\left(15b_k + 27h^2 b_k'' + 6h^2 c_k' + 2h^2 d_k\right)(2\mu_x \delta_x)u_k'$$

$$+ \frac{h^3}{180}\left(24b_k' - 3c_k + 5h^2 c_k'' + 2h^2 d_k'\right)(2\mu_x \delta_x)u_k$$

$$+ \frac{h^4}{90}\left(15f_k + h^2 f_k''\right), \quad k = 1(1)N.$$

$$(5.9)$$

Similarly, using the difference scheme (2.11), a fourth-order approximation for the derivative u' for the singular equation (5.2) in the compact form may be written as

$$-3(u_{k+1} - u_{k-1}) + h\left(u_{k+1}' + 4u_k' + u_{k-1}'\right)$$

$$= \frac{h}{60}(-24b_k)\delta_x^2 u_k + \frac{h^2}{60}(-3b_k')(2\mu_x \delta_x)u_k + \frac{h^3}{60}(2c_k + 3b_k')\delta_x^2 u_k'$$

$$+ \frac{h^2}{60}\left(12b_k + h^2(c_k' + d_k)\right)(2\mu_x \delta_x)u_k' + \frac{h^3}{60}\left(6b_k' + 2h^2 d_k'\right)u_k' + \frac{h^4}{60}(2hf_k'), \quad k = 1(1)N,$$

$$(5.10)$$

where $\delta_r u_k = u_{k+1/2} - u_{k-1/2}, \mu_r u_k = (u_{k+1/2} + u_{k-1/2})/2$.

Finite difference equations (5.9) and (5.10) along with the boundary conditions (5.4) gives a $2N \times 2N$ linear system of equations for the unknowns $u_1, u_2, \ldots, u_N, u_1', u_2', \ldots, u_N'$. The resulting block tri-diagonal system can be solved using the BGS method. The schemes (5.9) and (5.10) are free from the terms $1/(k \pm 1)$, hence very easily solved for $k = 1(1)N$ in the region (0,1).

Consider the coupled nonlinear singular equations

$$u^{IV} = a(r)\left[u'v'' + v'u''\right] + f(r), \quad 0 < r < 1,$$

$$v^{IV} = -a(r)u'u'' + g(r), \quad 0 < r < 1,$$

$$(5.11)$$

where $a(r) = 1/r$, with known boundary conditions $u(0)$, $v(0)$, $u'(0)$, $v'(0)$, $u(1)$, $v(1)$, $u'(1)$, and $v'(1)$. The coupled equations represent model equations of equilibrium for a load symmetrical about the centre (see [30]).

The second-order difference scheme for solving the system (5.11) is given by

$$-2(u_{k+1} - 2u_k + u_{k-1}) + h\left(u_{k+1}' - u_{k-1}'\right) = \frac{h^4}{6}\left[a_k(u_k'\overline{v}_k'' + v_k'\overline{u}_k'') + f_k\right],$$

$$-2(v_{k+1} - 2v_k + v_{k-1}) + h\left(v_{k+1}' - v_{k-1}'\right) = \frac{h^4}{6}\left[-a_k u_k'\overline{u}_k'' + g_k\right],$$

$$-3(u_{k+1} - u_{k-1}) + h(u'_{k+1} + 4u'_k + u'_{k-1}) = 0,$$

$$-3(v_{k+1} - v_{k-1}) + h(v'_{k+1} + 4v'_k + v'_{k-1}) = 0,$$

$$(5.12)$$

where $a_k = a(r_k)$, $f_k = f(r_k)$, and $g_k = g(r_k)$.

The fourth-order difference scheme for solving the system (5.11) for u, v, u', and v' is given by

$$-2(u_{k+1} - 2u_k + u_{k-1}) + h(u'_{k+1} - u'_{k-1})$$

$$= H_1\left[15f_k + h^2 f''_k\right]$$

$$+ H_1 a_{11}\left[u'_{k+1}\left\{\frac{-1}{2h^2}(11v_{k+1} - 16v_k + 5v_{k-1}) + \frac{1}{h}(4v'_{k+1} - v'_{k-1})\right\}\right.$$

$$\left. + v'_{k+1}\left\{\frac{-1}{2h^2}(11u_{k+1} - 16u_k + 5u_{k-1}) + \frac{1}{h}(4u'_{k+1} - u'_{k-1})\right\}\right]$$

$$+ H_1 a_{12}\left[u'_{k-1}\left\{\frac{-1}{2h^2}(11v_{k-1} - 16v_k + 5v_{k+1}) - \frac{1}{h}(4v'_{k-1} - v'_{k+1})\right\}\right.$$

$$\left. + v'_{k-1}\left\{\frac{-1}{2h^2}(11u_{k-1} - 16u_k + 5u_{k+1}) - \frac{1}{h}(4u'_{k-1} - u'_{k+1})\right\}\right]$$

$$+ H_1 a_{10}\left[u'_k\left\{\frac{2}{h^2}(v_{k+1} - 2v_k + v_{k-1}) - \frac{1}{2h}(v'_{k+1} - v'_{k-1})\right\}\right.$$

$$\left. + v'_k\left\{\frac{2}{h^2}(u_{k+1} - 2u_k + u_{k-1}) - \frac{1}{2h}(u'_{k+1} - u'_{k-1})\right\}\right],$$

$$-2(v_{k+1} - 2v_k + v_{k-1}) + h(v'_{k+1} - v'_{k-1})$$

$$= H_1\left[15g_k + h^2 g''_k\right] - H_1 a_{11} u'_{k+1}\left[\frac{-1}{2h^2}(11u_{k+1} - 16u_k + 5u_{k-1}) + \frac{1}{h}(4u'_{k+1} - u'_{k-1})\right]$$

$$- H_1 a_{12} u'_{k-1}\left[\frac{-1}{2h^2}(11u_{k-1} - 16u_k + 5u_{k+1}) - \frac{1}{h}(4u'_{k-1} - u'_{k+1})\right]$$

$$- 13 H_1 a_{10} u'_k\left[\frac{2}{h^2}(u_{k+1} - 2u_k + u_{k-1}) - \frac{1}{2h}(u'_{k+1} - u'_{k-1})\right],$$

$$- 3(u_{k+1} - u_{k-1}) + h(u'_{k+1} + 4u'_k + u'_{k-1})$$

$$= H_2(2h f'_k) + H_2 a_{11}\left[u'_{k+1}\left\{\frac{3v'_{k+1} - 4v'_k + v'_{k-1}}{2h}\right\} + v'_{k+1}\left\{\frac{3u'_{k+1} - 4u'_k + u'_{k-1}}{2h}\right\}\right]$$

$$- H_2 a_{12}\left[u'_{k-1}\left\{\frac{-3v'_{k-1} + 4v'_k - v'_{k+1}}{2h}\right\} + v'_{k-1}\left\{\frac{-3u'_{k-1} + 4u'_k - u'_{k+1}}{2h}\right\}\right],$$

$$- 3(v_{k+1} - v_{k-1}) + h(v'_{k+1} + 4v'_k + v'_{k-1})$$

$$= H_2(2hg'_k) - H_2 a_{11} u'_{k+1} \left\{ \frac{3u'_{k+1} - 4u'_k + u'_{k-1}}{2h} \right\} - H_2 a_{12} u'_{k-1} \left\{ \frac{-3u'_{k-1} + 4u'_k - u'_{k+1}}{2h} \right\},$$

$$(5.13)$$

where

$$a_k = a(r_k), \quad f_k = f(r_k), \quad g_k = g(r_k),$$

$$a_{11} = a_k + ha'_k + \frac{h^2}{2} a''_k, \qquad a_{12} = a_k - ha'_k + \frac{h^2}{2} a''_k, \qquad a_{10} = a_k.$$

$$(5.14)$$

The scheme (5.13) is free from the terms $1/k \pm 1$, hence very easily solved for $k = 1(1)N$ in the region $(0,1)$. The system (5.13) can be solved using the NBSOR method.

Consider the boundary value problem [31]

$$y^{iv} - \lambda y y'' = f(x),$$

$$y(0) = V_0, \qquad y(1) = V_1, \qquad y'(0) = 0, \qquad y'(1) = 0.$$

$$(5.15)$$

This arises from the time-dependent Navier-Strokes equations for axisymmetric flow of an incompressible fluid contained between infinite disks that occupy the planes $z = -d$ and $z = d$. The disks are porous and the fluid is injected or extracted normally with velocity V_0 at $z = -d$ and V_1 at $z = d$. Here, $=d/v$, where v is the kinematic viscosity.

6. Numerical Illustrations

To illustrate our method and to demonstrate computationally its convergence, we have solved the following linear problem using the BGS method (See [25, 32–35]), whose exact solution is known to us. We have taken $[0, 1]$ as our region of integration. The right-hand side functions and the boundary conditions are obtained using the exact solution. The iterations were stopped when the absolute error tolerance became $\leq 10^{-12}$. All computations were performed using double length arithmetic.

Problem 1. The problem is to solve (5.2) subject to the boundary conditions (5.4). The exact solution is $u = r^4 \sin r$. The root mean square errors (RMSEs) are tabulated in Table 1. The graph of the errors for $N = 32$ is given in Figure 1.

Problem 2. The boundary value problem is to solve (5.15). The exact solution is $(1-x^2) \exp(x)$. The maximum absolute errors (MAE) and RMSE are tabulated in Table 2.

Problem 3. The system of nonlinear equation (5.11) is to be solved subject to the natural boundary conditions. The exact solutions are $u = \cos(r)$ and $v = \exp(r)$. The MAE and RMSE are tabulated in Table 3.

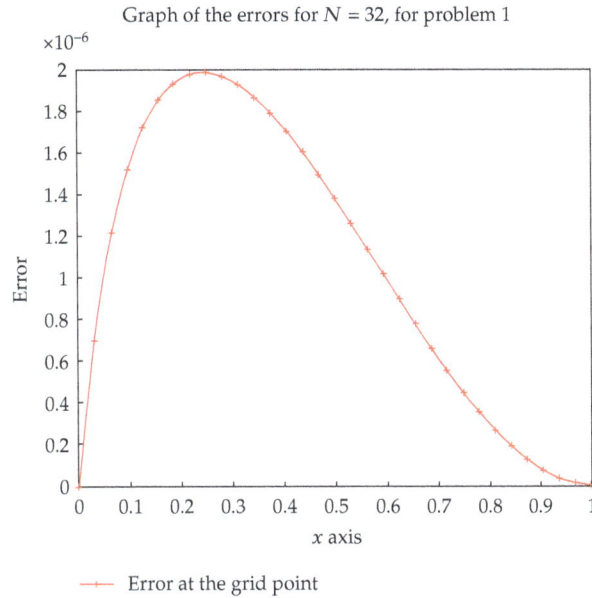

Figure 1

Table 1: The RMSE.

h		2nd-order method		4th-order method	
		MAE	RMSE	MAE	RMSE
1/8	u	0.9512 (−03)	0.6737 (−03)	0.6018 (−03)	0.4089 (−03)
	u'	0.3162 (−02)	0.2137 (−02)	0.2961 (−02)	0.1402 (−02)
1/16	u	0.2212 (−03)	0.1501 (−03)	0.3799 (−04)	0.2502 (−04)
	u'	0.7021 (−03)	0.4995 (−03)	0.3022 (−03)	0.9846 (−04)
1/32	u	0.5410 (−04)	0.3576 (−04)	0.1989 (−05)	0.1304 (−05)
	u'	0.1729 (−03)	0.1206 (−03)	0.2302 (−05)	0.5629 (−05)
1/64	u	0.1339 (−04)	0.8769 (−05)	0.9996 (−07)	0.6578 (−07)
	u'	0.4298 (−04)	0.2972 (−04)	0.1562 (−05)	0.2983 (−06)
1/128	u	0.3363 (−05)	0.2174 (−05)	0.4198 (−08)	0.3345 (−08)
	u'	0.1072 (−04)	0.7383 (−05)	0.9892 (−07)	0.1538 (−07)

Observe that for the linear singular problem for $\alpha = 1$, $h_1 = 1/64$, $h_2 = 1/128$, $\log(e_{h_1}/e_{h_2})/\log(h_1/h_2) = \log(0.6578(-07)/0.3345(-08))/\log 2 \approx 4$. Thus, we obtain fourth-order convergence for u. Similarly, we get fourth-order convergence for u'.

7. Concluding Remarks

The numerical results confirm that the proposed finite difference methods yield second- and fourth-order convergence for the solution and its derivative for the fourth-order ordinary differential equation. Difference formulas for mesh points near the boundary are obtained without using the fictitious points. The proposed method is applicable to problems in polar coordinates and the derivative of the solution is obtained as the by-product of the method. We

Table 2: The MAE and RMSE.

h		2nd-order method		4th-order method	
		MAE	RMSE	MAE	RMSE
$\lambda = 1.0$					
1/8	u	0.8435 (−04)	0.5652 (−04)	0.2744 (−05)	0.1880 (−05)
	u'	0.3552 (−03)	0.2322 (−03)	0.1058 (−04)	0.7178 (−05)
1/16	u	0.2292 (−04)	0.1499 (−04)	0.1854 (−06)	0.1217 (−06)
	u'	0.7762 (−04)	0.5391 (−04)	0.6695 (−06)	0.4351 (−06)
1/32	u	0.5844 (−05)	0.3771 (−05)	0.1151 (−07)	0.7446 (−08)
	u'	0.1885 (−04)	0.1322 (−04)	0.4115 (−07)	0.2618 (−07)
$\lambda = 100$					
1/8	u	0.1961 (−02)	0.1401 (−02)	0.5662 (−04)	0.3877 (−04)
	u'	0.6514 (−02)	0.4687 (−02)	0.3824 (−03)	0.1910 (−03)
1/16	u	0.4808 (−03)	0.3312 (−03)	0.3821 (−05)	0.2558 (−05)
	u'	0.1597 (−02)	0.1110 (−02)	0.1929 (−04)	0.9922 (−05)
1/32	u	0.1197 (−03)	0.8103 (−04)	0.2435 (−06)	0.1609 (−06)
	u'	0.4023 (−03)	0.2714 (−03)	0.1135 (−05)	0.5843 (−06)

Table 3: The MAE and RMSE.

h		2nd-order method		4th-order method	
		MAE	RMSE	MAE	RMSE
1/10	u	0.7141 (−05)	0.4859 (−05)	0.2673 (−05)	0.1818 (−05)
	v	0.3339 (−05)	0.2240 (−05)	0.6974 (−06)	0.4745 (−06)
	u'	0.2488 (−04)	0.1683 (−04)	0.1266 (−04)	0.6538 (−05)
	v'	0.1126 (−04)	0.8137 (−05)	0.2548 (−05)	0.1656 (−05)
1/20	u	0.1806 (−05)	0.1182 (−05)	0.2467 (−06)	0.1639 (−06)
	v	0.8550 (−06)	0.5595 (−06)	0.4846 (−07)	0.3212 (−07)
	u'	0.6486 (−05)	0.4128 (−05)	0.1167 (−05)	0.5946 (−06)
	v'	0.2659 (−05)	0.1957 (−05)	0.1795 (−06)	0.1119 (−06)
1/40	u	0.4514 (−06)	0.2917 (−06)	0.2048 (−07)	0.1346 (−07)
	v	0.2152 (−06)	0.1391 (−06)	0.3315 (−08)	0.2161 (−08)
	u'	0.1620 (−05)	0.1020 (−05)	0.1027 (−06)	0.4911 (−07)
	v'	0.6757 (−06)	0.4828 (−06)	0.1222 (−07)	0.7515 (−08)
1/80	u	0.1125 (−06)	0.7216 (−07)	0.1510 (−08)	0.9836 (−09)
	v	0.5307 (−07)	0.3442 (−07)	0.2931 (−09)	0.1883 (−09)
	u'	0.4038 (−06)	0.2525 (−06)	0.8190 (−08)	0.3662 (−08)
	v'	0.1686 (−06)	0.1192 (−06)	0.9925 (−09)	0.6537 (−09)
1/160	u	0.3050 (−07)	0.1946 (−07)	0.1213 (−09)	0.7945 (−10)
	v	0.1461 (−07)	0.9331 (−08)	0.1358 (−10)	0.8769 (−11)
	u'	0.1076 (−06)	0.6812 (−07)	0.6420 (−09)	0.2913 (−09)
	v'	0.4591 (−07)	0.3237 (−07)	0.5009 (−10)	0.3048 (−10)

employ the BGS method to solve the block matrix systems of the linear singular problem and the BSOR method to solve the nonlinear singular problem. We have solved here a physical problem that arises in the axisymmetric flow of an incompressible fluid.

Acknowledgments

This research was supported by "The Council of Scientific and Industrial Research" under Research Grant no. 09/045(0836)2009-EMR-I. The authors thank the reviewers for their valuable suggestions, which substantially improved the standard of the paper.

References

[1] S. P. Timoshenko, *Theory of Elastic Stability*, McGraw-Hill Book, New York, NY, USA, 2nd edition, 1961.

[2] A. R. Aftabizadeh, "Existence and uniqueness theorems for fourth-order boundary value problems," *Journal of Mathematical Analysis and Applications*, vol. 116, no. 2, pp. 415–426, 1986.

[3] R. P. Agarwal and P. R. Krishnamoorthy, "Boundary value problems for nth order ordinary differential equations," *Bulletin of the Institute of Mathematics. Academia Sinica*, vol. 7, no. 2, pp. 211–230, 1979.

[4] R. P. Agarwal, "Boundary value problems for higher order differential equations," *Bulletin of the Institute of Mathematics. Academia Sinica*, vol. 9, no. 1, pp. 47–61, 1981.

[5] R. P. Agarwal and G. Akrivis, "Boundary value problems occurring in plate deflection theory," *Journal of Computational and Applied Mathematics*, vol. 8, no. 3, pp. 145–154, 1982.

[6] Z. Bai, B. Huang, and W. Ge, "The iterative solutions for some fourth-order p-Laplace equation boundary value problems," *Applied Mathematics Letters*, vol. 19, no. 1, pp. 8–14, 2006.

[7] J. R. Graef and L. Kong, "A necessary and sufficient condition for existence of positive solutions of nonlinear boundary value problems," *Nonlinear Analysis*, vol. 66, no. 11, pp. 2389–2412, 2007.

[8] J. R. Graef, C. Qian, and B. Yang, "A three point boundary value problem for nonlinear fourth order differential equations," *Journal of Mathematical Analysis and Applications*, vol. 287, no. 1, pp. 217–233, 2003.

[9] M. D. Greenberg, *Differential Equations and Linear Algebra*, chapter 7, Prentice Hall, Engelwood Cliffs, NJ, USA, 2001.

[10] R. K. Mohanty, "A fourth-order finite difference method for the general one-dimensional nonlinear biharmonic problems of first kind," *Journal of Computational and Applied Mathematics*, vol. 114, no. 2, pp. 275–290, 2000.

[11] R. K. Nagle and E. B. Saff, *Fundamentals of Differential Equations*, chapter 6, The Benjamin/Cummingspp, 1986.

[12] M. A. Noor and S. T. Mohyud-Din, "An efficient method for fourth-order boundary value problems," *Computers & Mathematics with Applications*, vol. 54, no. 7-8, pp. 1101–1111, 2007.

[13] D. O'Regan, "Solvability of some fourth (and higher) order singular boundary value problems," *Journal of Mathematical Analysis and Applications*, vol. 161, no. 1, pp. 78–116, 1991.

[14] J. Schröder, "Numerical error bounds for fourth order boundary value problems, simultaneous estimation of $u(x)$ and $u''(x)$," *Numerische Mathematik*, vol. 44, no. 2, pp. 233–245, 1984.

[15] V. Shanthi and N. Ramanujam, "A numerical method for boundary value problems for singularly perturbed fourth-order ordinary differential equations," *Applied Mathematics and Computation*, vol. 129, no. 2-3, pp. 269–294, 2002.

[16] D. G. Zill and M. R. Cullen, *Differential Equations with Boundary-Value Problems*, chapter 5, Brooks Cole, NewYork, NY, USA, 1997.

[17] R. A. Usmani, "Finite difference methods for computing eigenvalues of fourth order boundary value problems," *International Journal of Mathematics and Mathematical Sciences*, vol. 9, no. 1, pp. 137–143, 1986.

[18] R. A. Usmani and M. Sakai, "Two new finite difference methods for computing eigenvalues of a fourth order linear boundary value problem," *International Journal of Mathematics and Mathematical Sciences*, vol. 10, no. 3, pp. 525–530, 1987.

[19] R. A. Usmani and P. J. Taylor, "Finite difference methods for solving $[p(x)y'']'' + q(x)y = r(x)$," *International Journal of Computer Mathematics*, vol. 14, no. 3-4, pp. 277–293, 1983.

[20] E. H. Twizell and S. I. A. Tirmizi, "Multiderivative methods for linear fourth order boundary value problems," Tech. Rep. TR/06/84, Department of Mathematics And Statistics, Brunel University, 1984.

[21] R. P. Agarwal and Y. M. Chow, "Iterative methods for a fourth order boundary value problem," *Journal of Computational and Applied Mathematics*, vol. 10, no. 2, pp. 203–217, 1984.

[22] A. Cabada, "The method of lower and upper solutions for second, third, fourth, and higher order boundary value problems," *Journal of Mathematical Analysis and Applications*, vol. 185, no. 2, pp. 302–320, 1994.

[23] D. Franco, D. O'Regan, and J. Perán, "Fourth-order problems with nonlinear boundary conditions," *Journal of Computational and Applied Mathematics*, vol. 174, no. 2, pp. 315–327, 2005.

[24] G. Han and F. Li, "Multiple solutions of some fourth-order boundary value problems," *Nonlinear Analysis*, vol. 66, no. 11, pp. 2591–2603, 2007.

[25] L. A. Hageman and D. M. Young, *Applied Iterative Methods*, Dover Publications, Mineola, NY, USA, 2004.

[26] R. K. Mohanty and D. J. Evans, "Block iterative methods for one-dimensional nonlinear biharmonic problems on a parallel computer," *Parallel Algorithms and Applications*, vol. 13, no. 3, pp. 239–263, 1999.

[27] M. M. Chawla, "A fourth-order tridiagonal finite difference method for general non-linear two-point boundary value problems with mixed boundary conditions," *Journal of the Institute of Mathematics and its Applications*, vol. 21, no. 1, pp. 83–93, 1978.

[28] R. K. Mohanty and D. J. Evans, "New algorithms for the numerical solution of one dimensional singular biharmonic problems of second kind," *International Journal of Computer Mathematics*, vol. 73, no. 1, pp. 105–124, 1999.

[29] D. J. Evans and R. K. Mohanty, "Alternating group explicit method for the numerical solution of non-linear singular two-point boundary value problems using a fourth order finite difference method," *International Journal of Computer Mathematics*, vol. 79, no. 10, pp. 1121–1133, 2002.

[30] J. Prescott, *Applied Elasticity*, Dover Publications Inc, New York, NY, USA, 1961.

[31] A. R. Elcrat, "On the radial flow of a viscous fluid between porous disks," *Archive for Rational Mechanics and Analysis*, vol. 61, no. 1, pp. 91–96, 1976.

[32] C. T. Kelly, *Iterative Methods for Linear and Non-Linear Equations*, SIAM Publication, Philadelphia, Pa, USA, 1995.

[33] Y. Saad, *Iterative Methods for Sparse Linear Systems*, Society for Industrial and Applied Mathematics, Philadelphia, Pa, USA, 2nd edition, 2003.

[34] R. S. Varga, *Matrix Iterative Analysis*, vol. 27 of *Springer Series in Computational Mathematics*, Springer, Berlin, Germany, 2000.

[35] D. M. Young, *Iterative Solution of Large Linear Systems*, Dover Publications, Mineola, NY, USA, 2003.

9

Interpreting the Phase Spectrum in Fourier Analysis of Partial Ranking Data

Ramakrishna Kakarala

School of Computer Engineering, Nanyang Technological University, Singapore 637665

Correspondence should be addressed to Ramakrishna Kakarala, ramakrishna@ntu.edu.sg

Academic Editor: Mustapha Ait Rami

Whenever ranking data are collected, such as in elections, surveys, and database searches, it is frequently the case that partial rankings are available instead of, or sometimes in addition to, full rankings. Statistical methods for partial rankings have been discussed in the literature. However, there has been relatively little published on their Fourier analysis, perhaps because the abstract nature of the transforms involved impede insight. This paper provides as its novel contributions an analysis of the Fourier transform for partial rankings, with particular attention to the first three ranks, while emphasizing on basic signal processing properties of transform magnitude and phase. It shows that the transform and its magnitude satisfy a projection invariance and analyzes the reconstruction of data from either magnitude or phase alone. The analysis is motivated by appealing to corresponding properties of the familiar DFT and by application to two real-world data sets.

1. Introduction

Ranking data, which arise in scenarios such as elections or database searches, describe how many times a given ordering of objects is chosen. It is frequently the case that when, ranking data are collected, partial ranking data are obtained in addition to, or perhaps instead of, full rankings. A partial or incomplete ranking only specifies the ordering of the top k out of n possibilities and usually indicates that the ranker is either unable to, or indifferent to, the ordering of the remaining $n - k$ items. Full ranking data are obviously a special case of partial ranking data. A classic approach is to treat full ranking data for n items as a function on the symmetric group S_n; for each permutation $p \in S_n$, the value of $x(p)$ is the number of times the ordering represented by that permutation is chosen [1]. For example, if 3 items are ranked, then $x([2,1,3])$ is the number of times the survey respondents chose to rank item 2 first, item 1 second, followed by 3. As discussed in more detail below, partial ranking data also form functions on S_n that are piecewise constant over cosets of the subgroup fixing the first k items.

The analysis of ranking data, including both full and partial rankings, is well established. Statistical methods exist both for data in the "time domain" (using signal processing terminology), which in this case is the permutation group S_n, and in the "frequency domain" that is obtained through Fourier analysis on the group. Recent papers by Lebanon and Mao [2] and Hall and Miller [3] explore, respectively, the nonparametric modeling and bootstrap analysis of partial ranking data in the time domain. Time domain analysis does not allow such interesting possibilities as using band-limited or "smooth" approximations to the data, on analyzing the strength of various components. Diaconis [1, 4] and Diaconis and Sturmfels [5] use the Fourier transform on S_n to analyze frequency components of both full and partial ranking data. Those papers, while addressing the fundamentals of Fourier analysis in terms of invariant subspaces, do not consider signal processing aspects as considered here. Other papers using the Fourier transform on S_n include Huang et al. [6] for inference on permutations of identities in tracking, and Kondor and Borgwardt [7] to provide labeling-invariant matching of graphs. Kakarala [8] shows that the Fourier transform on S_n may be interpreted in terms of signal processing concepts such as magnitude and phase, but the work is limited to full rankings. In this paper, we take a similar approach to analyze the properties of the Fourier transform on S_n for partial rankings, with particular emphasis on the role of phase in forming the top three ranks, $k \leq 3$.

Underlying our approach is the intuition that, in any frequency-domain approach, whether on the symmetric group S_n or on the more familiar discrete domain $\mathbb{Z}_N = \{0, 1, \ldots, N - 1\}$, the Fourier transform values may be separated into magnitudes, which indicate component strengths, and phases, which indicate relative component locations. Such a separation is basic to a signal processing approach, and is well understood in the ordinary discrete Fourier transform (DFT) on $\mathbb{Z}_N$, and also in two dimensions in the case of images. A familiar demonstration of the importance of phase is to combine the magnitude spectrum of image X with the phase spectrum of image Y and observe that, after inverse transform, the result appears very similar to Y [9]; in other words, phase is more important to our perception of image structure. Therefore, it seems appropriate to ask the following question: what is the role of phase in forming partial ranking data?

The problem of analyzing phase on S_n is not as straightforward as with the DFT on $\mathbb{Z}_N$, because the Fourier transform on S_n has matrix-valued coefficients, not scalars as with the DFT, making even such elementary concepts as "frequency" nonobvious. Though various papers describe the S_n transform in detail [6] and, code for computing a fast Fourier transform (FFT) on S_n has been published by Kondor [10], the level of abstraction required to understand the S_n transform is high. Therefore, this paper makes a concerted effort to reason from the familiar DFT to explore the relevant concepts on S_n. It shows that the coefficients of the Fourier transform for top k choice partial ranking data are invariant under projections that are determined by the subgroup S_{n-k}. The projection approach provides a relatively simple explanation of the roles of magnitude and phase for partial ranking. The explanation is tested on two real-world data sets.

It should be noted that the concept of partially measured ranking data has interpretations other than the one explored in this paper, which is top k out n choices data. For example, an "incomplete" ranking specifies a preference among a subset of the choices, not which is most preferred. Among choices A, B, and C, an incomplete ranking might simply say that A is preferred to C, but nothing about A versus B, or B versus C; mathematically, this may be modelled as a partial order on the choices [2]. Diaconis [4] describes other kinds of incomplete rankings: "committee selection," where one chooses the top k out of n choices but does not rank among the choices; "most and least desirable," where one chooses the most important and least important attributes among n choices but does not specify the order of the middle elements. What is common mathematically to the previous types of data is that they are constant on cosets of a suitably chosen subgroup H of S_n. The mathematical results of this paper concerning magnitude and phase apply to every coset space S_n/H. However, the results provided below on approximation by linear phase or unit-magnitude functions are limited to top k-choice data, whose domain in S_n/S_{n-k}. Though mathematically a special case of partially measured rankings, top k-choice data appears in sufficiently many scenarios to be worth analysis on its own.

2. Background Material

Fourier analysis on the symmetric group S_n is normally described in abstract terms involving group representation theory, which makes the subject difficult to understand for non-specialists. As mentioned in the Introduction, we use analogy to the better known DFT on $\mathbb{Z}_N$. The DFT is defined for data x by the familiar pair of equations for transform and inverse:

$$X[k] = \sum_{n=0}^{N-1} x[n]e^{-j2\pi kn/N}, \qquad x[n] = \frac{1}{N}\sum_{k=0}^{N-1} X[k]e^{j2\pi kn/N}. \qquad (2.1)$$

Each complex-valued DFT coefficient is expressed in terms of magnitude and phase by writing $X[k] = |X[k]|e^{j\phi(k)}$, where the absolute value determines the magnitude, and the angle $\phi(k)$ measures the starting value at $n = 0$ in the period of the constituent sinusoid $e^{j2\pi kn/N}$. The translation property of the DFT shows that the transform of the circularly shifted function $y[n] = x[n + t]$ has coefficients $Y[k] = X[k]e^{j2\pi kt/N}$, which shows that the magnitude does not change but the phase changes linearly, that is, $\phi(k) \mapsto \phi(k) + 2\pi kt/N$. Hence, phase is closely connected with location.

Suppose now that the data x has the additional symmetry of having a subperiod, that is, $x[n + M] = x[n]$ where M divides N. Then, it is well known that the DFT coefficients $X[k]$ are zero unless k is a multiple of N/M. For example, if $N = 128$ and $M = 4$, then, of the 128 possible DFT coefficients, only four are nonzero: $X[0]$, $X[32]$, $X[64]$, and $X[96]$. It is helpful to see the previous example in a different way to better understand the discussion of the symmetric group below. Suppose that we define $\tilde{x}$ as the data within one period, that is, $\tilde{x}[n] = x[n]$ for $n = 0,\ldots,3$, and $\tilde{x}[n] = 0$ otherwise. Let τ denote the periodic pulse train of Kronecker δ functions defined as follows:

$$\tau[n] = \sum_{m=0}^{31} \delta[n - 4m]. \qquad (2.2)$$

Then, $x = \tilde{x} * \tau$, where $*$ denotes circular convolution over 128 points. We have, therefore, by the convolution property of the DFT that $X[k] = \tilde{X}[k]T[k]$, where both $\tilde{X}$ and T are the respective DFTs on 128 points of $\tilde{x}$ and τ. It is easy to see that $T[k] = 32$ for $k = 0, 32, 64, 96$ but $T[k] = 0$ otherwise. We might consider the function $P[k] = T[k]/32$ a *projection* of the DFT coefficients; the term projection is appropriate because P takes values of either 0 or 1, and therefore $P[k]P[k] = P[k]$ for all k. With the projection so defined, we have that $X[k] = P[k]X[k]$, which shows that the data are invariant to the projection and therefore lie in its image. The projection approach helps considerably below in formulating the transform for partial rankings on the symmetric group.

The symmetric group S_n is the collection of all $n!$ possible permutations of the set $\{1, 2, \ldots, n\}$. If p and q represent two permutations in S_n, then the product pq denotes q applied first followed by p. For example, if $n = 4$ and $p = [2, 3, 4, 1]$, which indicates that $p(1) = 2$, $p(2) = 3$, $p(3) = 4$, $p(4) = 1$, and similarly $q = [3, 4, 1, 2]$, then $pq = [4, 1, 3, 2]$. With that product, S_n forms a group, with identity-denoted e and inverse p^{-1} being the unique permutation that exactly undoes the action of p, that is, $p^{-1}p = pp^{-1} = [1, 2, \ldots, n]$. For example, the inverse of $[2, 3, 4, 1]$ is $[4, 1, 3, 2]$.

Data consisting of full rankings form functions on S_n in the manner described in the Introduction. The same domain also serves for partial ranking data. If we have data where only the first k of the n items is ranked, then, for each $p \in S_n$, let us define the value of $x(p)$ to be the number of times the first k elements of p is chosen. The definition leads to piecewise constant functions on S_n. An example illustrates the approach. Suppose $n = 3$ items are to be ranked in an election given to 600 voters, but the respondents give only their top choices as follows: item 1 gets 100 votes, item 2 gets 200 votes, and item 3 gets 300 votes. Then, we construct x on S_3 by extending the votes to all permutations p based on first item, so that $x([1, 2, 3]) = x([1, 3, 2]) = 100$, and similarly for the other 4 choices of p. If we were to view the previous construction in group-theoretic terms, the function x is such that it is constant on left cosets of the subgroup S_2 fixing the first element, that is, $x(p) = x(ps)$ for all $s \in S_2$ where $s(1) = 1$ by definition, and $p(1) = t$ for the item t being chosen. Though the constant vote given to each coset is mathematically convenient, it does not capture certain effects that may be interesting; for example, if I choose oranges as my favorite fruit, I may be more likely to choose apples than durians as my next favorite, even if I am not required to state my next favorite. Nevertheless, due to its convenience, we use the constant on cosets approach in the remainder of this paper.

A detailed example helps to illustrate the model. In the famous American Psychological Association (APA) election data [1], which is available online (http://www .stat.ucla.edu/data/hand-daly-lunn-mcconway-ostrowski/ELECTION.DAT), 5,738 voters provided full rankings of each of 5 candidates for president. The full rankings form a function on S_5 and are shown plotted in Figure 1(a) against the $5! = 120$ elements of the group arranged in lexicographic order. In the same election, many voters chose not to submit full rankings but provided instead partial rankings. Specifically, 5,141 voters submitted only their top choice, 2,462 voters submitted only the their first and second choices in order, and 2,108 voters submitted only their top three choices in order. Consequently, there were a total of 9,711 voters giving only partial rankings, more than the 5,738 that gave full rankings. After forming piecewise constant functions as described above, the partial ranking data are displayed in Figures 1(b)–1(d).

An advantage of placing both full and partial rankings on the same domain is that we may apply the same Fourier transform in both cases. The Fourier transform on S_n, which is formally obtained from the theory of group representations, has important differences to the DFT. We review some basic facts from the literature [4]. First, the Fourier coefficients on S_n are matrix valued, unlike the scalar values of the DFT. Second, they are indexed by arithmetic partitions of n with nonincreasing elements, which are roughly analogous to the frequency index k of the DFT. For example, for $n = 5$, the seven such partitions are $(5), (4, 1), (3, 2), (3, 1^2) := (3, 1, 1), (2^2, 1), (2, 1^3)$, and (1^5). For every partition λ of n, the Fourier basis elements belonging to it are collected into a square-matrix-denoted D_λ whose dimensions n_λ are calculated using standard formulas [4]. For S_5, the seven partitions

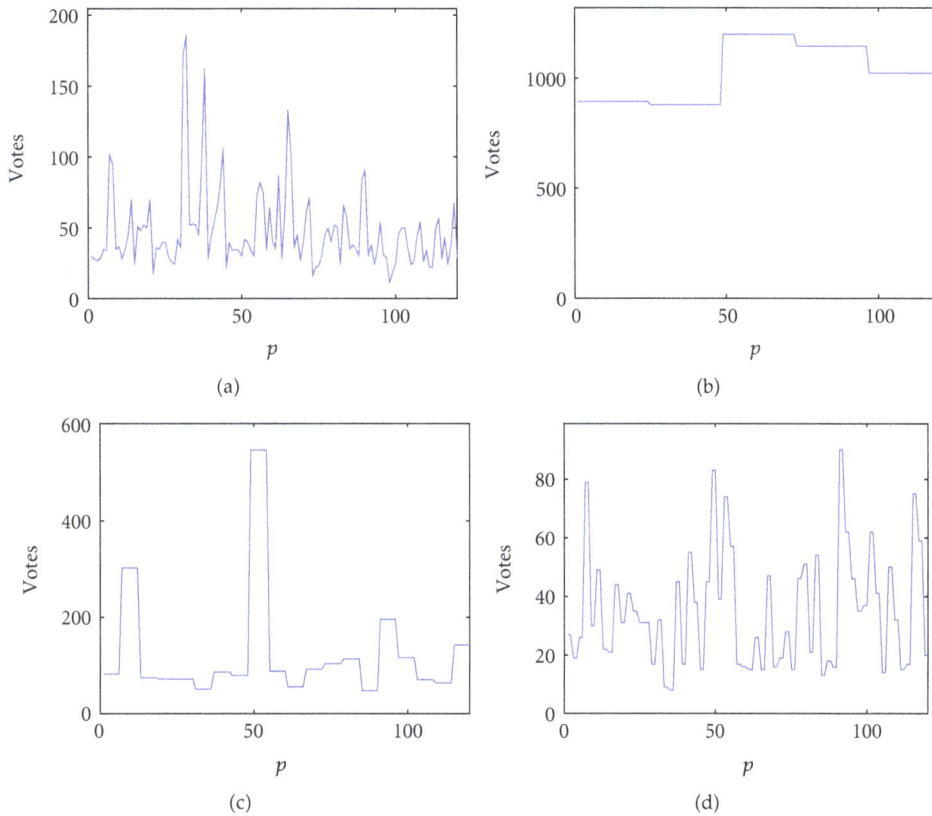

Figure 1: APA election data with both full and partial rankings are shown plotted in lexicographic order on S_5. The four subplots indicate as follows: (a) full ranking data; (b) votes where only the top candidate was given; (c) votes where the top two candidates in order were given; (d) votes where the top three were given. Note that, in each case, the votes peak at different locations: (a) the maximum votes (186) go to the ordering 23154, that is, candidate 2 has top preference, 3 second, followed by 1, 5, and 4 in decreasing preference; (b) the maximum votes (1,198) go to candidate 3; (c) the maximum votes (547) go to the ordering 31; (d) the maximum votes (90) go to the ordering 451. Note that, in (d), the second most popular ordering, getting 83 votes, is 312, which is more consistent with the result of (b) and (c).

described previously have square basis matrices D with respective dimensions 1, 4, 5, 6, 5, 4, 1, giving a total of 120 basis functions on S_5, where the number 120 is obtained by summing squares of dimensions. The basis may be constructed using real-valued functions, using the Young orthogonal representation (YOR). The Fourier transform and its inverse are, respectively, written

$$X(\lambda) = \sum_{p \in S_n} x(p) D_\lambda(p)^\top, \qquad x(p) = \sum_{\lambda \vdash n} \frac{n_\lambda}{n!} \mathrm{Trace} \left[X(\lambda) D_\lambda(p) \right]. \qquad (2.3)$$

The symbol $\lambda \vdash n$ on the right hand sum indicates a sum over all partitions for which D_λ is defined. Algorithms for constructing the D matrices are given in Huang et al. [6, Algs 3,4] and are used in obtaining the experimental results of this paper. In particular, we have $D_n(p) = 1$, so that $X(n)$ is a scalar containing the "d.c" value of the signal, and $D_{1^n}(p)$ is also scalar alternating between +1 and −1 in the manner similar to the Nyquist frequency $k = N/2$ in the DFT.

Two important properties of the Fourier transform are relevant to this paper: the Fourier basis matrices D_λ that are obtained from the YOR are orthogonal, $D_\lambda(p)D_\lambda(p)^\top = I$, which mimics the exponential unitarity $e^{j\omega}e^{-j\omega} = 1$ in the DFT; under a left translation of the data on S_n obtained by $x(p) \mapsto x(sp)$, the coefficients undergo the transformation $X(\lambda) \mapsto X(\lambda)D_\lambda(s)$, and, under a right translation $x(p) \mapsto x(sp)$, the coefficients transform as $X(\lambda) \mapsto D_\lambda(s)X(\lambda)$. Those two properties suggest an interpretation of the matrix-valued Fourier coefficients in terms of magnitude and phase [8]. The Fourier coefficient may be written using the matrix polar decomposition as $X(\lambda) = \hat{X}(\lambda)O(\lambda)$, where $\hat{X}$ representing magnitude is the positive semidefinite matrix obtained as the square root of $XX^\top$, and O is an orthogonal matrix representing phase. A standard result in matrix theory [11, page 190] shows that the magnitude $\hat{X}$ is unique, though the phase O needs not be unless X is nonsingular. Under left translation by s, the magnitude remains invariant while the phase changes by $O(\lambda) \mapsto D_\lambda(s)O(\lambda)$, which is analogous to the phase shift $\phi(k) \mapsto 2\pi kt/N + \phi(k)$ for the DFT. Note that both magnitude and phase may be computed using the singular value decomposition (SVD), $X = USV^\top$, by setting $\hat{X} = USU^\top$ and $O = UV^\top$. Below, we use the polar decomposition of magnitude and phase and analyze its properties for partial ranking data.

3. Fourier Analysis of Partial Rankings

In the previous section, we saw that translational symmetry in the DFT domain results in a projection invariance $X[k] = P]k]X[k]$ for the DFT coefficients. Inspired by that result and noting that our method of placing partial ranking data on S_n results in a kind of translational symmetry, we look for the relevant projection characteristics of the Fourier coefficient matrices on S_n. Finding the projection characteristics provides significant reduction in computational complexity and also shows the role of phase for partial ranking data as discussed below. For that purpose, define for each subgroup H of S_n and each $\lambda \vdash n$ the matrix

$$P_H(\lambda) = \frac{1}{|H|}\sum_{h \in H} D_\lambda(h). \tag{3.1}$$

Then, it is known [12, page 111] that $P_H = P_H^\top$ and $P_H P_H = P_H$, so that P_H is an orthogonal projection. The main result of this paper is now stated.

Theorem 3.1. *Let x denote a function on S_n that is piecewise constant with respect to a subgroup, that is, $x(p) = x(ph)$ for every $p \in S_n$ and h in the subgroup H with $|H|$ elements. Then, each Fourier coefficient of x is invariant under the corresponding projection: $X(\lambda) = P_H(\lambda)X(\lambda)$, and that is true of its magnitude as well $\hat{X}(\lambda) = P_H(\lambda)\hat{X}(\lambda)$.*

Proof. The projection invariance of X follows from the translational property of the S_n Fourier transform, from which $x(p) = x(ph)$ results in $X = D(h)X$, when averaged over all elements of H result in $X = P_H X$. (This fact has been shown in the literature; see [12] and Kondor [13, Section 5]). To prove that $\hat{X}$ is invariant, note that P_H being a projection means that there exists an orthogonal matrix U such that $P_H = UI_rU^\top$, where I_r is the identity matrix up to the first $r = \mathrm{Rank}[P_H]$ entries. Then, $\widehat{X}' = U^\top \widehat{X} U$ is the unique positive semidefinite square root of $X' = U^\top X U$. Since $X' = I_r X'$, $\widehat{X}' \widehat{X}' = X'(X')^\top$ implies that $\widehat{X}' = I_r \widehat{X}' I_r$, so that $\widehat{X}'$ is zero outside the upper left $r \times r$ subblock. Consequently, $\widehat{X}' = I_r \widehat{X}'$, and, therefore, $\widehat{X} = P_H \widehat{X}$.   □

The theorem may be applied to partial ranking data consisting of k out of n elements ranked by using the subgroup $H = S_{n-k}$ that fixes the first k elements and varies the remaining ones. Table 1 shows the ranks of the projections for the first three values of k. The reader may note that Diaconis [1] provides essentially the same numbers as in Table 1, though not obtained through projections. For $k = 1$, only two frequencies λ are involved, each with rank 1. The dimension of the representation $D_{(n-1,1)}$ is $n - 1$, and consequently the projection $P_H X$ has only $n - 1$ degrees of freedom. Therefore, the n degrees of freedom for first-choice-only data ($k = 1$) are divided between the one-dimensional "d.c." value obtained for frequency $\lambda = (n)$ and the $n - 1$ degrees of freedom for $\lambda = (n - 1, 1)$.

The theorem and table are illustrated with examples in the next section.

We examine the roles that magnitude and phase play in partial ranking data by appealing to the more familiar DFT for intuition. If X is the DFT of real-valued data x, with magnitude-phase decomposition $X = |X|e^{j\phi}$, then the inverse DFT of the magnitude $|X|$ alone is the zero-phase signal

$$x_{zp}[n] = \frac{1}{N} \sum_{k=0}^{N-1} |X[k]| e^{j2\pi kn/N}. \tag{3.2}$$

The zero-phase signal has certain properties: its peak value occurs at the origin since $x_{zp}[0] \geq |x_{zp}[n]|$; it is symmetric with respect to sign inversion, since $x_{zp}[-n] = x_{zp}[n]$. We may shift the peak of x_{zp} from 0 to any desired location q by applying the linear phase shift $\phi \mapsto \phi - 2\pi kq/N$. The resulting linear phase signal is

$$x_{lp}[n] = \frac{1}{N} \sum_{k=0}^{N-1} |X[k]| e^{-j2\pi kq/N} e^{j2\pi kn/N}. \tag{3.3}$$

The properties of the linear phase signal x_{lp} are now as follows: its peak value occurs at $n = q$; it is symmetric about q since $x_{lp}[q-n] = x_{lp}[q+n]$. In other words, we see that, in the absence of phase, the basic components add directly to peak at the starting point, and by shifting the starting point to any given location produces a linear phase version of the signal. Analogous to the zero-phase signal, we may define the unit-magnitude signal by applying the inverse DFT to only the phase:

$$x_{um}[n] = \frac{1}{N} \sum_{k=0}^{N-1} e^{j\phi(k)} e^{j2\pi kn/N}. \tag{3.4}$$

For the DFT, magnitude, and phase, each contains half the degrees of freedom of the original signal, and therefore both are equally important to exact numerical reconstruction. The concepts discussed also apply for the symmetric group S_n as we now show.

Table 1: Rank of projection matrices $P_{n-k}(\lambda)$ for various k and λ. All of the required λ for $k \leq 3$ are shown [1].

$k \setminus \lambda$	(n)	$(n-1,1)$	$(n-2,2)$	$(n-2,1^2)$	$(n-3,3)$	$(n-3,2,1)$	$(n-3,1^3)$
1	1	1	0	0	0	0	0
2	1	2	1	1	0	0	0
3	1	3	3	3	1	2	1

Using the inverse transform (2.3), we define the zero-phase signal x_{zp} on S_n corresponding to the data x as

$$x_{zp}(p) = \sum_{\lambda \vdash n} \frac{n_\lambda}{n!} \mathrm{Trace}\left[\widehat{X}(\lambda)D_\lambda(p)\right]. \tag{3.5}$$

Noting that $D_\lambda(e) = I$ for the identity permutation, we see that the positive semidefiniteness of $\widehat{X}$ implies that $\mathrm{Trace}[\widehat{X}U] \leq \mathrm{Trace}[\widehat{X}]$ for every orthogonal matrix U, as easily seen by using the eigen-decomposition $\widehat{X} = V\Lambda V^\top$ and applying the circular invariance of trace. Consequently, $x_{zp}(e) \geq x_{zp}(p)$ for all p. Furthermore, there is inversion symmetry since $x_{zp}(p^{-1}) = x_{zp}(p)$ due to the trace property $\mathrm{Trace}[\widehat{X}D^\top] = \mathrm{Trace}[D\widehat{X}^\top] = \mathrm{Trace}[\widehat{X}D]$. The properties of a zero-phase signal are formally similar to those of an "autocorrelation," which we define on S_n as follows:

$$a_x(s) = \sum_{p \in S_n} x(p)x(ps). \tag{3.6}$$

The connection between zero-phase signals and autocorrelations is made clear in a theorem stated below.

Reasoning as above, we see that we may shift the peak of the zero-phase signal to any given permutation q by the linear phase transformation $\widehat{X} \mapsto \widehat{X}D_\lambda(q^{-1})$, resulting in the linear-phase signal

$$x_{lp}(p) = \sum_{\lambda \vdash n} \frac{n_\lambda}{n!} \mathrm{Trace}\left[\widehat{X}(\lambda)D_\lambda(q)^\top D_\lambda(p)\right]. \tag{3.7}$$

Properties of the linear-phase signal are established in the following theorem, the proof of which is given in an earlier paper [8].

Theorem 3.2. *For every real-valued function x on S_n with Fourier transform X, we have the following.*

(a) *The transform X is symmetric with respect to matrix transpose if and only if x is symmetric with respect to inversion:*

$$\forall \lambda \vdash n \quad X(\lambda) = X(\lambda)^\top \iff x(p) = x(p^{-1}) \quad \forall p \in S_n. \tag{3.8}$$

(b) *$X(\lambda)$ is positive semidefinite for all λ if and only if there exists a function y such that x is the autocorrelation of y, that is, $x = a_y$ using the notation of (3.6).*

(c) *Symmetric functions are precisely those with linear-phase transforms: there exists $q \in S_n$ such that $X(\lambda) = S(\lambda)D_\alpha(q^{-1})$ with $S(\lambda) = S(\lambda)^\top$ if and only if $x(pq) = x(p^{-1}q)$ for all $p \in S_n$.*

The theorem shows that each linear-phase signal is inversion symmetric about its peak location q, that is, $x_{lp}(qp) = x_{lp}(qp^{-1})$. As above, we may define the unit-magnitude signal by using only the phase O in the polar decomposition $X = \widehat{X}O$ in the inverse DFT on S_n as follows:

$$x_{um}(p) = \sum_{\lambda \vdash n} \frac{n_\lambda}{n!} \text{Trace}\left[O(\lambda)D_\lambda(p)\right]. \tag{3.9}$$

Noting that the polar decomposition $X = \widehat{X}O$ of an $m \times m$ matrix places $m(m+1)/2$ degrees of freedom in the positive definite matrix $\widehat{X}$ and $m(m-1)/2$ in the orthogonal matrix O, we see that magnitude is slightly more important (by m) to numerically reconstructing full ranking data. However, the situation is much different when partial rank data is involved. By examining Table 1 and using Theorem 3.1, we show that the unit-magnitude signal is nearly complete in the case of first rank data.

Theorem 3.3. *If x is top choice only data ($k = 1$) on S_n, then there exist constants α and β such that $x = \alpha x_{um} + \beta$.*

The proof follows after noting that, by Theorem 3.1 and Table 1, the magnitude $\widehat{X}(n-1,1)$ is a scalar, so that $\alpha = \widehat{X}(n-1,1)$ and $\beta = (n!)^{-1}X(n)(1 - \alpha/|X(n)|)$.

4. Examples

Consider the group S_5 used for the APA data shown in Figure 1. For the top two choice data ($k = 2$), the ranks of the projections in Table 1 show that the $5 \times 4 = 20$ degrees of freedom are allocated as follows: 1 in the d.c. term $X(5)$; 8 in the term $X(4,1)$; the remaining 11 degrees of freedom allocated as 5 and 6, respectively, in each of the Fourier coefficients for $(3,2)$ and $(3,1^2)$. By choosing a basis in which $P(4,1) = I_2$, we obtain the following for the nonzero entries of the Fourier coefficient and its magnitude (rounded to integers):

$$X(4,1) = \begin{pmatrix} -729 & 1452 & 986 & 505 \\ -2808 & 237 & -1885 & -59 \end{pmatrix}, \quad \widehat{X}(4,1) = \begin{pmatrix} 1964 & 94 \\ 94 & 3389 \end{pmatrix}. \tag{4.1}$$

Each matrix is actually 4×4, and the zero entries are not shown.

To illustrate the properties of phase for partial ranking data on S_n, we reconstruct each of the partial rank signals in Figure 1 using only zero and linear phase and show the results in Figure 2. In Figure 2(d), we see a strikingly good fit between the partial rank data with two preferences and its linear phase approximation: numerically, we have $\|x - x_{lp}\| \cdot \|x + x_{lp}\|^{-1} = 0.08$, where $\|\cdot\|$ is the l_2 norm. This suggests that the phase structure of the two-preference data is relatively simple, and the inversion symmetry property indicates that voters are equally content with transposing the order of the two top preferences moving away from the peak. The result is made more interesting by noting that, of the 20 degrees of freedom in top-two preference data, only 6 are constrained by the magnitude spectrum given by the $\widehat{X}$ matrices;

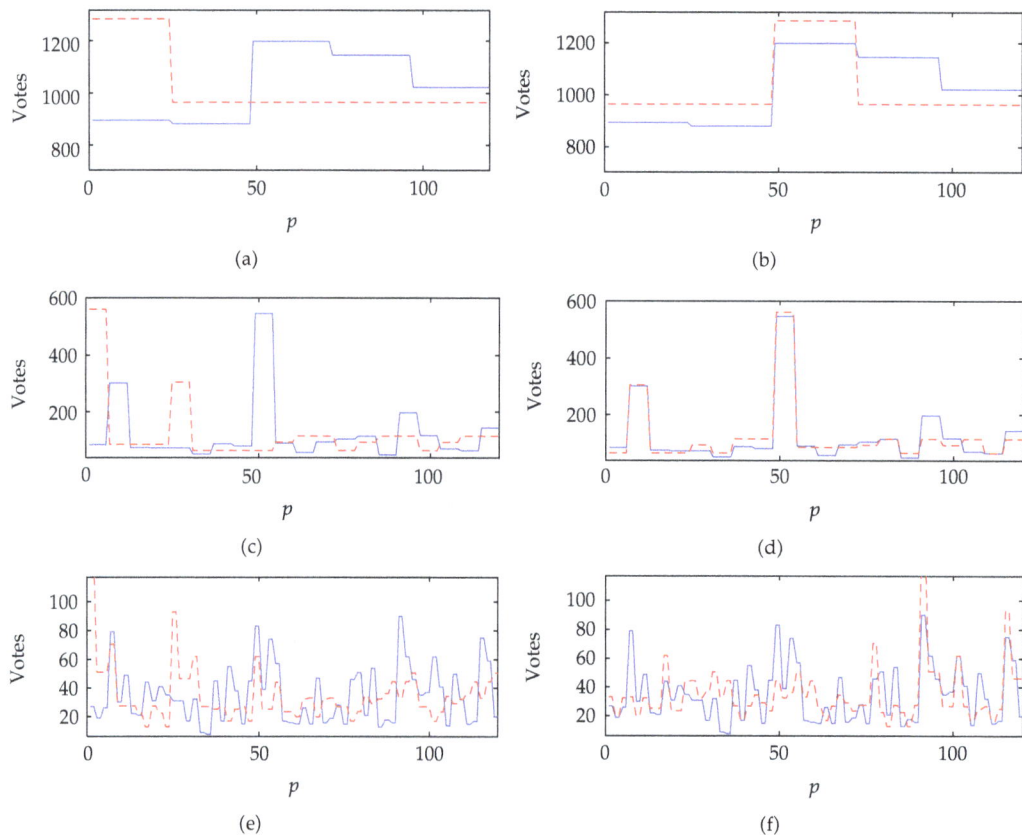

Figure 2: Partial ranking data from the APA election shown plotted in lexicographic order on S_5, with comparisons to the zero-phase (magnitude only) approximations in the left column, and the linear-phase approximations in the right. The linear-phase approximations are in each case adjusted so that the location of their peak value q matches that of the original data. The six subplots indicate as follows: (a) first preference only (blue) compared to zero phase (red dashed line); (b) first preference only with linear-phase approximation; (c) and (d) the same for the top two preferences; (e) and (f) the same for the top three preferences. Note that the two-preference data in (d) is well fit by the linear-phase approximation.

hence, adding only the linear phase term necessary to shift the peak should not be sufficient to reconstruct 92% of the signal, but it is.

The different levels of fit between the partial rank data and its linear-phase approximations may be understood also by considering the degrees of freedom involved. On the domain S_5 for the APA data, first preference data has 5 degrees of freedom. From Table 1, we see that are two frequencies involved, both with rank 1. As discussed above, $\widehat{X}(n-1,1)$ is a scalar. Consequently, the magnitude spectrum constrains 2 out of the 5 degrees of freedom. The case $k = 2$ is discussed above, and, for $k = 3$, we have that 24 out of the 60 degrees of freedom are constrained by magnitudes. However, as n increases, the degrees of freedom for the magnitude spectrum do not increase, because the ranks of the projection matrices are independent of n. For example, for $n = 50$, the magnitude spectrum for top three choices data ($k = 3$) constrains only 24 out of the 117,600 degrees of freedom. Consequently, for three choices data with large n, the phase spectrum by far exceeds the magnitude component in constraining data.

To illustrate the role of phase for top-choice data for large n, we examine the college rankings from 2009 by US News and World Report that is available online (http://supportingadvancement.com/potpourri/us_news_and_world_report/us_news_rankings.htm). In this data, $n = 65$ American universities are ranked on 17 numerical categories, including acceptance rate, percentage of classes with fewer than 20 students, and alumni giving rate. We consider each category as a voter giving a vote to only the university having the top category value. In the event of ties, which happens only in one category—the percentage of need met for full time students, where 23 universities met 100% of the need—all of the universities having the top value were given a vote. Figure 3 shows the data is poorly fit by with zero phase, as expected, but the shape of the data is well fit up to a scale factor by the unit-magnitude signal as expected from Theorem 3.3.

5. Discussion

We have seen in the previous section that the fit between partial ranking data and its linear phase approximation can be surprisingly good, especially in the case of the APA data for $k = 2$. The quality of linear phase fit is not limited to partial rank data. Full ranking data, which are discussed in [8], may also show a good linear phase approximation. Consider the German survey data, which consists of full rankings of four items by 2,262 voters [14]. Figure 4(a) shows that the data is well reconstructed by a linear-phase approximation; in fact, the linear-phase approximation reproduces 93% of the original signal as measured by $\|x - x_{lp}\| \cdot \|x + x_{lp}\|^{-1}$. Similarly, Figure 4(b) shows that the full ranking data for the APA election is well approximated (78%) by its linear-phase version. However, with full ranking data, the magnitude spectrum dominates: on S_4, as with the German survey data, we obtain 17 out of the 24 degrees of freedom in the full-ranking data from the magnitude spectrum, while, on S_5, we obtain 73 of the 120 or 62.5% of the d.o.f. from the magnitude spectrum of full rankings. Therefore, with full ranking data, we should not be as surprised by the quality of fit by linear-phase approximation as we might be with partial rank data.

It is reasonable to wonder what we gain by approximating data that we already have in exact form. Diaconis [1] states a general principle in analyzing data: "if you've found some structure, take it out, and look at what's left." The results in this and the previous section show cases where linear-phase structure exists in full rank and, more surprisingly, given the degrees of freedom argument, in partial-rank data. The high level of fit in the cases we have analyzed suggest that, once we remove the linear phase structure, there is little left. it would be interesting to apply linear phase approximation to a larger variety of data sets to see whether such symmetry is common. Also, a potential application of the linear-phase formulation is that it provides a way of reasoning about ranking data with reduced complexity, where phase is essentially eliminated except for a single component. It would be interesting to apply the linear phase approximation as a simplifying means to compare graphs up to relabeling of data [7].

5.1. Complexity

One of the limitations of ranking data is that the size of the domain S_n increases as $n!$, making it impractical to capture a complete set of fully ranked data for n much larger than 10. Furthermore, the complexity of the group theoretic FFT for S_n is $\mathcal{O}(n! \log^2 n!)$, as shown in Maslen [15, Theorem 1.1]. This is very difficult to compute for $n > 10$. However, partial ranking data and their spectral analysis allow data for much larger n to be analyzed. For

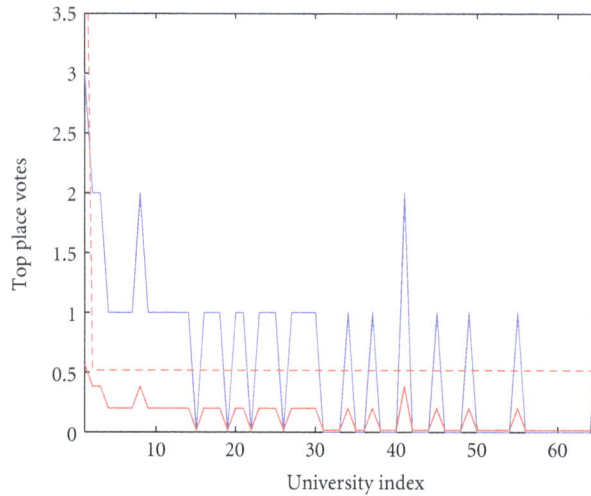

Figure 3: Top place votes for 65 American universities are shown plotted in the solid blue line. The dashed red line shows the reconstruction using the zero-phase version x_{zp}, and the solid red line shows the result using x_{um}, the unit-magnitude signal.

example, the number of data points for the top 3 out of n choices is $n!/(n-3)! \approx n^3$, which remains tractable for n up to 100. Maslen [15] showed that the group-theoretic FFT on S_n when adapted for $k = 3$ has $\mathcal{O}(n^4)$ complexity; in comparison, the ordinary FFT on $\mathbb{Z}_m$ for $m = (100)^4$ can be completed in 3 seconds on a 2.6 GHz quad-core Xeon processor. Therefore, we see that processing only partial rank data allows a capability of roughly an order-of-magnitude increase in n over fully-ranked data. If we restrict to only top choice data ($k = 1$), then there is a linear-time algorithm for computing the Fourier transform [16].

Knowing the complexity of the transform helps to determine the complexity of either the zero-phase (3.5) or the unit-magnitude (3.9) approximations. Each of those approximations requires the following three steps: computing the forward transform, separating each coefficient matrix into magnitude and phase components, and computing the inverse transform. The inverse transform has the same complexity as the forward transform. The magnitude-phase separation requires performing an SVD of each matrix coefficient, followed by two matrix multiplications for the magnitude, or one for the phase. The cost of each SVD is $\mathcal{O}(n_\lambda^3)$, where n_λ is the size of each representation λ. Unfortunately, there are no simple, closed-form, expressions for n_λ. However, when using partial rank data, the number of coefficients involved is relatively small due to the projection property. From Table 1, we see that there are only 7 coefficient matrices for top-three choices data ($k = 3$), the largest of which has rank 3. Note that the ranks listed in the Table are independent of n. We may use reduced SVDs for these 7 coefficients, resulting in efficient calculation of the magnitude-phase separation due to their low ranks. Consequently, for large n, the cost of either the zero-phase or the unit-magnitude approximation is dominated by the cost of the forward and inverse transforms, which are each $\mathcal{O}(n^4)$ for top-three choice data.

5.2. Approximation and Compression

It is reasonable to wonder whether we may obtain signal compression by approximating partial rank data by either (3.5) or (3.9). Clearly, for large n and small k, the zero-phase

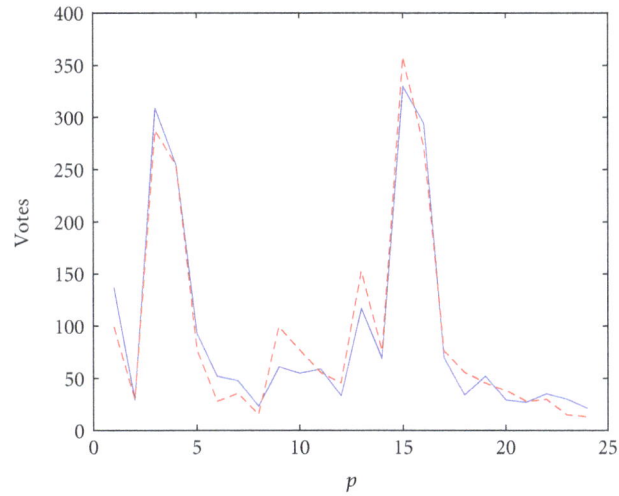

(a) Full ranking German survey data (blue) and its linear phase approximation (dashed red)

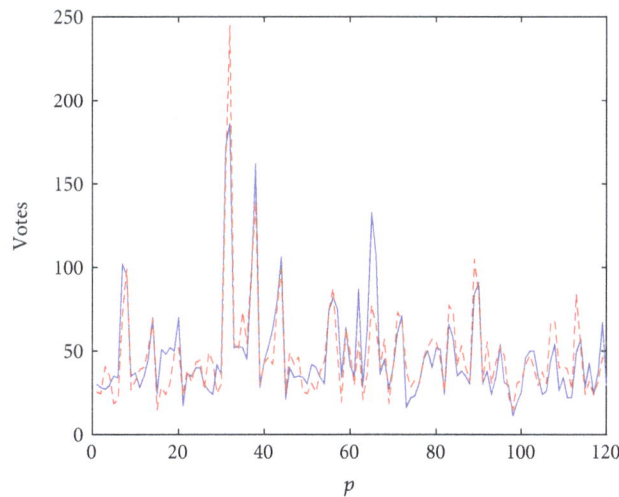

(b) Full ranking APA data (blue) and its linear phase approximation (dashed red)

Figure 4: Two examples illustrating linear-phase approximation for full ranking data. Lexicographic ordering is used for permutations on the horizontal axes of both graphs. Note that the data in (a) has domain S_4, while (b) has domain S_5.

approximation (3.5) is poor because magnitude constrains only a small number of degrees of freedom, as described in the previous section. Conversely, the phase spectrum constrains much of the data; as discussed previously, phase constrains all but 24 of the 117,600 degrees of freedom for $n = 50$, meaning that it is really a very minor compression. To summarize using (3.5) to replace the signal is too much compression, while using (3.9) is too little compression.

The error in approximating data with either its zero-phase version (3.5) or unit-magnitude (3.9) may be determined as follows. Considering the inverse transform on S_n is determined by Fourier coefficients $X(\lambda)$, we see that the error in zero-phase approximation is governed by $\|X(\lambda) - \widehat{X}(\lambda)\|^2$, where the norm means the sum of squared entries. Consequently, due to the submultiplicative property of the matrix norm, we estimate a relative error at each λ of

$$\frac{\left\|X(\lambda) - \widehat{X}(\lambda)\right\|^2}{\|X(\lambda)\|^2} \leq \|O(\lambda) - I\|^2 \leq 2n_\lambda. \tag{5.1}$$

Here, n_λ is the dimension of $O(\lambda)$, and we used the identity $\|\widehat{X}\|^2 = \|X\|^2$. A similar calculation for unit-magnitude approximation shows that the error at each λ is

$$\frac{\|X(\lambda) - O(\lambda)\|^2}{\|X(\lambda)\|^2} \leq 2n_\lambda. \tag{5.2}$$

These are weak upper bounds, and it would be desirable to improve on them in future work.

6. Summary

This paper analyzes the properties of the Fourier spectrum for partial ranking data and shows that the transform coefficients satisfy a projection invariance. The coefficients may be converted to magnitude and phase components, with the magnitude also showing projection invariance. We show that first rank data is essentially determined by its phase spectrum, but that as n increases, the phase dominates magnitude in forming partial rank data.

Acknowledgment

The author thanks the anonymous reviewers for their comments, which greatly improved the paper.

References

[1] P. Diaconis, "A generalization of spectral analysis with application to ranked data," *The Annals of Statistics*, vol. 17, no. 3, pp. 949–979, 1989.

[2] G. Lebanon and Y. Mao, "Non-parametric modeling of partially ranked data," *Journal of Machine Learning Research*, vol. 9, pp. 2401–2429, 2008.

[3] P. Hall and H. Miller, "Modeling the variability of rankings," *The Annals of Statistics*, vol. 38, no. 5, pp. 2652–2677, 2010.

[4] P. Diaconis, *Group Representations in Probability and Statistics*, Institute of Mathematical Statistics, Hayward, Calif, USA, 1988.

[5] P. Diaconis and B. Sturmfels, "Algebraic algorithms for sampling from conditional distributions," *The Annals of Statistics*, vol. 26, no. 1, pp. 363–397, 1998.

[6] J. Huang, C. Guestrin, and L. Guibas, "Fourier theoretic probabilistic inference over permutations," *Journal of Machine Learning Research*, vol. 10, pp. 997–1070, 2009.

[7] R. Kondor and K. Borgwardt, "The skew spectrum of graphs," in *Proceedings of the International Conference on Machine Learning (ICML)*, A. McCallum and S. Roweis, Eds., pp. 496–503, Omnipress, 2008.

[8] R. Kakarala, "A signal processing approach to Fourier analysis of ranking data: the importance of phase," *IEEE Transactions on Signal Processing*, vol. 59, no. 4, pp. 1518–1527, 2011.

[9] A. V. Oppenheim, A. S. Willsky, and S. H. Nawab, *Sigals and Systems*, Prentice-Hall, 2nd edition, 1996.

[10] R. Kondor, "$S_n$ob: a C++ library for fast Fourier transforms on the symmetric group," 2006, http://www.its.caltech.edu/~risi/index.html.

[11] P. Lancaster and M. Tismenetsky, *The Theory of Matrices*, Computer Science and Applied Mathematics, Academic Press, Orlando, Fla, USA, 2nd edition, 1985.

[12] E. Hewitt and K. A. Ross, *Abstract Harmonic Analysis. Vol. II: Structure and Analysis for Compact Groups. Analysis on Locally Compact Abelian Groups*, Springer, New York, NY, USA, 1970.

[13] R. Kondor, "The skew spectrum of functions on finite groups and their homogeneous spaces," *Representation Theory*. In press, http://arxiv.org/abs/0712.4259.

[14] M. Croon, "Latent class models for the analysis of rankings," in *New Developments in Psychological Choice Modeling*, G. D. Solte, H. Feger, and K. C. Klauer, Eds., pp. 99–121, North-Holland, 1989.

[15] D. K. Maslen, "The efficient computation of Fourier transforms on the symmetric group," *Mathematics of Computation*, vol. 67, no. 223, pp. 1121–1147, 1998.

[16] M. Clausen and R. Kakarala, "Computing Fourier transforms and convolutions of $S_n - 1$-invariant signals on S_n in time linear in n," *Applied Mathematics Letters*, vol. 23, no. 2, pp. 183–187, 2010.

10

Signorini Cylindrical Waves and Shannon Wavelets

Carlo Cattani

Department of Mathematics, University of Salerno, 84084 Fisciano (SA), Italy

Correspondence should be addressed to Carlo Cattani, ccattani@unisa.it

Academic Editor: Doron Levy

Hyperelastic materials based on Signorini's strain energy density are studied by using Shannon wavelets. Cylindrical waves propagating in a nonlinear elastic material from the circular cylindrical cavity along the radius are analyzed in the following by focusing both on the main nonlinear effects and on the method of solution for the corresponding nonlinear differential equation. Cylindrical waves' solution of the resulting equations can be easily represented in terms of this family of wavelets. It will be shown that Hankel functions can be linked with Shannon wavelets, so that wavelets can have some physical meaning being a good approximation of cylindrical waves. The nonlinearity is introduced by Signorini elastic energy density and corresponds to the quadratic nonlinearity relative to displacements. The configuration state of elastic medium is defined through cylindrical coordinates but the deformation is considered as functionally depending only on the radial coordinate. The physical and geometrical nonlinearities arising from the wave propagation are discussed from the point of view of wavelet analysis.

1. Introduction

In this paper, cylindrical waves arising from the nonlinear equation of hyperelastic Signorini materials [1–6] are studied. In particular, it will be shown that cylindrical waves can be easily given in terms of Shannon wavelets.

Hyperelastic materials based on Signorini's strain energy density [7, 8] were recently investigated [1–6, 9, 10], because of the simple form of the Signorini potential, which has the main advantage to be dependent only on three constants, including the two classical Lamé constants (λ, μ). Hyperelastic materials and composites are interesting for the many recent advances both in theoretical approaches and in practical discoveries of new composites, having extreme behaviors under deformation [9].

However, Signorini hyperelastic materials, as a drawback, lead to some nonlinear equations, to be studied in cylindrical coordinates [11–13]. The starting point, for searching the solution of these equations, is the Weber equation, which is classically solved by the special functions of Bessel type. Thus the main advantage of three parameters' potential is counterbalanced by the Bessel function approximation. It has been recently shown [14] that Bessel functions locally coincide with Shannon wavelets, thus enabling us to represent cylindrical waves by the multiscale approach [9, 15, 16] of Shannon wavelets [14, 17–20]. In this way, Shannon wavelets might have some physical meaning through the cylindrical waves propagation.

In recent years wavelets have been successfully applied to the wavelet representation of integrodifferential operators [16–24], thus giving rise to the so-called wavelet solutions of PDE (see, e.g., [16, 21, 22]) and integral equations (see, e.g., [20, 23, 24]).

In fact, wavelets enjoy many interesting features such as the localization, the multiscale representation, and the fast decay to zero (either in space or in frequency domain), which are a useful tool in many different applications, (see, e.g., [17–20] and references therein).

Usually wavelets have been used only as any other kind of orthogonal functions, with some additional features but seldom they have shown to have also some physical meanings [15, 25].

We will see that Shannon wavelets can approximate very well the Bessel functions, thus being the most suitable tool for investigating cylindrical waves. Shannon wavelets are analytically defined functions, infinitely differentiable, and sharply bounded in the frequency domain. Their derivatives can be defined to any order by a simple analytical function [17–20], thus enabling us to approximate a function and its derivatives and easily performing the projection of differential operators.

This paper is organized as follows. Section 2 deals with some preliminary remarks on the elastic materials in generalized coordinates. In Section 3, Signorini density energy is defined and the basic equations in cylindrical coordinates for wave propagation in materials are given. The main properties of Shannon wavelets, reconstruction of a function, and connection coefficients are shortly described in Section 4. In Section 5 the similarities and distinctions between Bessel functions and Shannon wavelets are given. Section 6 deals with some remarks on perturbation method. In the same section the Shannon wavelet solution of the nonlinear wave propagation is given and the corresponding nonlinear effects are commented.

2. Preliminary Remarks

Let $(\theta^1, \theta^2, \theta^3)$, be the (Lagrangian) cylindrical coordinate system $\theta^1 = r$, $\theta^2 = \vartheta$, $\theta^3 = z$, and $(ds)^2 = g_{ik} d\theta^i d\theta^k = (dr)^2 + r^2 (d\vartheta)^2 + (dz)^2$ with

$$\|g_{ik}\| = \begin{Vmatrix} 1 & 0 & 0 \\ 0 & r^2 & 0 \\ 0 & 0 & 1 \end{Vmatrix}, \qquad \|g^{ik}\| = \begin{Vmatrix} 1 & 0 & 0 \\ 0 & \dfrac{1}{r^2} & 0 \\ 0 & 0 & 1 \end{Vmatrix}, \tag{2.1}$$

being the corresponding vector length and metric.

The Cauchy-Green strain tensor is defined as

$$\varepsilon_{ik} = \frac{1}{2} \left(\nabla_i u_k + \nabla_k u_i + \nabla_i u_j \nabla_k u^j \right), \tag{2.2}$$

with $\vec{u} = \{u_i\}$ being the displacement vector (in each point of the continuum).

The covariant derivatives of a vector $\{v_i\}$ are

$$\nabla_i v^k = \frac{\partial v^k}{\partial \theta^i} + v^j \Gamma^k_{ji}, \qquad \nabla_i v_j = \frac{\partial v_j}{\partial \theta^i} - v_k \Gamma^k_{ji} \tag{2.3}$$

and can be easily computed by means of the Christoffel's symbols

$$\Gamma^m_{ki} = \frac{1}{2} g^{mn} \left(\frac{\partial g_{kn}}{\partial \theta^i} + \frac{\partial g_{in}}{\partial \theta^k} - \frac{\partial g_{ki}}{\partial \theta^n} \right) \tag{2.4}$$

and the metric values (2.1). Thanks to (2.1) the only nonvanishing components of these symbols are

$$\Gamma^1_{22} = -r, \qquad \Gamma^2_{12} = \Gamma^2_{21} = \left(\frac{1}{r} \right). \tag{2.5}$$

Concerning the deformation, it can be classified according to the nonvanishing components of the displacement vector. We have cylindrical waves [4, 9, 10, 26–29] when

$$\vec{u}\left(\theta^1, \theta^2, \theta^3 \right) = \vec{u}(r, \vartheta, z) = \{ u_1 = u_r(r), u_2 = r \cdot u_\vartheta = u_3 = u_z = 0 \}. \tag{2.6}$$

When the components of the Cauchy-Green tensor are known, we can easily evaluate the three invariants:

$$I_1(\varepsilon_{ik}) = \varepsilon_{ik} g^{ik} = \varepsilon_{11} \cdot 1 + \varepsilon_{22} \cdot \frac{1}{r^2} + \varepsilon_{33} \cdot 1,$$

$$I_2(\varepsilon_{ik}) = \varepsilon_{im} \varepsilon_{nk} g^{ik} g^{nm}$$

$$= (\varepsilon_{11} \cdot 1)^2 + \left(\varepsilon_{22} \cdot \frac{1}{r^2} \right)^2 + (\varepsilon_{33} \cdot 1)^2 + \left(\varepsilon_{12} \cdot \frac{1}{r} \right)^2 + \left(\varepsilon_{23} \cdot \frac{1}{r} \right)^2 + (\varepsilon_{13} \cdot 1)^2,$$

$$I_3(\varepsilon_{ik}) = \varepsilon_{pm} \varepsilon_{in} \varepsilon_{kq} g^{im} g^{pq} g^{kn}$$

$$= (\varepsilon_{11})^3 + \left(\varepsilon_{22} \frac{1}{r^2} \right)^3 + (\varepsilon_{33})^3 + (\varepsilon_{13} \cdot 1) \left(\varepsilon_{13} \varepsilon_{11} + \varepsilon_{23} \varepsilon_{12} \frac{1}{r^2} + \varepsilon_{13} \varepsilon_{33} \right)$$

$$+ \left(\varepsilon_{12} \cdot \frac{1}{r^2} \right) \left(\varepsilon_{12} \varepsilon_{11} + \varepsilon_{12} \varepsilon_{22} \frac{1}{r^2} + \varepsilon_{13} \varepsilon_{23} \right) + \left(\varepsilon_{23} \cdot \frac{1}{r^2} \right) \left(\varepsilon_{12} \varepsilon_{13} + \varepsilon_{23} \varepsilon_{22} \frac{1}{r^2} + \varepsilon_{23} \varepsilon_{33} \right), \tag{2.7}$$

which, in dealing with hyperelastic materials, enable us to compute the potential.

In the case of cylindrical waves (2.6), by taking into account (2.2), the only nonzero components of the strain tensor are

$$\varepsilon_{11} = \varepsilon_{rr} = u_{r,r} + \frac{1}{2}(u_{r,r})^2, \qquad \varepsilon_{22} = r^2 \varepsilon_{\vartheta\vartheta} = r u_r + \frac{1}{2}(u_r)^2, \tag{2.8}$$

$$I_1(\varepsilon_{ik}) = \varepsilon_{ik}g^{ik} = \varepsilon_{11} + \varepsilon_{22}$$

$$= u_{r,r} + \frac{1}{2}(u_{r,r})^2 + ru_r + \frac{1}{2}(u_r)^2,$$

$$I_2(\varepsilon_{ik}) = \varepsilon_{im}\varepsilon_{nk}g^{ik}g^{nm} = \varepsilon_{11}^2 + \frac{1}{r^4}\varepsilon_{22}^2$$

$$= (u_{r,r})^2 + (u_{r,r})^3 + \frac{1}{r^2}(u_r)^2 + \frac{1}{r^3}(u_r)^3 + \frac{1}{4}(u_{r,r})^4 + \frac{1}{4r^4}(u_r)^4, \tag{2.9}$$

$$I_3(\varepsilon_{ik}) = \varepsilon_{pm}\varepsilon_{in}\varepsilon_{kq}g^{im}g^{pq}g^{kn} = \varepsilon_{11}^3 + \frac{1}{r^6}\varepsilon_{22}^3 = (u_{r,r})^3 + \frac{1}{r^3}(u_r)^3$$

$$+ \frac{3}{2}\left[(u_{r,r})^4 + \frac{1}{r^4}(u_r)^4\right] + \frac{3}{4}\left[(u_{r,r})^5 + \frac{1}{r^5}(u_r)^5\right] + \frac{1}{8}\left[(u_{r,r})^6 + \frac{1}{r^6}(u_r)^6\right],$$

so that, by neglecting displacements of order higher than three, we have

$$I_1(\varepsilon_{ik}) = \varepsilon_{ik}g^{ik} = \varepsilon_{11} + \varepsilon_{22}$$

$$= u_{r,r} + \frac{1}{2}(u_{r,r})^2 + ru_r + \frac{1}{2}(u_r)^2,$$

$$I_2(\varepsilon_{ik}) = \varepsilon_{im}\varepsilon_{nk}g^{ik}g^{nm} = \varepsilon_{11}^2 + \frac{1}{r^4}\varepsilon_{22}^2 \cong (u_{r,r})^2 + (u_{r,r})^3 + \frac{1}{r^2}(u_r)^2 + \frac{1}{r^3}(u_r)^3, \tag{2.10}$$

$$I_3(\varepsilon_{ik}) = \varepsilon_{pm}\varepsilon_{in}\varepsilon_{kq}g^{im}g^{pq}g^{kn} = \varepsilon_{11}^3 + \frac{1}{r^6}\varepsilon_{22}^3 \cong (u_{r,r})^3 + \frac{1}{r^3}(u_r)^3.$$

Signorini potential, which belongs to the polynomial hyperelastic model (also called generalized Rivlin model) [26–29], is defined as [1–9, 14]

$$W = \left(\frac{1}{\sqrt{I_{A3}}}\right)\left[cI_{A2} + \left(\frac{1}{2}\right)\left(\lambda + \mu - \left(\frac{c}{2}\right)\right)(I_{A1})^2 + \left(\lambda + \left(\frac{c}{2}\right)\right)(1 - I_{A1})\right] - \left(\mu + \left(\frac{c}{2}\right)\right) \tag{2.11}$$

with

$$I_{A1} = \frac{I_1 + 2(I_1)^2 - 2I_2 + 2(I_1)^3 - 6I_1I_2 + 4I_3}{1 + 2I_1 + 2(I_1)^2 - 2I_2 + (4/3)(I_1)^3 - 4I_1I_2 + (8/3)I_3},$$

$$I_{A2} = \frac{1}{2}(I_1)^2 - \frac{(1/2)(I_1)^2 - (1/2)I_2 + (I_1)^3 - 3I_1I_2 + 2I_3}{1 + 2I_1 + 2(I_1)^2 - 2I_2 + (4/3)(I_1)^3 - 4I_1I_2 + (8/3)I_3}, \tag{2.12}$$

$$I_{A3} = \frac{2}{3}I_1I_2 - \frac{1}{4\sqrt{3}}(I_1)^3 \frac{(I_1)^3 - I_1I_2 + 2I_3}{1 + 2I_1 + 2(I_1)^2 - 2I_2 + (4/3)(I_1)^3 - 4I_1I_2 + (8/3)I_3}.$$

Therefore from the previous equation, by taking into account (2.10), we have the approximation

$$
\begin{aligned}
I_{A1} &\cong (\varepsilon_{11} + \varepsilon_{22}) + 2(\varepsilon_{11} + \varepsilon_{22})^2 - 2\left(\varepsilon_{11}^2 + \frac{1}{r^4}\varepsilon_{22}^2\right) + 2(\varepsilon_{11} + \varepsilon_{22})^3 \\
&\quad - 6(\varepsilon_{11} + \varepsilon_{22})\left(\varepsilon_{11}^2 + \frac{1}{r^4}\varepsilon_{22}^2\right) + 4\left(\varepsilon_{11}^3 + \frac{1}{r^6}\varepsilon_{22}^3\right) \\
&= (\varepsilon_{11} + \varepsilon_{22}) + 2\varepsilon_{11}\varepsilon_{22}, \\
I_{A2} &\cong -\frac{1}{2}\left(\varepsilon_{11}^2 + \varepsilon_{22}^2\right) + (\varepsilon_{11} + \varepsilon_{22})^3 - 3(\varepsilon_{11} + \varepsilon_{22})\left(\varepsilon_{11}^2 + \frac{1}{r^4}\varepsilon_{22}^2\right) + 2\left(\varepsilon_{11}^3 + \frac{1}{r^6}\varepsilon_{22}^3\right) \\
&= -\frac{1}{2}\left(\varepsilon_{11}^2 + \frac{1}{r^4}\varepsilon_{22}^2\right), \\
I_{A3} &\cong \frac{2}{3}(\varepsilon_{11} + \varepsilon_{22})\left(\varepsilon_{11}^2 + \frac{1}{r^4}\varepsilon_{22}^2\right),
\end{aligned} \tag{2.13}
$$

that is

$$
\begin{aligned}
I_{A1} &\cong (\varepsilon_{11} + \varepsilon_{22}) + 2\varepsilon_{11}\varepsilon_{22}, \\
I_{A2} &\cong -\frac{1}{2}\left(\varepsilon_{11}^2 + \frac{1}{r^4}\varepsilon_{22}^2\right), \\
I_{A3} &\cong \frac{2}{3}(\varepsilon_{11} + \varepsilon_{22})\left(\varepsilon_{11}^2 + \frac{1}{r^4}\varepsilon_{22}^2\right),
\end{aligned} \tag{2.14}
$$

and, according to (2.8),

$$
\begin{aligned}
I_{A1} &\cong \left(u_{r,r} + \frac{1}{2}(u_{r,r})^2 + ru_r + \frac{1}{2}(u_r)^2\right) + 2\left[u_{r,r} + \frac{1}{2}(u_{r,r})^2\right]\left[ru_r + \frac{1}{2}(u_r)^2\right], \\
I_{A2} &\cong -\frac{1}{2}\left(\left[u_{r,r} + \frac{1}{2}(u_{r,r})^2\right]^2 + \frac{1}{r^4}\left[ru_r + \frac{1}{2}(u_r)^2\right]^2\right), \\
I_{A3} &\cong \frac{2}{3}\left(u_{r,r} + \frac{1}{2}(u_{r,r})^2 + ru_r + \frac{1}{2}(u_r)^2\right)\left(\left[u_{r,r} + \frac{1}{2}(u_{r,r})^2\right]^2 + \frac{1}{r^4}\left[ru_r + \frac{1}{2}(u_r)^2\right]^2\right).
\end{aligned} \tag{2.15}
$$

3. Cylindrical Waves Equation

The basic equations of motion are [1–5, 9, 10, 14, 26–29]

$$
\nabla_i T^{ik} - \rho \nabla_i \varepsilon^{ik} = \frac{\partial^2 u^k}{\partial t^2}, \tag{3.1}
$$

where T^{ik} is the Piola-Kirchoff stress tensor. For hyperelastic materials it is $T^{ik} = (\partial W / \partial \varepsilon_{ik})$ where W is given by (2.11), for Signorini's materials.

Taking into account that

$$\frac{\partial W}{\partial \varepsilon_{ik}} = \sum_{h=1}^{3} \frac{\partial W}{\partial I_{Ah}} \frac{\partial I_{Ah}}{\partial \varepsilon_{ik}}, \qquad (3.2)$$

and, according to (2.11), it is

$$\frac{\partial W}{\partial I_{A1}} = \left(\frac{1}{\sqrt{I_{A3}}}\right)\left[\left(\lambda + \mu - \left(\frac{c}{2}\right)\right)I_{A1} - \left(\lambda + \left(\frac{c}{2}\right)\right)\right],$$

$$\frac{\partial W}{\partial I_{A2}} = \frac{c}{2\sqrt{I_{A3}}}, \qquad (3.3)$$

$$\frac{\partial W}{\partial I_{A3}} = \frac{1}{2\sqrt{I_{A3}}}\left[cI_{A2} + \left(\frac{1}{2}\right)\left(\lambda + \mu - \left(\frac{c}{2}\right)\right)(I_{A1})^2 + \left(\lambda + \left(\frac{c}{2}\right)\right)(1 - I_{A1})\right],$$

and, by (2.14), the only unvanishing derivatives are

$$\frac{\partial I_{A1}}{\partial \varepsilon_{11}} = 1 + 2\varepsilon_{22}, \qquad \frac{\partial I_{A1}}{\partial \varepsilon_{22}} = 1 + 2\varepsilon_{11},$$

$$\frac{\partial I_{A2}}{\partial \varepsilon_{11}} = -\varepsilon_{11}, \qquad \frac{\partial I_{A2}}{\partial \varepsilon_{22}} = -\frac{1}{r^4}\varepsilon_{22}, \qquad (3.4)$$

$$\frac{\partial I_{A3}}{\partial \varepsilon_{11}} = 2\varepsilon_{11} + \frac{2}{3r^4}\varepsilon_{22}^2 + \frac{4}{3}\varepsilon_{11}\varepsilon_{22}, \qquad \frac{\partial I_{A3}}{\partial \varepsilon_{22}} = \frac{2}{3}\varepsilon_{11}^2 + \frac{2}{r^4}\varepsilon_{22}^2 + \frac{4}{3}\varepsilon_{11}\varepsilon_{22}.$$

The Piola-Kirchoff tensor for the Signorini model (see also [3–5, 9]) is

$$T^{ik} = \left[\lambda I_{A1} + cI_{A2} + \frac{1}{2}\left(\lambda + \mu - \frac{c}{2}\right)(I_{A1})^2\right]g^{ik} + 2\left[\mu - \left(\lambda + \mu + \frac{c}{2}\right)I_{A1}\right]\varepsilon^{ik} + 2c\left(\varepsilon^{ij}\varepsilon_j^k\right). \quad (3.5)$$

In the strain components, we will neglect those terms with order higher than 3, so that the only unvanishing components of T^{ik} are

$$T^{11} = \left[\lambda I_{A1} + cI_{A2} + \frac{1}{2}\left(\lambda + \mu - \frac{c}{2}\right)(I_{A1})^2\right] + 2\left[\mu - \left(\lambda + \mu + \frac{c}{2}\right)I_{A1}\right]\varepsilon_{11} + 2c(\varepsilon_{11})^2,$$

$$T^{22} = \frac{1}{r^2}\left[\lambda I_{A1} + cI_{A2} + \frac{1}{2}\left(\lambda + \mu - \frac{c}{2}\right)(I_{A1})^2\right] + 2\left[\mu - \left(\lambda + \mu + \frac{c}{2}\right)I_{A1}\right]\frac{\varepsilon_{22}}{r^2} + 2c\left(\frac{\varepsilon_{22}}{r^2}\right)^2,$$

$$T^{33} = \left[\lambda I_{A1} + cI_{A2} + \frac{1}{2}\left(\lambda + \mu - \frac{c}{2}\right)(I_{A1})^2\right].$$

$$(3.6)$$

By using (2.8),(2.10), and (3.6) we finally get the Kirchoff tensor in terms of displacements:

$$T^{11} = T^{rr}$$

$$= (\lambda + 2\mu)u_{r,r} + \lambda\frac{u_r}{r} + \frac{1}{4}(-10\lambda - 4\mu + 5c)(u_{r,r})^2 + \frac{1}{2}(2\lambda - 2\mu - 5c)\frac{1}{r}\,u_r u_{r,r}$$

$$+ \frac{1}{4}(6\lambda + 2\mu + c)\frac{1}{r^2}(u_r)^2 + \frac{1}{2}(6\lambda + 13c)(u_{r,r})^3 + \frac{1}{4}(70\lambda - 18\mu + c)\frac{1}{r}\,u_r u_{r,r}$$

$$+ \frac{1}{4}(-42\lambda + 10\mu + 15c)\frac{1}{r^2}(u_r)^2 + \frac{1}{2}(4\lambda - 2\mu + 3c)\frac{1}{r^3}(u_r)^3,$$

$$r^2 T^{22} = T_{\vartheta\vartheta}$$

$$= (\lambda + 2\mu)\frac{u_r}{r} + \lambda u_{r,r} + \frac{1}{4}(-2\lambda + 2\mu + c)(u_{r,r})^2 + \frac{1}{2}(2\lambda - 2\mu - 5c)\frac{1}{r}\,u_r u_{r,r}$$

$$+ \frac{1}{4}(-2\lambda - 4\mu + 5c)\frac{1}{r^2}(u_r)^2 + (2\lambda - \mu + 4c)\,(u_{r,r})^3 + \frac{1}{4}(-42\lambda + 10\mu + 15c)\frac{1}{r}u_r u_{r,r}$$

$$+ \frac{1}{4}(70\lambda - 18\mu + c)\frac{1}{r^2}(u_r)^2 + (-\lambda + 4c)\frac{1}{r^3}(u_r)^3.$$

$$(3.7)$$

From (3.1) the only nontrivial equation is the first one:

$$(\lambda + 2\mu)\left(u_{r,rr} + \frac{u_{r,r}}{r} + u_r - \frac{u_r}{r^2}\right) - \rho\ddot{u}_r = S_1 u_{r,rr}u_{r,r} + S_2\frac{1}{r}u_{r,rr}u_r + S_3\frac{1}{r}(u_{r,r})^2 + S_4\frac{1}{r^2}u_{r,r}u_r$$

$$+ S_5\frac{1}{r^3}(u_r)^2 + S_6 u_{r,rr}(u_{r,r})^2 + S_7\frac{1}{r^3}u_{r,rr}(u_r)^2$$

$$+ S_8\frac{1}{r}u_{r,rr}u_{r,r}u_r + S_9\frac{1}{r}(u_{r,r})^3 + S_{10}\frac{1}{r^4}(u_r)^3$$

$$+ S_{11}\frac{1}{r^2}(u_{r,r})^2 u_r + S_{12}\frac{1}{r^3}u_{r,r}(u_r)^2,$$

$$(3.8)$$

where the coefficients $S_1, S_2, \ldots, S_{12}$ depend on Signorini parameters $\lambda, \mu,$ and c:

$$S_1 = \frac{1}{2}(-6\lambda + 4\mu + 5c), \qquad S_2 = \frac{1}{2}(4\lambda - 2\mu - 5c), \qquad S_3 = \frac{1}{2}(2\lambda - \mu - 3c),$$

$$S_4 = \frac{1}{2}(2\mu - 5c), \qquad S_5 = \frac{1}{2}(5\mu - 3c), \qquad S_6 = \frac{1}{4}(9\lambda - 12\mu + 93c),$$

$$S_7 = \frac{1}{2}(24\lambda - 4\mu - 7c), \qquad S_8 = 36\lambda - 10\mu - 2c, \qquad S_9 = \frac{1}{2}(32\lambda - 13\mu - 2c),$$

$$S_{10} = -\frac{1}{4}(10\lambda + c), \qquad S_{11} = \frac{1}{4}(-74\lambda + 26\mu + 33c), \qquad S_{12} = \frac{1}{4}(22\lambda - 18\mu + 7c).$$

$$(3.9)$$

In the following, we will search solutions in the following form:

$$u_r = e^{i\omega t}u(r), \tag{3.10}$$

where time-harmonic waves $e^{i\omega t}$ are separated by the longitudinal waves $u(r)$, so that

$$\ddot{u}_r = -\omega^2 u_r, \qquad \omega = \sqrt{\frac{1 - (\lambda + 2\mu)}{\rho}}, \tag{3.11}$$

and $u(r)$ is the solution of the following equation:

$$
\begin{aligned}
\left(u_{,rr} + \frac{u_{,r}}{r} + u - \frac{u}{r^2} \right) &= a_1 u_{,rr} u_{,r} + a_2 \frac{1}{r} u_{,rr} u + a_3 \frac{1}{r}(u_{,r})^2 + a_4 \frac{1}{r^2} u_{,r} u + a_5 \frac{1}{r^3}(u)^2 \\
&\quad + a_6 u_{,rr}(u_{,r})^2 + a_7 \frac{1}{r^3} u_{,rr}(u)^2 + a_8 \frac{1}{r} u_{,rr} u_{,r} u \\
&\quad + a_9 \frac{1}{r}(u_{,r})^3 + a_{10} \frac{1}{r^4}(u)^3 + a_{11} \frac{1}{r^2}(u_{,r})^2 u + a_{12} \frac{1}{r^3} u_{,r}(u)^2,
\end{aligned}
\tag{3.12}
$$

with $a_i = S_i/(\lambda + 2\mu)$, $i = 1, \ldots, 12$.

Equation (3.12) gives the more general model of cylindrical wave propagation for Signorini hyperelastic materials. At the r.h.s. there appear nonlinear terms up to the third order in $u, u_{,r}$, and $u_{,rr}$ while the coefficients depend on both inverse r up to the 4th power and the physical parameters λ, μ, and c. In the following we will search the Shannon wavelet solution of (3.12), by neglecting $O(r^{-1})$ terms in the r.h.s., by showing that Shannon wavelets are linked with Bessel functions.

3.1. Linear Equation

If we neglect the nonlinear terms of the right-hand side, from (3.12) we simply get the linear equation:

$$\left(u_{,rr} + \frac{u_{,r}}{r} + u - \frac{u}{r^2} \right) = 0, \tag{3.12'}$$

which is the (homogeneous) Weber equation [30, 31], classically solved by Bessel functions.

In fact, Bessel function $J_n(x)$ of order n is defined as the solution of the Weber equation:

$$x^2 y'' + x y' + \left(x^2 - n^2 \right) y = 0, \quad n \in \mathbb{C}. \tag{3.13}$$

In particular, when $n = 1$, the more general solution of

$$x^2 y'' + x y' + \left(x^2 - 1 \right) y = 0 \tag{3.14}$$

is

$$y(x) = c_1 J_1(x) + c_2 J_2(x). \tag{3.15}$$

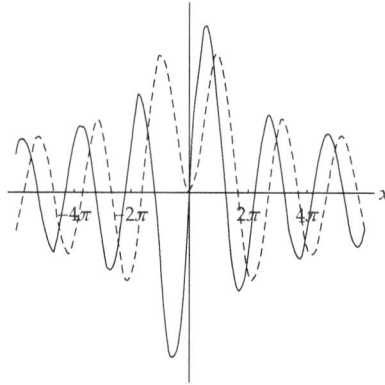

Figure 1: Bessel Functions $J_1(x)$ (bold) and $J_2(x)$ (dashed).

The Taylor series for Bessel function is

$$J_n(x) = \sum_{k=0}^{\infty} \frac{(-1)^k}{k!\Gamma(n+k+1)} \left(\frac{1}{2}x\right)^{2k+n}, \quad x \in (-\varepsilon, \varepsilon) \tag{3.16}$$

with $\Gamma(n)$ being gamma function.

So, for integer values of n, being $\Gamma(n+1) = n!$, there result

$$\begin{aligned}
J_1(x) &= \sum_{k=0}^{\infty} \frac{(-1)^k}{k!(k+1)!} \left(\frac{1}{2}x\right)^{2k+1} \\
&= \left(\frac{1}{2}x\right) - \frac{1}{2!}\left(\frac{1}{2}x\right)^3 + \frac{1}{2!3!}\left(\frac{1}{2}x\right)^5 - \frac{1}{3!4!}\left(\frac{1}{2}x\right)^7 \cdots, \\
J_2(x) &= \sum_{k=0}^{\infty} \frac{(-1)^k}{k!(k+2)!} \left(\frac{1}{2}x\right)^{2k+2} \\
&= \frac{1}{2!}\left(\frac{1}{2}x\right)^2 - \frac{1}{3!}\left(\frac{1}{2}x\right)^4 + \frac{1}{2!4!}\left(\frac{1}{2}x\right)^6 + \cdots.
\end{aligned} \tag{3.17}$$

It can be easily seen that $J_{2n}(x)$ $(n \in \mathbb{N})$ are even functions and $J_{2n+1}(x)$ $(n \in \mathbb{N})$ are odd functions, while both are localized functions with some decay to zero (Figure 1).

A good approximation of $J_1(x)$ in the interval $[-\pi/2, \pi/2]$ can be already obtained by the third-order polynomial, while with the 7th power polynomial we can have a good approximation in $[-\pi, \pi]$:

$$J_1(x) \cong \left(\frac{1}{2}x\right) - \frac{1}{2!}\left(\frac{1}{2}x\right)^3 + \frac{1}{2!3!}\left(\frac{1}{2}x\right)^5 - \frac{1}{3!4!}\left(\frac{1}{2}x\right)^7. \tag{3.16'}$$

Analogously a good approximation of $J_2(x)$ is obtained in the interval $[-\pi/2, \pi/2]$ by a second order polynomial, whereas with the 6th power polynomial we have a good approximation in $[-\pi, \pi]$

$$J_2(x) \cong \frac{1}{2!}\left(\frac{1}{2}x\right)^2 - \frac{1}{3!}\left(\frac{1}{2}x\right)^4 + \frac{1}{2!4!}\left(\frac{1}{2}x\right)^6. \tag{3.16''}$$

3.2. Second-Order Equation

Equation (3.12) gives rise to many interesting nonlinear equations for cylindrical waves. In fact, up to the second-order nonlinearities, it becomes

$$\left(u_{,rr} + \frac{u_{,r}}{r} + u - \frac{u}{r^2}\right) = u_{,r}\left(a_1 u_{,rr} + a_3 \frac{u_{,r}}{r} - a_4 \frac{u}{r^2}\right) + \frac{1}{r}u\left(a_2 u_{,rr} - a_5 \frac{u}{r^2}\right). \tag{3.18}$$

So, by keeping only the first term of the right-hand side, which is equivalent to neglect terms $O(1/r)$, we have

$$\left(u_{,rr} + \frac{u_{,r}}{r} + u - \frac{u}{r^2}\right) = a_1 u_{,r} u_{,rr}. \tag{3.12''}$$

3.3. Third-Order Equation

Up to the third-order nonlinearities, and neglecting all terms $O(1/r)$, (3.12) gives

$$\left(u_{,rr} + \frac{u_{,r}}{r} + u - \frac{u}{r^2}\right) = u_{,rr}\left[a_1 u_{,r} + a_6(u_{,r})^2\right]. \tag{3.12'''}$$

We will give the solutions of (3.12'), (3.12''), and (3.12''') by using Shannon wavelets. In order to do so, we need first to show that, in a sufficient large neighborhood of zero, Shannon wavelets are equivalent to the Bessel function. We can also see that at the same approximation the Taylor polynomial for Shannon wavelets is one order lower than the Taylor polynomial for the corresponding Bessel function, so that Shannon wavelets are more efficient from computational point of view.

4. Shannon Wavelet

In this section Shannon wavelets and their differential properties are shortly summarized (for further readings and explicit computations see, e.g., [14, 17–20] and references therein).

Shannon scaling function $\varphi(x)$ and wavelet function $\psi(x)$ are localized functions with some decay to zero (like Bessel functions), defined as

$$\varphi(x) = \operatorname{sinc} x = \frac{\sin \pi x}{\pi x} = \frac{e^{\pi i x} - e^{-\pi i x}}{2\pi i x},$$

$$\psi(x) = \frac{\sin \pi(x - 1/2) - \sin 2\pi(x - 1/2)}{\pi(x - 1/2)} \tag{4.1}$$

$$= \frac{e^{-2i\pi x}\left(-i + e^{i\pi x} + e^{3\,i\pi x} + i\,e^{4\,i\pi x}\right)}{(\pi - 2\pi x)}.$$

The corresponding families of translated and dilated instances wavelet [17–20], on which is based the multiscale analysis, are

$$\varphi_k^n(x) = 2^{n/2}\varphi(2^n x - k) = 2^{n/2}\frac{\sin \pi(2^n x - k)}{\pi(2^n x - k)}$$

$$= 2^{n/2}\frac{e^{\pi i(2^n x - k)} - e^{-\pi i(2^n x - k)}}{2\pi i(2^n x - k)},$$

$$\psi_k^n(x) = 2^{n/2}\frac{\sin \pi(2^n x - k - 1/2) - \sin 2\pi(2^n x - k - 1/2)}{\pi(2^n x - k - 1/2)}, \tag{4.2}$$

$$= \frac{2^{n/2}}{2\pi(2^n x - k + 1/2)}\sum_{s=1}^{2} i^{1+s} e^{s\pi i(2^n x - k)} - i^{1-s} e^{-s\pi i(2^n x - k)},$$

with $\varphi_k^0(x) = \varphi_k(x)$ and $\psi_k^0(x) = \psi_k(x)$. In the following, we will denote

$$\varphi_0^0(x) = \varphi(x), \qquad \psi_0^0(x) = \psi(x). \tag{4.3}$$

Both families of Shannon scaling and wavelet are $L_2(\mathbb{R})$-functions, with a slow decay to zero, so that

$$\lim_{x \to \pm\infty} \varphi_k^n(x) = 0, \qquad \lim_{x \to \pm\infty} \psi_k^n(x) = 0. \tag{4.4}$$

For each $f(x) \in L_2(\mathbb{R})$ and $g(x) \in L_2(\mathbb{R})$, the inner product is defined as

$$\langle f, g \rangle = \int_{-\infty}^{\infty} f(x)\overline{g(x)}dx, \tag{4.5}$$

where the bar stands for the complex conjugate.

With respect to this inner product, Shannon wavelets are orthogonal functions so that [18–20]

$$\langle \psi_k^n(x), \psi_h^m(x) \rangle = \delta^{nm}\delta_{hk},$$

$$\left\langle \varphi_k^0(x), \varphi_h^0(x) \right\rangle = \delta_{kh}, \tag{4.6}$$

$$\left\langle \varphi_k^0(x), \psi_h^m(x) \right\rangle = 0, \quad m \geq 0,$$

with δ^{nm} and δ_{hk} being the Kroenecker symbols.

Let $f(x) \in L_2(\mathbb{R})$ be a function such that the integrals

$$\alpha_k = \left\langle f(x), \varphi_k^0(x) \right\rangle = \int_{-\infty}^{\infty} f(x)\overline{\varphi_k^0(x)}dx,$$

$$\beta_k^n = \left\langle f(x), \psi_k^n(x) \right\rangle = \int_{-\infty}^{\infty} f(x)\overline{\psi_k^n(x)}dx \tag{4.7}$$

exist and have finite values; it can be shown that the series

$$f(x) = \sum_{h=-\infty}^{\infty} \alpha_h \varphi_h^0(x) + \sum_{n=0}^{\infty} \sum_{k=-\infty}^{\infty} \beta_k^n \psi_k^n(x) \tag{4.8}$$

on the right side converges to $f(x)$. For a fixed upper bound we simply have the approximation (for the error estimate see [20])

$$f(x) \cong \sum_{h=-K}^{K} \alpha_h \varphi_h^0(x) + \sum_{n=0}^{N} \sum_{k=-S}^{S} \beta_k^n \psi_k^n(x). \tag{4.9}$$

4.1. Differentiable Properties of Shannon Wavelets

The derivatives of the Shannon wavelets are [19, 20]

$$\frac{d^\ell}{d\,x^\ell}\varphi_h^0(x) = \sum_{k=-\infty}^{\infty} \lambda_{hk}^{(\ell)}\ \varphi_k^0(x),$$

$$\frac{d^\ell}{dx^\ell}\psi_h^m(x) = \sum_{n=0}^{\infty} \sum_{k=-\infty}^{\infty} \gamma_{hk}^{(\ell)mn}\psi_k^n(x)\ , \tag{4.10}$$

with

$$\lambda_{kh}^{(\ell)} \stackrel{\text{def}}{=} \left\langle \frac{d^\ell}{dx^\ell}\varphi_k^0(x), \varphi_h^0(x) \right\rangle, \qquad \gamma_{hk}^{(\ell)mn} \stackrel{\text{def}}{=} \left\langle \frac{d^\ell}{dx^\ell}\psi_k^n(x), \psi_h^m(x) \right\rangle \tag{4.11}$$

being the connection coefficients [18–20].

It has been shown [19, 20] that

$$\lambda_{kh}^{(\ell)} = \begin{cases} (-1)^{k-h}\dfrac{i^\ell}{2\pi}\displaystyle\sum_{s=1}^{\ell}\dfrac{\ell!\pi^s}{s![i(k-h)]^{\ell-s+1}}\left[(-1)^s - 1\right], & k \neq h, \\[20pt] \dfrac{i^\ell \pi^{\ell+1}}{2\pi(\ell+1)}\left[1 + (-1)^\ell\right], & k = h, \end{cases} \tag{4.12}$$

when $\ell \geq 1$, and $\lambda_{kh}^{(0)} = \delta_{kh}$, and

$$
\gamma^{(\ell)mn}_{hk} = \mu(h-k)\delta^{nm}\left\{ \sum_{s=1}^{\ell+1} (-1)^{[1+\mu(h-k)](2\ell-s+1)/2} \frac{\ell!\,i^{\ell-s}\,\pi^{\ell-s}}{(\ell-s+1)!|h-k|^s}(-1)^{-s-2(h+k)}2^{n\ell-s-1} \right.
$$
$$
\left. \times\left\{2^{\ell+1}\left[(-1)^{4h+s}+(-1)^{4k+\ell}\right]-2^s\left[(-1)^{3k+h+\ell}+(-1)^{3h+k+s}\right]\right\}\right\}, \quad k \neq h,
$$
$$
\gamma^{(\ell)mn}_{hk} = \delta^{nm}\left[i^\ell \frac{\pi^\ell 2^{n\ell-1}}{\ell+1}\left(2^{\ell+1}-1\right)\left(1+(-1)^\ell\right)\right], \quad k = h,
$$

$$(4.13)$$

with

$$
\mu(m) = \text{sign}(m) = \begin{cases} 1, & m > 0, \\ -1, & m < 0, \\ 0, & m = 0. \end{cases} \tag{4.14}
$$

According to (4.10) the Taylor series of the scaling and Shannon wavelet, nearby the origin, are

$$
\varphi_h^0(x) = \sum_{\ell=0}^{\infty} \frac{1}{\ell!}\lambda_{h0}^{(\ell)}x^\ell, \quad |x| < \varepsilon,
$$
$$
\psi_h^m(x) = \sum_{\ell=0}^{\infty} \frac{1}{\ell!}\gamma^{(\ell)mn}_{h0}2^{m/2}\left(x-2^{-m-1}\right)^\ell, \quad \left|x-2^{-m-1}\right| < \varepsilon. \tag{4.15}
$$

5. Similarities between Bessel Functions and Shannon Wavelets

Since Bessel functions are $L_2(\mathbb{R})$, they can be easily represented in terms of Shannon wavelets as follows:

$$
J_1(x) \cong -\frac{1}{2}\psi\left(\frac{x}{3\sqrt{2}}+\frac{1}{5}\right)-0.08, \quad x \in \left(-\frac{\pi}{2},\frac{\pi}{2}\right),
$$
$$
J_2(x) \cong -\frac{1}{2}\varphi\left(\frac{x}{2\sqrt{2}}\right)+\frac{1}{2}, \quad x \in (-\pi,\pi). \tag{5.1}
$$

In particular, around $x = 0$ they nearly coincide with the Shannon scaling and wavelet, so that the even $J_{2n}(x)$, $(n \in \mathbb{N})$ can be well approximated by the scaling Shannon functions (Figure 2), while the odd Bessel functions $J_{2n+1}(x)$ and $(n \in \mathbb{N})$ can be approximated by the Shannon wavelets (Figure 3).

Although this approximation for both is restricted to an interval, we can assume that in the interval $|\varepsilon| \leq \pi/2$, where the perturbation method is applied, Bessel functions substantially coincide with the Shannon wavelet families; in other words, Shannon scaling functions and Shannon wavelets are solution of the Weber equation in the interval $|\varepsilon| \leq \pi/2$.

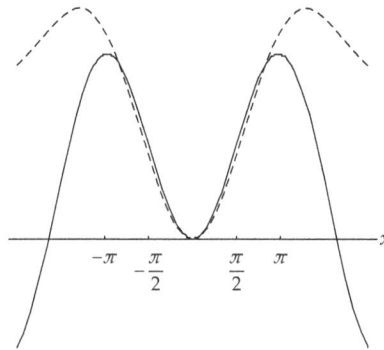

Figure 2: Bessel function $J_2(x)$ and (dashed) the Shannon scaling function $-(1/2)\varphi(x/2\sqrt{2}) + 1/2$.

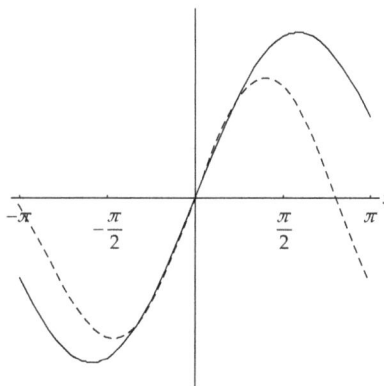

Figure 3: Bessel function $J_1(x)$ and (dashed) the Shannon wavelet function $-(1/2)\psi(x/3\sqrt{2} + 1/5) - 0.08$.

According to (4.13) and (4.15) the Taylor expansion (in $x = 0$) for the scaling wavelet is

$$\varphi(x) = \sum_{k=0}^{\infty} (-1)^k \frac{\pi^{2k} x^{2k}}{(2k+1)!'} \tag{5.2}$$

so that at the sixth order

$$\varphi(x) = 1 - \frac{\pi^2 x^2}{3!} + \frac{\pi^4 x^4}{5!} - \frac{\pi^6 x^6}{7!} \cdots. \tag{5.3}$$

Analogously, for the Shannon wavelet $\psi(x)$, in $x = 0$, it is up to the sixth order;

$$\psi(x) = -1 + \frac{1}{2!}\frac{7}{3}\pi^2\left(x - \frac{1}{2}\right)^2 - \frac{1}{4!}\frac{31}{5}\pi^4\left(x - \frac{1}{2}\right)^4 + \frac{1}{6!}\frac{127}{7}\pi^6\left(x - \frac{1}{2}\right)^6. \tag{5.4}$$

By comparing the Taylor expansion for Bessel functions, as given by (3.16′), and (3.16″) and the Taylor expansion of Shannon wavelets (5.3) and (5.4), we can see that a good

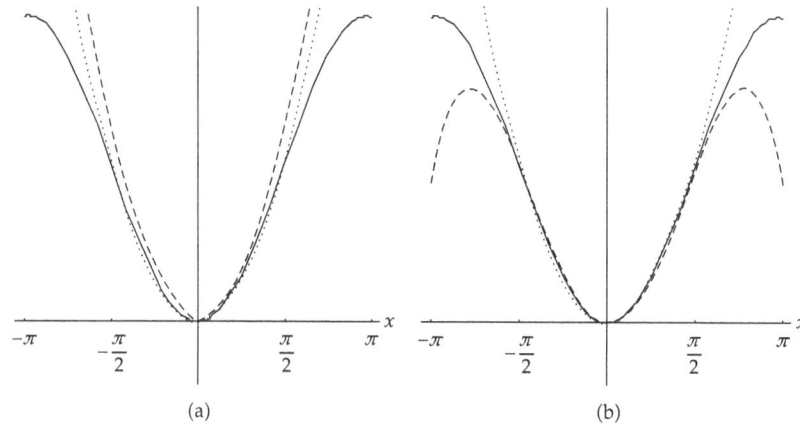

Figure 4: Approximation of $J_2(x)$ with a second-order polynomial from (5.3) (dotted) and, dashed, the Taylor polynomial (3.16″) at the second order (a) and fouth order (b).

approximation of the Bessel can be obtained by a lower-order polynomial approximation of the Shannon wavelet (Figures 4 and 5).

Taking into account (4.15) and (5.2) it can be easily shown that, for $x \in (-\pi, \pi)$, the error of the approximation in (5.1) tends to zero for $k \to \infty$. For instance, it is

$$\left| J_2(x) + \frac{1}{2}\varphi\left(\frac{x}{2\sqrt{2}}\right) - \frac{1}{2} \right| \le \left| J_2(x) + \frac{1}{2}\varphi\left(\frac{x}{2\sqrt{2}}\right) \right|,$$

$$\left| J_2(x) + \frac{1}{2}\varphi\left(\frac{x}{2\sqrt{2}}\right) \right| \overset{(4.15),(5.2)}{=} \sum_{k=0}^{\infty}(-1)^k \frac{x^{2k}}{2^{2k+1}}\left(\frac{x^2}{2k!(k+2)!} + \frac{\pi^{2k}}{2^k(2k+1)!} \right), \tag{5.5}$$

so that for $|x| \le \pi$ it is

$$\left| J_2(x) + \frac{1}{2}\varphi\left(\frac{x}{2\sqrt{2}}\right) \right| \le \sum_{k=0}^{\infty}(-1)^k \frac{\pi^{2k}}{2^{2k+1}}\left(\frac{\pi^2}{2k!(k+2)!} + \frac{\pi^{2k}}{2^k(2k+1)!} \right). \tag{5.6}$$

The series at the r.h.s is an alternating series which converges to zero, since, according to Leibniz rule, it is

$$\lim_{k \to \infty} \frac{\pi^{2k}}{2^{2k+1}}\left(\frac{\pi^2}{2k!(k+2)!} + \frac{\pi^{2k}}{2^k(2k+1)!} \right) = 0. \tag{5.7}$$

Analogously, we can show the same result for the wavelet approximation $(4.15)_1$ of the Bessel function $J_1(x)$.

By using the approximation (5.1) we can assume as solution of the Weber equation $(3.12')$ the Shannon wavelet

$$-\frac{1}{2}\psi\left(\frac{x}{3\sqrt{2}} + \frac{1}{5}\right) - 0.08, \quad x \in \left(-\frac{\pi}{2}, \frac{\pi}{2}\right). \tag{5.8}$$

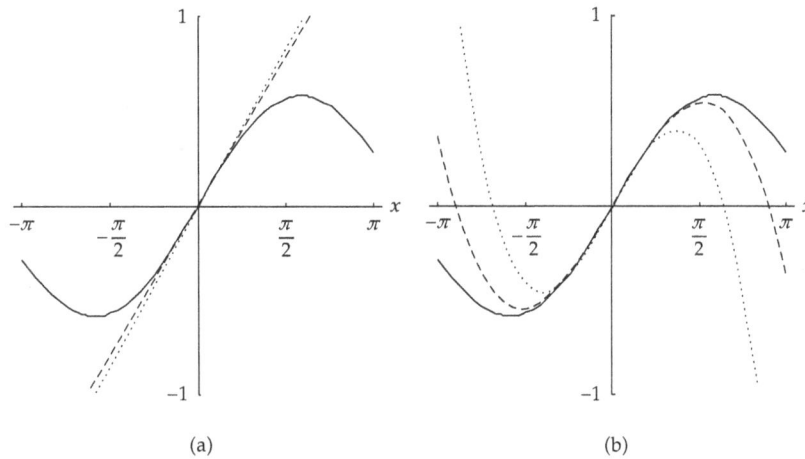

(a)   (b)

Figure 5: Approximation of $J_1(x)$ with a first-order (a) and third-order (b) polynomial from (5.4) (dotted) and, dashed, the Taylor polynomial (3.16′).

The derivatives of this function, according to (4.10), are

$$\frac{d^\ell}{dx^\ell}\left[-\frac{1}{2}\psi\left(\frac{x}{3\sqrt{2}}+\frac{1}{5}\right)-0.08\right]=-\frac{1}{2\left(3\sqrt{2}\right)^\ell}\sum_{k=-\infty}^{\infty}\gamma^{(\ell)00}{}_{0k}\psi_k^0\left(\frac{x}{3\sqrt{2}}+\frac{1}{5}\right), \qquad (5.9)$$

and up to the second order,

$$\frac{d}{dx}\left[-\frac{1}{2}\psi\left(\frac{x}{3\sqrt{2}}+\frac{1}{5}\right)-0.08\right]\cong-\frac{1}{2\left(3\sqrt{2}\right)}\left[\frac{1}{4}\psi_1^0\left(\frac{x}{3\sqrt{2}}+\frac{1}{5}\right)\right],$$

$$\frac{d^2}{dx^2}\left[-\frac{1}{2}\psi\left(\frac{x}{3\sqrt{2}}+\frac{1}{5}\right)-0.08\right]\cong-\frac{1}{2\left(3\sqrt{2}\right)^2}\left[-\frac{7}{3}\psi_0^0\left(\frac{x}{3\sqrt{2}}+\frac{1}{5}\right)+\frac{1}{8}\psi_1^0\left(\frac{x}{3\sqrt{2}}+\frac{1}{5}\right)\right],$$

$$(5.10)$$

where, the explicit values of the connection coefficients are (4.13):

$$\gamma^{(1)00}{}_{00}=0, \qquad \gamma^{(1)00}{}_{01}=\frac{1}{4}, \qquad \gamma^{(1)00}{}_{02}=\frac{1}{8},\dots,$$

$$\gamma^{(2)00}{}_{00}=-\frac{7}{3}, \qquad \gamma^{(2)00}{}_{01}=\frac{1}{8}, \qquad \gamma^{(2)00}{}_{02}=\frac{1}{32},\dots. \qquad (4.13')$$

The derivatives (5.10) have two components along two orthogonal functions, so that the projection with respect to $\psi(x/3\sqrt{2} + 1/5)$ gives

$$\frac{d}{dx}\left[-\frac{1}{2}\psi\left(\frac{x}{3\sqrt{2}} + \frac{1}{5}\right) - 0.08\right] \cong 0,$$

$$\frac{d^2}{dx^2}\left[-\frac{1}{2}\psi\left(\frac{x}{3\sqrt{2}} + \frac{1}{5}\right) - 0.08\right] \cong \frac{7}{\left(18\sqrt{2}\right)}\psi\left(\frac{x}{3\sqrt{2}} + \frac{1}{5}\right), \qquad (5.11)$$

while, with respect to $\psi_1(x/3\sqrt{2} + 1/5)$, we get

$$\frac{d}{dx}\left[-\frac{1}{2}\psi\left(\frac{x}{3\sqrt{2}} + \frac{1}{5}\right) - 0.08\right] \cong -\frac{1}{24\sqrt{2}}\left[\psi_1^0\left(\frac{x}{3\sqrt{2}} + \frac{1}{5}\right)\right],$$

$$\frac{d^2}{dx^2}\left[-\frac{1}{2}\psi\left(\frac{x}{3\sqrt{2}} + \frac{1}{5}\right) - 0.08\right] \cong -\frac{1}{288}\left[\psi_1^0\left(\frac{x}{3\sqrt{2}} + \frac{1}{5}\right)\right]. \qquad (5.12)$$

It can be easily shown by a direct computation that it is also

$$\frac{d}{dx}\left[\psi\left(\frac{x}{3\sqrt{2}} + \frac{1}{5}\right) - 0.08\right] = \frac{1}{12\sqrt{2}}\left[\psi_1^0\left(\frac{x}{3\sqrt{2}} + \frac{1}{5}\right)\right],$$

$$\frac{d}{dx}\psi_1^0\left(\frac{x}{3\sqrt{2}} + \frac{1}{5}\right) = \frac{1}{6\sqrt{2}}\psi_1^0\left(\frac{x}{3\sqrt{2}} + \frac{1}{5}\right), \qquad (5.10')$$

$$\frac{d^2}{dx^2}\psi_1^0\left(\frac{x}{3\sqrt{2}} + \frac{1}{5}\right) = \frac{1}{72}\psi_1^0\left(\frac{x}{3\sqrt{2}} + \frac{1}{5}\right).$$

6. Perturbation Method

In order to compute the cylindrical waves solution of the nonlinear equations (3.12'') and (3.12''') we will consider the perturbation method [9]. This method is based on the assumption that the solution of the nonlinear problem

$$Lu(x) = Nu(x), \qquad (6.1)$$

with L and N being the linear and nonlinear parts of the differential operator, can be expressed as a converging series, which depends on a small parameter $0 \leq \varepsilon \leq 1$:

$$u(x, \varepsilon) = \sum_{n=0}^{\infty} \varepsilon^n u^{(n)}(x) \qquad (6.2)$$

such that $u^{(0)}(x)$ is the solution of the linear problem:

$$Lu^{(0)}(x) = 0. \qquad (6.3)$$

The other terms of the series are computed by solving the recursive set of (of linear) equations:

$$Lu^{(n+1)}(x) = Nu^{(n)}(x), \quad n \geq 0. \tag{6.4}$$

6.1. Second-Order Nonlinearity

Let us search the solution of the second-order nonlinear equation (3.12') (where for convenience $r \to x$):

$$\left(u_{,xx} + \frac{u_{,x}}{x} + u - \frac{u}{x^2} \right) = a_1 u_{,xx} u_{,x} \tag{6.5}$$

by assuming that

$$u(x) = u^{(0)}(x) + \varepsilon u^{(1)}(x), \tag{6.6}$$

where $u^{(0)}(x)$ is the solution of the linear equation:

$$\left(u_{,xx} + \frac{u_{,x}}{x} + u - \frac{u}{x^2} \right) = 0. \tag{6.7}$$

When $u^{(0)}(x)$ is known, $u^{(1)}(x)$ is computed as the solution of

$$\left(u_{,xx} + \frac{u_{,x}}{x} + u - \frac{u}{x^2} \right) = a_1 u^{(0)}_{,xx} u^{(0)}_{,x}. \tag{6.8}$$

Moreover as initial condition is taken, $u(x,0) = u^{(0)}(x)$ and the perturbation is on time so that the small parameter can be identified with time $\varepsilon \to t$ and the solution of (3.12'') can be written as

$$u(x,t) = e^{-i\omega t} \left[u^{(0)}(x) + t u^{(1)}(x) \right]. \tag{6.9}$$

The general solution of (6.8) implies some cumbersome hypergeometric series and Laguerre polynomials (see, e.g., [14]); however, it should be noticed that since the r.h.s. of (6.8) is obtained from (3.12), by neglecting all terms $O(1/r)$ we can approximate also the l.h.s with the same hypotheses so that $u^{(1)}(x)$ can be searched as solution of

$$u_{,xx} + u = a_1 u^{(0)}_{,xx} u^{(0)}_{,x}. \tag{6.10}$$

The solution of (6.7) is (5.8), so that by inserting this wavelet function in the right-hand side of (6.8) and taking into account (5.9) and (5.10'), the function $u^{(1)}(x)$ will be obtained by solving

$$
\begin{aligned}
u_{,xx} + u &= \frac{a_1}{4\left(3\sqrt{2}\right)^3} \sum_{k=-\infty}^{\infty} \gamma^{(2)00}{}_{0k}\psi_k^0\left(\frac{x}{3\sqrt{2}}+\frac{1}{5}\right) \sum_{k=-\infty}^{\infty} \gamma^{(1)00}{}_{0k}\psi_k^0\left(\frac{x}{3\sqrt{2}}+\frac{1}{5}\right) \\
&= \frac{a_1}{4\left(3\sqrt{2}\right)^3}\left[\gamma^{(2)00}{}_{00}\psi_0^0\left(\frac{x}{3\sqrt{2}}+\frac{1}{5}\right) + \gamma^{(2)00}{}_{01}\psi_1^0\left(\frac{x}{3\sqrt{2}}+\frac{1}{5}\right) + \cdots\right] \\
&\times \left[\gamma^{(1)00}{}_{00}\psi_0^0\left(\frac{x}{3\sqrt{2}}+\frac{1}{5}\right) + \gamma^{(1)00}{}_{01}\psi_1^0\left(\frac{x}{3\sqrt{2}}+\frac{1}{5}\right) + \cdots\right].
\end{aligned}
\tag{6.11}
$$

By using the values of the connection coefficients (4.13'), for $-1 \le k \le 1$ and the orthogonality property of wavelets, we have

$$
u_{,xx} + u = \frac{a_1}{4\left(3\sqrt{2}\right)^3}\frac{1}{32}\left[\psi_1^0\left(\frac{x}{3\sqrt{2}}+\frac{1}{5}\right)\right]^2.
\tag{6.12}
$$

The solution $u^{(1)}(x)$ of (6.12) is searched in the form

$$
u^{(1)}(x) = f(x)\left[\psi_1^0\left(\frac{x}{3\sqrt{2}}+\frac{1}{5}\right)\right]^2.
\tag{6.13}
$$

By deriving and taking into account (5.10'),

$$
\begin{aligned}
u_{,x}^{(1)}(x) &= \left(f_{,x}+\frac{1}{3\sqrt{2}}f\right)\left[\psi_1^0\left(\frac{x}{3\sqrt{2}}+\frac{1}{5}\right)\right]^2, \\
u_{,xx}^{(1)}(x) &= \left[f_{,xx}+\frac{2}{3\sqrt{2}}f_{,x}+\frac{1}{18}f\right]\left[\psi_1^0\left(\frac{x}{3\sqrt{2}}+\frac{1}{5}\right)\right]^2.
\end{aligned}
\tag{6.14}
$$

Equation (6.12) becomes

$$
\left[f_{,xx}+\frac{2}{3\sqrt{2}}f_{,x}+\frac{19}{18}f\right]\left[\psi_1^0\left(\frac{x}{3\sqrt{2}}+\frac{1}{5}\right)\right]^2 = \frac{a_1}{4\left(3\sqrt{2}\right)^3}\frac{1}{32}\left[\psi_1^0\left(\frac{x}{3\sqrt{2}}+\frac{1}{5}\right)\right]^2,
\tag{6.15}
$$

that is,

$$
f_{,xx}+\frac{2}{3\sqrt{2}}f_{,x}+\frac{19}{18}f = \frac{a_1}{128\left(3\sqrt{2}\right)^3}.
\tag{6.16}
$$

The solution is

$$f(x) = \frac{a_1}{7296\sqrt{2}} + e^{-x/(3\sqrt{2})}(c_1 \cos x + c_2 \sin x) \qquad (6.17)$$

so that

$$u^{(1)}(x) = \left[\frac{a_1}{7296\sqrt{2}} + e^{-x/3\sqrt{2}}(c_1 \cos x + c_2 \sin x)\right]\left[\psi_1^0\left(\frac{x}{3\sqrt{2}} + \frac{1}{5}\right)\right]^2. \qquad (6.18)$$

If we assume that at the initial time $t = 0$, the nonlinear effect is neglectable, in a such a way that $u^{(1)}(0) = 0$ so that

$$0 = \left[\frac{a_1}{7296\sqrt{2}} + c_1\right]\left[\psi_1^0\left(\frac{1}{5}\right)\right]^2, \qquad (6.19)$$

which simplifies the previous form of $u^{(1)}(x)$ into

$$u^{(1)}(x) = \frac{a_1}{7296\sqrt{2}}\left(1 - e^{-x/(3\sqrt{2})}\cos x\right)\left[\psi_1^0\left(\frac{x}{3\sqrt{2}} + \frac{1}{5}\right)\right]^2. \qquad (6.20)$$

There follows that the general solution of (6.5) is

$$u(x) = -\frac{1}{2}\psi\left(\frac{x}{3\sqrt{2}} + \frac{1}{5}\right) - 0.08 + \frac{a_1}{7296\sqrt{2}}\left(1 - e^{-x/(3\sqrt{2})}\cos x\right)\left[\psi_1^0\left(\frac{x}{3\sqrt{2}} + \frac{1}{5}\right)\right]^2, \qquad (6.21)$$

and the explicit solution of (3.12″) becomes (see Figure 6)

$$u(x,t) = \left\{-\frac{1}{2}\psi\left(\frac{x}{3\sqrt{2}} + \frac{1}{5}\right) - 0.08 + t\frac{a_1}{7296\sqrt{2}}\left(1 - e^{-x/(3\sqrt{2})}\cos x\right)\left[\psi_1^0\left(\frac{x}{3\sqrt{2}} + \frac{1}{5}\right)\right]^2\right\}e^{-i\omega t}. \qquad (6.22)$$

As expected the evolution of the initial profile (Figure 6) shows the main nonlinear effect of large (increasing) amplitude. The initial profile is deformed by showing the increasing amplitude.

6.2. Third-Order Nonlinearity

Let us search the solution of the third-order nonlinear equation (3.12″) (where $r \to x$):

$$\left(u_{,xx} + \frac{u_{,x}}{x} + u - \frac{u}{x^2}\right) = a_6 u_{,xx}\,(u_{,x})^2. \qquad (6.23)$$

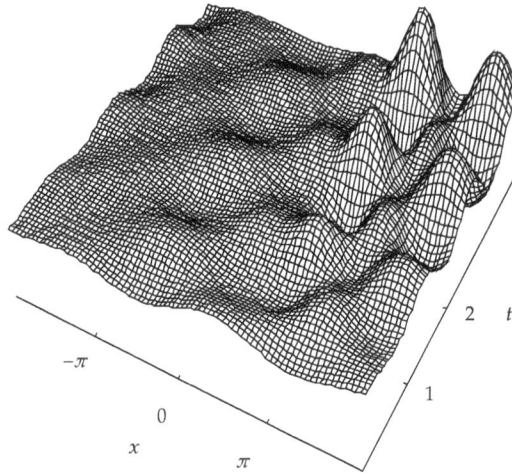

Figure 6: Wave solution with $a_1 = 10^4$ and $\omega = 5$.

The solution of (6.12) can be written as

$$u(x,t) = e^{-i\omega t}\left[u^{(0)}(x) + t u^{(1)}(x)\right], \tag{6.24}$$

where $u^{(0)}(x)$ is given by (5.8). Inserting this wavelet function in the right-hand side of (6.23), with the same approximation as in the previous case, and taking into account (5.9) and (5.10′), the function $u^{(1)}(x)$ will be obtained by solving

$$
\begin{aligned}
u_{,xx} + u &= -\frac{a_6}{2592}\sum_{k=-\infty}^{\infty}\gamma^{(2)00}{}_{0k}\psi_k^0\left(\frac{x}{3\sqrt{2}}+\frac{1}{5}\right)\left[\sum_{k=-\infty}^{\infty}\gamma^{(1)00}{}_{0k}\psi_k^0\left(\frac{x}{3\sqrt{2}}+\frac{1}{5}\right)\right]^2 \\
&= -\frac{a_6}{2592}\left[\gamma^{(2)00}{}_{00}\psi_0^0\left(\frac{x}{3\sqrt{2}}+\frac{1}{5}\right) + \gamma^{(2)00}{}_{01}\psi_1^0\left(\frac{x}{3\sqrt{2}}+\frac{1}{5}\right) + \cdots\right] \\
&\quad \times \left[\gamma^{(1)00}{}_{00}\psi_0^0\left(\frac{x}{3\sqrt{2}}+\frac{1}{5}\right) + \gamma^{(1)00}{}_{01}\psi_1^0\left(\frac{x}{3\sqrt{2}}+\frac{1}{5}\right) + \cdots\right],
\end{aligned}
\tag{6.25}
$$

that is,

$$u_{,xx} + u = -\frac{a_6}{2592}\gamma^{(2)00}{}_{01}\left[\gamma^{(1)00}{}_{01}\right]^2\left[\psi_1^0\left(\frac{x}{3\sqrt{2}}+\frac{1}{5}\right)\right]^3. \tag{6.26}$$

By taking into account the values of the connection coefficients (4.13′), we have

$$u_{,xx} + u = -\frac{a_6}{331776}\left[\psi_1^0\left(\frac{x}{3\sqrt{2}}+\frac{1}{5}\right)\right]^3. \tag{6.27}$$

The solution $u^{(1)}(x)$ of (6.23) is searched in the form

$$u^{(1)}(x) = f(x) \left[\psi_1^0 \left(\frac{x}{3\sqrt{2}} + \frac{1}{5} \right) \right]^3.$$

(6.28)

By deriving and taking into account (5.10'),

$$u_{,x}^{(1)}(x) = \left(f_{,x} + \frac{1}{2\sqrt{2}} f \right) \left[\psi_1^0 \left(\frac{x}{3\sqrt{2}} + \frac{1}{5} \right) \right]^3,$$

$$u_{,xx}^{(1)}(x) = \left[f_{,xx} + \frac{3}{2\sqrt{2}} f_{,x} + \frac{1}{8} f \right] \left[\psi_1^0 \left(\frac{x}{3\sqrt{2}} + \frac{1}{5} \right) \right]^3.$$

(6.29)

Equation (6.27) becomes

$$\left[f_{,xx} + \frac{3}{2\sqrt{2}} f_{,x} + \frac{9}{8} f \right] \left[\psi_1^0 \left(\frac{x}{3\sqrt{2}} + \frac{1}{5} \right) \right]^3 = -\frac{a_6}{331776} \left[\psi_1^0 \left(\frac{x}{3\sqrt{2}} + \frac{1}{5} \right) \right]^3.$$

(6.30)

So, by assuming the same hypotheses of the previous quadratic case, and with the same computations, we have

$$u^{(1)}(x) = -\frac{a_6}{373248} \left(1 - e^{-3x/(4\sqrt{2})} \cos \frac{3}{4} \sqrt{\frac{3}{2}} x \right) \left[\psi_1^0 \left(\frac{x}{3\sqrt{2}} + \frac{1}{5} \right) \right]^3.$$

(6.31)

The general solution of (6.23) is

$$u(x) = -\frac{1}{2} \psi \left(\frac{x}{3\sqrt{2}} + \frac{1}{5} \right) - 0.08 - \frac{a_6}{373248} \left(1 - e^{-3x/(4\sqrt{2})} \cos \frac{3}{4} \sqrt{\frac{3}{2}} x \right) \left[\psi_1^0 \left(\frac{x}{3\sqrt{2}} + \frac{1}{5} \right) \right]^3,$$

(6.32)

and the explicit solution of (3.12''') becomes (Figure 7)

$$u(x,t) \left\{ -\frac{1}{2} \psi \left(\frac{x}{3\sqrt{2}} + \frac{1}{5} \right) - 0.08 - \frac{a_6}{373248} t \right.$$

$$\left. \times \left(1 - e^{-3x/(4\sqrt{2})} \cos \frac{3}{4} \sqrt{\frac{3}{2}} x \right) \left[\psi_1^0 \left(\frac{x}{3\sqrt{2}} + \frac{1}{5} \right) \right]^3 \right\} e^{-i\omega t}.$$

(6.33)

As in the previous case we can observe the rapid growing of the amplitude, together with a splitting of the peak.

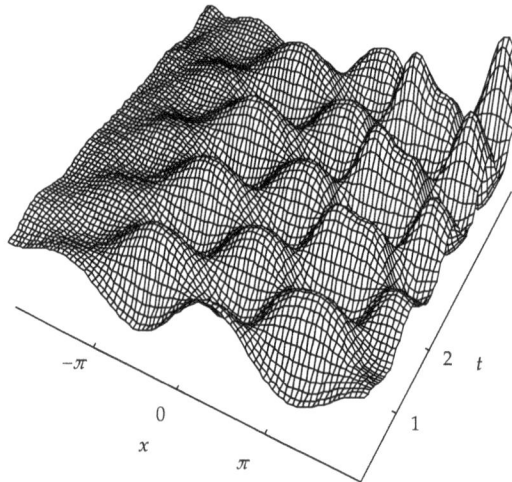

Figure 7: Wave solution with $a_6 = 10^5$ and $\omega = 5$.

7. Conclusions

It has been shown that cylindrical waves in a quadratic nonlinear Signorini structural model can be easily investigated by using Shannon wavelets. The initial profile, (solution of the linear equation) can be represented by Shannon wavelets and the evolution in time is described by the deforming wavelet profile thus giving a physical meaning to these kinds of wavelets. Shannon wavelets are equivalent to the Bessel function, at least in a quite sufficiently large neighborhood of 0. We have also noticed that, at the same approximation of the cylindrical wave, the Taylor polynomial for Shannon wavelets is one order lower than the Taylor polynomial for the corresponding Bessel function, so that Shannon wavelets are more efficient from computational point of view. It should be also noticed that Shannon wavelets are only the real part of the Newland harmonic wavelets [9, 17], so that also the Hankel functions, which are obtained by complex combination of Bessel functions, might have the same good approximation by harmonic wavelets.

References

[1] C. Cattani, "Waves in nonlinear Signorini structural model," *Journal of Mathematics*, vol. 1, no. 2, pp. 97–106, 2008.

[2] C. Cattani and E. Nosova, "Transversal waves in nonlinear Signorini model," *Lecture Notes in Computer Science*, vol. 5072, no. 1, pp. 1181–1190, 2008.

[3] C. Cattani and J. J. Rushchitsky, "Nonlinear plane waves in Signorini's hyperelastic material," *International Applied Mechanics*, vol. 42, no. 8, pp. 895–903, 2006.

[4] C. Cattani and J. J. Rushchitsky, "Nonlinear cylindrical waves in Signorini's hyperelastic material," *International Applied Mechanics*, vol. 42, no. 7, pp. 765–774, 2006.

[5] C. Cattani and J. J. Rushchitsky, "Similarities and differences between the Murnaghan and Signorini descriptions of the evolution of quadratically nonlinear hyperelastic plane waves," *International Applied Mechanics*, vol. 42, no. 9, pp. 997–1010, 2006.

[6] C. Cattani, J. J. Rushchitsky, and J. Symchuk, "Nonlinear plane waves in hyperelastic medium deforming by Signorini law. Derivation of basic equations and identification of Signorini constant," *International Applied Mechanics*, vol. 42, no. 10, pp. 58–67, 2006.

[7] A. Signorini, "Trasformazioni termoelastiche finite," *Annali di Matematica Pura ed Applicata, Series 4*, vol. 22, no. 1, pp. 33–143, 1943.

[8] A. Signorini, "Trasformazioni termoelastiche finite," *Annali di Matematica Pura ed Applicata, Series 4*, vol. 30, pp. 1–72, 1948.

[9] C. Cattani and J. J. Rushchitsky, *Wavelet and Wave Analysis as Applied to Materials with Micro or Nanostructure*, vol. 74 of *Series on Advances in Mathematics for Applied Sciences*, World Scientific, Singapore, 2007.

[10] C. Cattani and J. J. Rushchitsky, "Volterra's distortions in nonlinear hyperelastic media," *International Journal of Applied Mathematics and Mechanics*, vol. 1, no. 3, pp. 100–118, 2005.

[11] M. A. Biot, "Propagation of elastic waves in a cylindrical bore containing a fluid," *Journal of Applied Physics*, vol. 23, no. 9, pp. 997–1005, 1952.

[12] H. Demiray, "Wave propagation through a viscous fluid contained in a prestressed thin elastic tube," *International Journal of Engineering Science*, vol. 30, no. 11, pp. 1607–1620, 1992.

[13] H. H. Dai, "Model equations for nonlinear dispersive waves in a compressible Mooney-Rivlin rod," *Acta Mechanica*, vol. 127, no. 1–4, pp. 193–207, 1998.

[14] C. Cattani, "Shannon wavelet in nonlinear cylindrical waves ," submitted to *Ukrainian Mathematical Journal*.

[15] G. Kaiser, *A Friendly Guide to Wavelets*, Birkhäuser, 2011.

[16] K. Amaratunga, J. R. Williams, S. Qian, and J. Weiss, "Wavelet-Galerkin solutions for one-dimensional partial differential equations," *International Journal for Numerical Methods in Engineering*, vol. 37, no. 16, pp. 2703–2716, 1994.

[17] C. Cattani, "Harmonic wavelets towards the solution of nonlinear PDE," *Computers & Mathematics with Applications*, vol. 50, no. 8-9, pp. 1191–1210, 2005.

[18] C. Cattani, "Connection coefficients of Shannon wavelets," *Mathematical Modelling and Analysis*, vol. 11, no. 2, pp. 1–16, 2006.

[19] C. Cattani, "Shannon wavelets theory," *Mathematical Problems in Engineering*, vol. 2008, Article ID 164808, 24 pages, 2008.

[20] C. Cattani, "Shannon wavelets for the solution of integrodifferential equations," *Mathematical Problems in Engineering*, vol. 2010, Article ID 408418, 22 pages, 2010.

[21] G. Beylkin and J. M. Keiser, "On the adaptive numerical solution of nonlinear partial differential equations in wavelet bases," *Journal of Computational Physics*, vol. 132, no. 2, pp. 233–259, 1997.

[22] S. Bertoluzza and G. Naldi, "A wavelet collocation method for the numerical solution of partial differential equations," *Applied and Computational Harmonic Analysis*, vol. 3, no. 1, pp. 1–9, 1996.

[23] B. Alpert, G. Beylkin, R. Coifman, and V. Rokhlin, "Wavelet-like bases for the fast solution of second-kind integral equations," *SIAM Journal on Scientific Computing*, vol. 14, pp. 159–184, 1993.

[24] W. Dahmen, S. Prössdorf, and R. Schneider, "Wavelet approximation methods for pseudodifferential equations: I Stability and convergence," *Mathematische Zeitschrift*, vol. 215, no. 1, pp. 583–620, 1994.

[25] C. Cattani and Y. Y. Rushchitskii, "Solitary elastic waves and elastic wavelets," *International Applied Mechanics*, vol. 39, no. 6, pp. 741–752, 2003.

[26] J. D. Achenbach, *Wave Propagation in Elastic Solids*, North-Holland , 1973.

[27] R. W. Ogden, *Non-Linear Elastic Deformations*, Dover, 1974.

[28] C. W. Macosko, *Rheology: Principles, Measurement and Applications*, VCH, 1994.

[29] A. Bower, *Applied Mechanics of Solids*, CRC, 2009.

[30] D. Zwillinger, *Handbook of Differential Equations*, Academic Press, Boston, Mass, USA, 3rd edition, 1997.

[31] I. S. Gradshteyn and I. M. Ryzhik, *Tables of Integrals, Series, and Products*, Academic Press, San Diego, Calif, USA, 6th edition, 2000.

Permissions

The contributors of this book come from diverse backgrounds, making this book a truly international effort. This book will bring forth new frontiers with its revolutionizing research information and detailed analysis of the nascent developments around the world.

We would like to thank all the contributing authors for lending their expertise to make the book truly unique. They have played a crucial role in the development of this book. Without their invaluable contributions this book wouldn't have been possible. They have made vital efforts to compile up to date information on the varied aspects of this subject to make this book a valuable addition to the collection of many professionals and students.

This book was conceptualized with the vision of imparting up-to-date information and advanced data in this field. To ensure the same, a matchless editorial board was set up. Every individual on the board went through rigorous rounds of assessment to prove their worth. After which they invested a large part of their time researching and compiling the most relevant data for our readers. Conferences and sessions were held from time to time between the editorial board and the contributing authors to present the data in the most comprehensible form. The editorial team has worked tirelessly to provide valuable and valid information to help people across the globe.

Every chapter published in this book has been scrutinized by our experts. Their significance has been extensively debated. The topics covered herein carry significant findings which will fuel the growth of the discipline. They may even be implemented as practical applications or may be referred to as a beginning point for another development. Chapters in this book were first published by Hindawi Publishing Corporation; hereby published with permission under the Creative Commons Attribution License or equivalent.

The editorial board has been involved in producing this book since its inception. They have spent rigorous hours researching and exploring the diverse topics which have resulted in the successful publishing of this book. They have passed on their knowledge of decades through this book. To expedite this challenging task, the publisher supported the team at every step. A small team of assistant editors was also appointed to further simplify the editing procedure and attain best results for the readers.

Our editorial team has been hand-picked from every corner of the world. Their multi-ethnicity adds dynamic inputs to the discussions which result in innovative outcomes. These outcomes are then further discussed with the researchers and contributors who give their valuable feedback and opinion regarding the same. The feedback is then collaborated with the researches and they are edited in a comprehensive manner to aid the understanding of the subject.

Apart from the editorial board, the designing team has also invested a significant amount of their time in understanding the subject and creating the most relevant covers. They scrutinized every image to scout for the most suitable representation of the subject and create an appropriate cover for the book.

The publishing team has been involved in this book since its early stages. They were actively engaged in every process, be it collecting the data, connecting with the contributors or procuring relevant information. The team has been an ardent support to the editorial, designing and production team. Their endless efforts to recruit the best for this project, has resulted in the accomplishment of this book. They are a veteran in the field of academics and their pool of knowledge is as vast as their experience in printing. Their expertise and guidance has proved useful at every step. Their uncompromising quality standards have made this book an exceptional effort. Their encouragement from time to time has been an inspiration for everyone.

The publisher and the editorial board hope that this book will prove to be a valuable piece of knowledge for researchers, students, practitioners and scholars across the globe..

List of Contributors

R. Ingram
Department of Mathematics, University of Pittsburgh, PA 15260, USA

C. C. Manica
Departmento de Matemática Pura e Aplicada, Universidade Federal do Rio Grande do Sul, Porto Alegre 91509-900, RS, Brazil

N. Mays
Department of Mathematics, Wheeling Jesuit University, WV 26003, USA

I. Stanculescu
Farquhar College of Arts and Sciences, Nova Southeastern University, FL 33314, USA

Hamid Reza Marzban and Sayyed Mohammad Hoseini
Department of Mathematical Sciences, Isfahan University of Technology, P.O. Box 8415683111, Isfahan, Iran

Guoliang He
School of Mathematical Science, University of Electronic Science and Technology, Chengdu 610054, China

Jian Su
Faculty of Science, Xi'an Jiaotong University, Xi'an 710049, China

Wenqiang Dai
School of Management, University of Electronic Science and Technology, Chengdu 610054, China

Rajni Sharma
Department of Applied Sciences, DAV Institute of Engineering and Technology, Kabirnagar 144008, India

Janak Raj Sharma
Department of Mathematics, Sant Longowal Institute of Engineering and Technology, Longowal 148106, India

Anas Rachid
E´cole Nationale Supe´rieure des Arts et Me´tiers-Casablanca, Universite´ Hassan II, B.P. 150, Mohammedia, Morocco

Mohamed Bahaj and Noureddine Ayoub
Department of Mathematics and Computing Sciences, Faculty of Sciences and Technology, University Hassan 1st, B.P. 577, Settat, Morocco

Randhir Singh, Jitendra Kumar and Gnaneshwar Nelakanti
Department of Mathematics, Indian Institute of Technology Kharagpur, Kharagpur 721302, India

R. K. Pandey and G. K. Gupta
Department of Mathematics, Indian Institute of Technology, Kharagpur 721302, India

Jyoti Talwar and R. K. Mohanty
Department of Mathematics, Faculty of Mathematical Sciences, University of Delhi, 110007 Delhi, India

Ramakrishna Kakarala
School of Computer Engineering, Nanyang Technological University, Singapore 637665

Carlo Cattani
Department of Mathematics, University of Salerno, 84084 Fisciano (SA), Italy